Cell Death in Biology and Diseases

Series Editors

Xiao-Ming Yin
Zheng Dong

More information about this series at http://www.springer.com/series/8908

Wen-Xing Ding • Xiao-Ming Yin
Editors

Molecules, Systems and Signaling in Liver Injury

Springer

Editors
Wen-Xing Ding
Department of Pharmacology
Toxicology and Therapy
The University of Kansas Medical Center
Kansas City, KS, USA

Xiao-Ming Yin
Health Pathology Laboratory
Indiana University
Indianapolis, IN, USA

Cell Death in Biology and Diseases
ISBN 978-3-319-58105-7   ISBN 978-3-319-58106-4 (eBook)
DOI 10.1007/978-3-319-58106-4

Library of Congress Control Number: 2017945288

Printed on acid-free paper

This Springer imprint is published by Springer Nature
The registered company is Springer International Publishing AG
The registered company address is: Gewerbestrasse 11, 6330 Cham, Switzerland

Series Preface

Cell death, or conversely cell survival, is a major biological phenomenon. Just as with cell proliferation and cell differentiation, cell death is a choice that a cell has to make, sometimes voluntarily, other times accidentally. As such, cell death serves a purpose in the biology of a multicellular organism. The machinery of cell death and that of cell protection are evolutionarily conserved, and their elements can even be found in single-celled organism. The disruption of cell death mechanisms can often cause developmental abnormalities. Factors that can trigger cell death are diverse, and the cell death process is intricately connected with other biological processes. Cell death directly contributes to the pathogenesis of many diseases, including cancer, neurodegenerative diseases, and tissue injury in organ failure.

The study of cell death and cell survival has become a multidisciplinary subject, which requires expertise from all fields of the modern biology. Exploring the role of cell death in disease development and the modulation of cell death for the prevention and treatment of devastating disease demands constant updating of our knowledge through the broadest interactions among all investigators, basic and clinical. The rapid expansion of our knowledge in this field has gone beyond what could be summarized in a single book. Thus, this timely series Cell Death in Biology and Diseases summarizes new developments in different areas of cell death research in an elaborate and systemic way. Each volume of this series addresses a particular topic of cell death that either has a broad impact on the field or has an in-depth development in a unique direction. As a whole, this series provides a current and encyclopedic view of cell death.

We would like to sincerely thank the editors of each volume in the series and the authors of each chapter in these volumes for their strong commitment and great effort towards making this mission possible. We are also grateful to our team of professional Springer editors. They have worked with us diligently and creatively from the initiation and are continuing this on the development and production of each volume of the series. Finally, we hope the readers will enjoy the reading, find the content helpful to their work, and consider this series an invaluable resource.

Indianapolis, IN, USA — Xiao-Ming Yin
Augusta, GA, USA — Zheng Dong

Preface

Cell death is a fundamental biological phenomenon. It is evolutionarily conserved but can assume different forms under different conditions. While apoptosis, necrosis, and necroptosis are perhaps best studied, less known forms of cell death such as pyroptosis, ferroptosis, parthanatos, and entosis are increasingly found in various conditions or organisms. In multicellular organisms, cell death is important for development, shaping organ size, reforming tissue architecture, promoting functional differentiation, and determining mitochondrial inheritance. In the post--development stage, cell death determines the severity in tissue injury and the degree of subsequent response in inflammation, fibrosis, repair, regeneration, and tumorigenesis. The pathological changes in a complex organ, such as the liver, can be greatly affected by the death program in its cellular components. As the major organ for metabolism and detoxification, the liver is constantly under challenges from both internal and external sources. Viral infection and xenobiotics are the two major environmental stimuli that can cause significant hepatocyte death. Metabolic disturbance (such as in autophagy function) and special food components (such as lipids, ethanol, cholesterol, sugars) are the major internal stress that can also take a significant token on the hepatocytes that leads to cell death. Furthermore, the liver may experience traumatic injury as in cholestasis or ischemia, which often leads to tissue injury to various degrees. The liver is composed of hepatocytes (the majority), cholangiocytes, stellate cells, fibroblasts, macrophages/Kupffer cells, sinusoidal cells, and many more. They respond to these insults with different sensitivities and contribute to the overall liver pathology in various ways. Thus different signaling pathways may be involved in a cell type-specific and/or in a stimulus-specific way, which triggers subcellular organelle (mitochondria and endoplasmic reticulum as well as lysosomes) stress to induce cell death. Significant progresses have been made in the past decades to characterize these cell death pathways and the cells that are involved in liver injury. We are thus preparing two books devoted to the mechanisms of cell death in the liver and liver diseases. While the first volume mainly discusses liver injury and cell death caused by the various external and internal stimuli, the second volume discusses in great detail the type of cells and intracellular organelles involved in cell death as well as several major signaling pathways involved. We have paid particular attention to the interaction of different cells and aimed to provide a global view of the liver pathology that connects injury/cell death to other pathological changes, such as inflammation, fibrosis, and tumor development. Notably we will also discuss the interaction between the liver and the intestine where microbiota can both influence and be influenced by the liver pathology. Authors who are invited to contribute to the two volumes are respected experts in the subject, which should greatly enhance the authority of the chapter and the book. The volume editors are thus indebted to the authors for their outstanding contributions that make these two books so informative and thought-provoking.

Kansas City, KS, USA Wen-Xing Ding
Indianapolis, IN, USA Xiao-Ming Yin

Contents

Contributors

Gianfranco Alpini Department of Internal Medicine, College of Medicine Texas A&M Health Science Center, Central Texas Veterans Health Care System, Temple, TX, USA

Kevin E. Behrns Department of Surgery, University of Florida, Gainesville, FL, USA

Joseph G. Casian Department of Biopharmaceutical Sciences, School of Pharmacy, Keck Graduate Institute, Claremont, CA, USA

Carl W. Decker Department of Biopharmaceutical Sciences, School of Pharmacy, Keck Graduate Institute, Claremont, CA, USA

Laurent Ehrlich Department of Internal Medicine, College of Medicine Texas A&M Health Science Center, Central Texas Veterans Health Care System, Temple, TX, USA

Wenke Feng Department of Medicine, University of Louisville School of Medicine, Louisville, KY, USA

Scott L. Friedman, MD Division of Liver Diseases, Department of Medicine, Icahn School of Medicine at Mount Sinai New York, NY, USA

Shannon S. Glaser Department of Internal Medicine, College of Medicine Texas A&M Health Science Center, Central Texas Veterans Health Care System, Temple, TX, USA

Kristina L. Go Department of Surgery, University of Florida, Gainesville, FL, USA

Chad Hall Department of Internal Medicine, College of Medicine Texas A&M Health Science Center, Central Texas Veterans Health Care System, Temple, TX, USA

Derick Han Department of Biopharmaceutical Sciences, School of Pharmacy, Keck Graduate Institute, Claremont, CA, USA

Hui Han Department of Medicine, Keck School of Medicine of USC, University of Southern California, Los Angeles, CA, USA

Edward N. Harris Department of Biochemistry, University of Nebraska Lincoln, Lincoln, NE, USA

Hayato Hikita Department of Gastroenterology and Hepatology, Osaka University Graduate School of Medicine, Osaka, Japan

Luke A. Horton W.M. Keck Science Department, Claremont McKenna College, Claremont, CA, USA

Cheng Ji Department of Medicine, Keck School of Medicine of USC, University of Southern California, Los Angeles, CA, USA

Cynthia Ju Skaggs School of Pharmacy and Pharmaceutical Sciences, University of Colorado Anschutz Medical Campus, Aurora, CO, USA

Rui Kang Department of Surgery, University of Pittsburgh, Pittsburgh, PA, USA

Neil Kaplowitz Department of Medicine, Keck School of Medicine of USC, University of Southern California, Los Angeles, CA, USA

Jae-Sung Kim Department of Surgery, University of Florida, Gainesville, FL, USA

Sooyeon Lee Department of Surgery, University of Florida, Gainesville, FL, USA

John Marentette Skaggs School of Pharmacy and Pharmaceutical Sciences, University of Colorado Anschutz Medical Campus, Aurora, CO, USA

Craig McClain Department of Medicine, University of Louisville School of Medicine, Louisville, KY, USA

Fanyin Meng Department of Internal Medicine, College of Medicine Texas A&M Health Science Center, Central Texas Veterans Health Care System, Temple, TX, USA

Satdarshan Pal Singh Monga Department of Pathology, University of Pittsburgh, School of Medicine, Pittsburgh, PA, USA

Justin L. Mott Department of Biochemistry and Molecular Biology, Fred & Pamela Buffett Cancer Center, University of Nebraska Medical Center, Omaha, NE, USA

Kari Nichole Nejak-Bowen Department of Pathology, University of Pittsburgh, School of Medicine, Pittsburgh, PA, USA

Kim Tho Nguyen Department of Biopharmaceutical Sciences, School of Pharmacy, Keck Graduate Institute, Claremont, CA, USA

April O'Brien Department of Internal Medicine, College of Medicine Texas A&M Health Science Center, Central Texas Veterans Health Care System, Temple, TX, USA

Fatemeh P. Parvin-Nejad Division of Liver Diseases, Department of Medicine, Icahn School of Medicine at Mount Sinai, New York, NY, USA

Madhuri P. Rao W.M. Keck Science Department, Claremont McKenna College, Claremont, CA, USA

Kai H. Silkwood Department of Chemistry and Biochemistry, University of California San Diego, La Jolla, San Diego, CA, USA

Tetsuo Takehara Department of Gastroenterology and Hepatology, Osaka University Graduate School of Medicine, Osaka, Japan

Daolin Tang Department of Surgery, University of Pittsburgh, Pittsburgh, PA, USA

Tianhao Zhou Department of Internal Medicine, College of Medicine Texas A&M Health Science Center, Central Texas Veterans Health Care System, Temple, TX, USA [illegible] New York, NY, USA

[illegible]

[illegible]

[illegible]

[illegible]

[illegible]

About the Editors

Dr. Wen-Xing Ding is currently an associate professor of the Department of Pharmacology, Toxicology and Therapeutics at the University of Kansas Medical Center. He received his PhD from the National University of Singapore. He did his postdoctoral training at the University of Pittsburgh. In the past decade, Dr. Ding's research has been focused on the molecular mechanisms of cell death and autophagy in alcohol- and drug-induced liver injury as well as liver tumorigenesis. Dr. Ding is a member of the American Association for the Study of Liver Diseases (AASLD), Society of Toxicology (SOT), and Research Society on Alcoholism. Dr. Ding was a Liver Scholar of American Liver Foundation and a recipient of the University Scholarly Achievement Award (University of Kansas). Dr. Ding has published more than 100 papers in peer-reviewed journals, and his research work is currently supported by NIAAA and NIDDK.

Dr. Xiao-Ming Yin is a Louis Y. Mazzini Professor of pathology at the Department of Pathology and Laboratory Medicine, Indiana University School of Medicine. He obtained his medical degree from Shanghai Medical University and his PhD degree from the University of Texas Southwestern Medical Center at Dallas. He completed his postdoctoral fellowship training and his pathology residency training at Washington University in St. Louis. Dr. Yin is a member of several scientific societies and is a fellow of the American Association for the Study of Liver Diseases (AASLD). He won the Outstanding Investigator Award from the American Society for Investigative Pathology (ASIP). His research interest is cell death and survival in tissue injury and cancer biology, particularly in the liver system.

Chapter 1
Lipotoxicity in Non-parenchymal Liver Cells

Edward N. Harris and Justin L. Mott

Abbreviations

α-SMA	Alpha-smooth muscle actin
DR5	TRAIL death receptor 5
eNOS	Endothelial nitric oxide synthase
FasL	Fas ligand
GGT	Gamma-glutamyl transpeptidase
Hh	Hedgehog
HSCs	Hepatic stellate cells
KCs	Kupffer cells
LSECs	Liver sinusoidal endothelial cells
LXRs	Liver X receptors
NAFLD	Nonalcoholic fatty liver disease
NASH	Nonalcoholic steatohepatitis
PD-1	Programmed death 1 receptor
PD-L1/2	Programmed death 1 receptor ligands
PNPLA3	Patatin-like phosholipase domain-containing 3

Financial Support This chapter was supported in part by the National Institutes of Health and National Institute of General Medical Science grant P20GM104320 (to JLM) and the Fred & Pamela Buffett Cancer Center. The contents of this manuscript are solely the responsibility of the authors.

E.N. Harris
Department of Biochemistry, University of Nebraska Lincoln, Lincoln, NE, USA

J.L. Mott, MD, PhD (✉)
Department of Biochemistry and Molecular Biology, Fred & Pamela Buffett Cancer Center, University of Nebraska Medical Center, Omaha, NE 68198-5870, USA
e-mail: justin.mott@unmc.edu

W.-X. Ding, X.-M. Yin (eds.), *Molecules, Systems and Signaling in Liver Injury*, Cell Death in Biology and Diseases, DOI 10.1007/978-3-319-58106-4_1

X.-M. Yin and Z. Dong (Series eds.), *Cell Death in Biology and Diseases*

TGF-β	Transforming growth factor-β
TNFR	Tumor necrosis factor receptors
VEGF	Vascular endothelial growth factor

1.1 Introduction

Liver injury in the presence of metabolic alterations is becoming a major cause of morbidity and mortality. Nonalcoholic fatty liver disease (NAFLD) includes simple steatosis and the more severe NASH (Goh and McCullough 2016; Demir et al. 2015). NASH-associated cirrhosis and end-stage liver disease are currently the second leading indication for liver transplantation and are the fastest-growing cause for transplantation in the USA (Wong et al. 2015; Charlton et al. 2011). Altered metabolic states associated with NAFLD and NASH are obesity, insulin resistance, metabolic syndrome, and type 2 diabetes. Obviously, there is considerable overlap among these conditions. Development of metabolic syndrome is commonly associated with obesity, and increasing rates of obesity have paralleled increased NASH incidence. It is estimated that 34–40% of the adult population in the USA is obese (body mass index over 30 kg/m^2). Over 9 million adults in the USA have NASH, and over 100 million have NAFLD (Angulo 2002). The burden of this disease is hard to overstate.

Histologic findings of NASH include lipid droplet accumulation in hepatocytes (steatosis), hepatocellular injury and ballooning, and lobular inflammation. While not required for the diagnosis of NASH, fibrosis is the best predictor of disease outcomes (Angulo et al. 2015; Ekstedt et al. 2015). In stage 2 and higher fibrosis, there is portal fibrosis in addition to perisinusoidal lobular fibrosis. Another unscored histologic finding is ductular reaction with proliferation of hepatic progenitor cells in the lobule expressing markers of biliary epithelial cells and ductular reaction with a related expansion of biliary epithelial cells in the portal tract (Richardson et al. 2007). The degree of ductular reaction correlated with fibrosis stage in NASH with the greatest area of ductular reaction in stage 4 fibrotic NASH livers (Richardson et al. 2007; Gadd et al. 2014). Portal inflammation with CD68-positive macrophages and CD3- or CD8-positive lymphocytes also correlated with stage of fibrosis (Gadd et al. 2014). Ultrastructurally, liver sinusoidal endothelial cells (LSECs) lose their fenestrae and become capillarized.

Metabolic changes associated with NASH suggest a role for lipid signaling in disease. Indeed, genome-wide association studies have demonstrated that the lipase patatin-like phosholipase domain-containing 3 (PNPLA3) is associated with NAFLD (Romeo et al. 2008). Circulating free fatty acids are increased in NASH patients, and patients with stage 3 or 4 fibrosis had significantly higher free fatty acid levels than patients with stage 0–2 fibrosis. Interestingly, serum free fatty acids did not correlate with the degree of steatosis (Nehra et al. 2001). The predominant circulating fatty acid is palmitic acid, a 16-carbon saturated fatty acid. Saturated free fatty acids such as palmitic and stearic acid are cytotoxic to several cell types, termed lipoapoptosis. Monounsaturated free fatty acids such as oleate or palmitoleate are generally not toxic and even offer protection from lipoapoptosis (Welters et al. 2004; Akazawa et al. 2010; Mei et al. 2011; Ahn et al. 2013). Monounsaturated free fatty acids still promote lipid droplet accumulation, consistent with the concept that steatosis per se is not cytotoxic.

Here, we will discuss pathologic changes in NASH associated with lipid mediators. For some liver cell types, this includes lipoapoptosis; for others the lipotoxicity is manifested by altered function. The various cell types in the liver each contribute to NASH, and these roles will be discussed.

1.2 Hepatocytes

Apoptosis of hepatocytes in NASH is covered in depth elsewhere in this volume and will not be covered here at length. We note that hepatocyte apoptosis is a hallmark of NASH (Feldstein et al. 2003; Wieckowska et al. 2006) and free fatty acids can induce lipoapoptosis in cultured hepatocyte-derived cells (Ji et al. 2005; Malhi et al. 2006a; Cazanave et al. 2014). The term lipoapoptosis is used here to indicate apoptotic cell death caused by a lipid mediator and is not necessarily dependent on cellular accumulation of lipid droplets, a condition called steatosis. Steatosis is histologically striking in NAFLD and NASH, though may be absent in later stages of NASH, resulting in the historical term "cryptogenic cirrhosis" (Brunt 2005). On the other hand, apoptotic cell death is not readily apparent on general stains (such as H&E staining). Apoptotic hepatocytes can be identified on occasion in histologic sections, and their prevalence is better estimated using special staining such as TUNEL to detect cleaved DNA ends or cleaved caspase-3 to identify cells with activated caspases. In NASH, hepatocyte apoptosis is generally in the range of 2–5%, representing about a fivefold increase over normal liver (Feldstein et al. 2003).

1.3 Cholangiocytes

Evidence for biliary involvement in NASH is currently circumstantial. Pathological assessment of NASH is scored by steatosis, lobular inflammation, and hepatocyte ballooning, as well as unscored histologic findings such as fibrosis (Brunt et al. 2011). Biliary changes are not part of the histologic criteria for NAFLD or NASH. Bile duct epithelial cells do not accumulate histologically significant amounts of triglycerides in the form of lipid droplets. Because cholangiocytes do not become steatotic, it is tempting to overlook their potential role in nonalcoholic fatty liver disease. However, storing neutral lipids in the form of triglycerides and cholesterol esters in lipid droplets is not required to sense and respond to circulating lipids like free fatty acids.

Though not part of the scoring criteria, there are biliary histologic changes in NASH. For example, a common finding in NASH is the occurrence of the ductular reaction, a marked expansion of biliary epithelial cells (Roskams et al. 2003; Richardson et al. 2007; Gadd et al. 2014). Ductular reaction in NASH was most commonly associated with worse fibrosis (Richardson et al. 2007; Gadd et al. 2014).

Increasing fibrosis stages correlated with increasing area of ductular reaction (Richardson et al. 2007). Interpreting the significance of the ductular reaction is complicated by the observation that ductular reaction may be present from different mechanisms. For instance, ductular reaction may indicate hepatocyte injury in combination with impaired hepatocyte proliferation (Roskams et al. 2003; Falkowski et al. 2003; Clouston et al. 2005). Alternatively, ductular reaction may indicate a proliferative response to direct biliary damage (Xu et al. 2004; Munshi et al. 2011). Some contribution from each is also feasible.

Is there evidence of direct biliary damage? The biliary marker gamma-glutamyl transpeptidase (GGT) is released into the serum in biliary injury. Elevated serum GGT is correlated with metabolic syndrome (Rantala et al. 2000), risk of developing type 2 diabetes (Perry et al. 1998), histologic fibrosis (Tahan et al. 2008), and likelihood of NASH (Neuschwander-Tetri et al. 2010). However, GGT is not specific for injury to bile duct epithelial cells and can be found at canaliculi of hepatocytes as well (Hanigan and Frierson 1996; Irie et al. 2007). Patients with a cholestatic phenotype of NAFLD (elevated GGT, elevated alkaline phosphatase) were matched to patients with non-cholestatic NAFLD for body mass index and markers of hepatocyte injury (alanine and aspartate aminotransferases) and were assessed for microscopic evidence of biliary injury. Histologic assessment demonstrated bile duct swelling, duct loss, ductular proliferation (ductular reaction), and portal fibrosis. Notably, those patients with increased markers of biliary injury were more likely to have worse fibrosis or even cirrhosis than the matched cohort (Sorrentino et al. 2005). To date, these markers of injury may still reflect either primary biliary injury or secondary biliary damage due to signaling in other cells in the liver. Notably, primary biliary cholangitis was worsened by concomitant overweight, with more steatosis, higher inflammatory grade, and worse bile duct damage found in overweight patients suggesting that lipotoxicity may contribute in the biliary tree, although other metabolic traits were not correlated with severity (Hindi et al. 2013). Additionally, the same single nucleotide polymorphism in PNPLA3 (I148M) that increases NAFLD risk was found to be associated with worse transplant-free survival in patients with primary sclerosing cholangitis. This effect was found in men with severe fibrotic disease manifested by a dominant stenosis (Friedrich et al. 2013). The molecular mechanism behind worsening of disease is unknown. The communication between cells that is necessary for normal liver function means that perturbation in one cell type can, and often does, alter another cell type.

The location of fibrosis in early NASH is generally pericentral, suggesting that cells in the portal tract may contribute less to disease (Brunt et al. 2011). Certainly, hepatocyte injury, altered endothelial function, and activation of Kupffer and hepatic stellate cells result in fibrosis. Still, in pediatric NASH, fibrosis is commonly found predominantly in the portal tract, sometimes referred to as type 2 NASH (Schwimmer et al. 2005; Takahashi et al. 2011). Sequential liver biopsies have demonstrated that patients with periportal fibrosis (stage 2 or 3) can transition to a pattern of pericentral fibrosis (Adams et al. 2005), demonstrating that the region of fibrosis may not be constant or may be different depending on sampling of the liver by biopsy. Whether bile duct epithelial injury promotes portal fibrosis in NASH remains an open question, but biliary fibrosis in other conditions implies that the cholangiocyte could contribute.

Developmentally, hepatocytes and cholangiocytes arise from a bipotential common precursor cell. The cells generally are few in number in the healthy adult liver but likely contribute to the ductular reaction under conditions of injury. Because of this shared developmental origin, it is likely hepatocytes and cholangiocytes retain common signaling pathways. Also because of this shared development, and the fact that the albumin promoter becomes active in precursor cells, it is important to carefully interpret studies in which the albumin promoter has been used to drive liver-specific genetic changes. For example, it is common to employ a cell-type-specific promoter to express Cre recombinase allowing for genetic rearrangement at loxP sites in a subset of somatic cells, but not in all cells. These conditional genetic systems can be used to selectively delete genes that have floxed exons only in cells expressing Cre. Alternatively, the Cre system can be used to selectively induce expression of a constitutively silent transgene that contains a lox-stop-lox cassette upstream of the transcriptional start site. Recombination in Cre-positive cells then removes the stop cassette and allows expression in a cell-type-specific manner. In the liver, either the albumin promoter (Alb-Cre) or the albumin promoter and alpha-fetoprotein enhancers (Alfp-Cre) have been used. These systems are liver specific (Postic et al. 1999; Kellendonk et al. 2000) but should not be considered hepatocyte specific. Both Alb-Cre and Alfp-Cre result in recombination in cholangiocytes as well as the expected hepatocytes (Tchorz et al. 2009; Kellendonk et al. 2000), i.e., in both epithelial cell types in the liver.

Given these caveats, deletion of the apoptosis-related TRAIL death receptor (DR5) in liver epithelial cells reduced NASH injury in a mouse model (Idrissova et al. 2015). The TRIAL death receptor is upstream of the initiator caspase-8. Cre-mediated deletion of caspase-8 in liver epithelial cells also decreased injury and fibrosis in methionine–/choline-deficient mice (Hatting et al. 2013). These findings implicate death receptor signaling in lipoapoptosis and severity of liver injury. In hepatocyte-derived cell lines, interfering with mRNA for DR5, caspase-8, or FADD reduced lipoapoptosis induced by treating hepatocytes with saturated free fatty acids (Cazanave et al. 2011). The same study also found that human cholangiocyte-derived cell lines underwent apoptosis upon exposure to saturated free fatty acids and that inhibition of DR5 or caspase-8 prevented lipoapoptosis (Cazanave et al. 2011).

Based on this study, and the shared developmental origin of cholangiocytes and hepatocytes, we have begun to study cholangiocyte lipoapoptosis. Cultured cholangiocyte-derived cell lines undergo caspase-dependent lipoapoptosis upon treatment with stearate or palmitate. Apoptosis increased with increasing concentration of these fatty acids and was partially dependent upon activation of the transcription factor FoxO3 (Natarajan et al. 2014). In contrast to hepatocyte lipoapoptosis (Malhi et al. 2006a), cholangiocytes did not require JNK activation for cell death. This cell culture model employed cultured cells and individual fatty acids at relatively high concentrations. The serum free fatty acid content is increased in obesity and metabolic syndrome as well as in NASH patients, so the free fatty acid concentrations were chosen to mimic pathophysiologic, not physiologic, levels. These limitations should be recognized in considering the potential role of cholangiocyte lipoapoptosis in NASH. We hypothesize that cholangiocyte lipoapoptosis, in combination with activation of the surviving cholangiocytes, can promote or worsen fibrogenesis, inflammation, and NASH progression.

1.4 Kupffer Cells

Kupffer cell apoptosis does not appear to cause worsening of NASH. However, it is apparent Kupffer cells are an active cell type determining the severity of inflammation and injury. Specifically, in a series of needle biopsy samples, Kupffer cells were found to be aggregated in NASH but not in healthy liver samples or livers with steatosis (Lefkowitch et al. 2002). There was even sporadic evidence of intracellular lipids in Kupffer cells, which may have represented Kupffer cell lipid droplets or phagocytosed lipid-laden hepatocytes (Lefkowitch et al. 2002).

Experimental depletion of Kupffer cells, preferentially large periportal Kupffer cells, can be accomplished by treating animals with gadolinium chloride which forms aggregates above pH 6 and is taken up by KCs causing their death (Hardonk et al. 1992). Alternatively, the hydrophilic bisphosphonate clodronate can be packaged into liposomes facilitating uptake into macrophages by phagocytosis where it induced apoptosis (Van Rooijen and Sanders 1994). Using these tools revealed that Kupffer cells are necessary for short-term steatosis and hepatic insulin resistance. Depletion of Kupffer cells for 1 week prevented hepatic steatosis due to high-fat feeding (Stienstra et al. 2010). An acute, 3-day high-fat diet induced hepatic insulin resistance that was prevented by Kupffer cell depletion (Lanthier et al. 2010). Gadolinium chloride-mediated depletion of Kupffer cells prevented steatosis induced by 2 weeks of either high-fat or high-sucrose feeding (Huang et al. 2010). Additionally, Kupffer cell depletion prevented hepatic insulin resistance tested by insulin clamp. During this short dietary intervention, there was no change in fasting insulin or fasting glucose levels (Huang et al. 2010). Weekly Kupffer cell depletion with liposomal clodronate prevented glucose intolerance during a 4-week high-fat diet (Lanthier et al. 2011). Similarly, twice-weekly gadolinium improved systemic insulin resistance and hepatic steatosis in mice fed with a high-fat diet for 3 weeks (Neyrinck et al. 2009). Weekly clodronate-mediated depletion in mice fed with a methionine–/choline-deficient diet for 3 weeks that causes NASH-like liver changes without insulin resistance or obesity reduced steatosis and markers of fibrosis and inflammation (Rivera et al. 2007). However, a 16-week dietary challenge with high fat did not benefit from Kupffer cell depletion during the last 10 days (Lanthier et al. 2011). This suggests that once metabolic changes are present, Kupffer cells are not required to maintain these alterations.

In summary, lipid-induced apoptosis of liver macrophages is unlikely to contribute to NASH, while prevention of Kupffer cell activation (by killing off these cells) has a beneficial, preventative effect. Therapeutic benefit from Kupffer cell depletion after metabolic changes lacks experimental support.

1.5 Liver Sinusoidal Endothelial Cells

Liver sinusoidal endothelial cells (LSECs) are unique endothelial-type cells lining the capillary bed of the liver and serve as the blood barrier between general circulation and the space of Disse which allows access to stellate cells and hepatocytes. The three defining characteristics that make LSECs unique from other endothelia are (1) the lack of an underlying basal lamina (Wisse et al. 1985), (2) numerous cytoplasmic holes or fenestrae organized as sieve plates, and (3) a very high endocytic (fluid phase) capacity. Each fenestra is 100–175 nm in diameter and allows plasma components including small chylomicron and chylomicron remnants to pass through this single-cell layer and come in direct contact with hepatocytes (Braet and Wisse 2002; Wisse 1970). Formation of fenestrae occurs by an actin-dependent fusion between opposite sheets of the plasma membrane when there is a lack of intramembrane particles (Taira 1994). In addition, formation and overall numbers of fenestrae are maintained by adequate levels of vascular endothelial growth factor (VEGF), and loss of VEGF reduces fenestrae (Xie et al. 2012). Due to the dynamic nature of fenestrae, various external stimuli cause the number of fenestra to increase, including hypoxia (Frenzel et al. 1976) and pressure (Nopanitaya et al. 1976; Fraser et al. 1980), or decrease including ethanol (Mak et al. 1984; Jacobs et al. 2009; Sarphie et al. 1997) and nicotine (Fraser et al. 1988). It has been observed that normal aging decreases the size and number of fenestrae, termed pseudo-capillarization, and affects lipid metabolism due to reduced chylomicron contact with the hepatocytes (Le Couteur et al. 2001; McLean et al. 2003; Warren et al. 2005). LSECs are very efficient in the internalization of extracellular matrix and waste materials from blood. Rapid ligand internalization and receptor recycling are owed in part to a net-like distribution of clathrin heavy chains associated with clathrin-coated vesicles in the endocytic processes of internalization, vesicle maturation, and ligand trafficking (Falkowska-Hansen et al. 2007). It is believed that few cells of the mammalian body can match or exceed the endocytic capacity of LSECs.

Tumor necrosis factor receptors (TNFRs) are very low to absent in the normal liver, but in lipotoxic conditions, TNFR expression is increased in hepatocytes, LSECs, cholangiocytes, and KCs (Volpes et al. 1992). Fas receptor, of the TNFR family, is a 45-kDa receptor on the plasma membrane belonging to the tumor necrosis factor family. Fas ligand (FasL) is expressed on activated immune cells (T cells, monocytes, and neutrophils) inducing apoptosis in Fas-expressing cells (Depraetere and Golstein 1997). Normally, the expression of Fas on LSECs is quite low, but that increases under conditions of NASH or lipotoxicity (Malhi et al. 2006b; Muschen et al. 1998). Despite the increase in Fas expression, LSECs are more resistant to apoptosis in the presence of FasL than hepatocytes and stellate cells, though higher physiological doses of FasL did, eventually, induce apoptosis in LSECs (Fig. 1.1). Low doses of FasL or anti-Fas antibodies during in vitro studies have demonstrated an increased expression of ICAM-1 and VCAM-1 instead of apoptosis suggesting that alternative pathways stimulated by Fas are operative in LSECs (Cardier et al. 1999). It is speculated that an increase in adhesion molecules may be a signal for

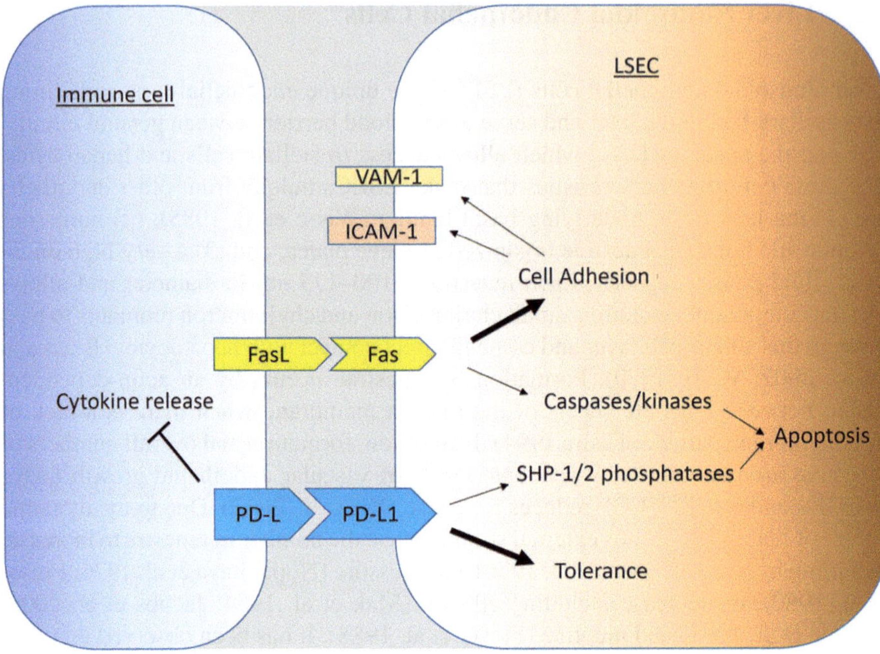

Fig. 1.1 Interaction of activated immune cells and LSECs. LSECs interact with both quiescent and activated immune cells. LSECs express Fas and PD-L1 receptors which induce parallel pathways that induce apoptosis. However, the dominant outcomes for each pathway are illustrated by the thicker arrow, and apoptosis is only induced after an ill-defined threshold of stimulation has been met. Overall, LSECs tend to dampen the inflammatory response in many environments including NAFLD/NASH

recruitment for liver-specific Kupffer cells (KCs). Clearance of cell debris and other particles is primarily performed by KCs as these cells are professional phagocytes, whereas LSECs internalize material through rapid pinocytosis. It is estimated that macromolecules suitable for LSEC internalization are less than 0.23 μm, which shifts the burden of phagocytosis to KCs (Shiratori et al. 1993; Park et al. 2008). KCs and LSECs coordinate controlled and tempered responses in chronic inflammatory responses that prove to be hepatoprotective in which both cell types are required to keep tissue damage in check. For instance, depletion of KCs prior to acetaminophen-induced liver injury results in exacerbated liver injury, particularly to the hepatic endothelium (Holt et al. 2010). Similarly in a sepsis model, LSEC apoptosis and Fas expression were increased after a depletion of KCs with the use of clodronate liposomes (Hutchins et al. 2013a). Both of these experimental examples show that KCs are necessary for the viability and well-being of LSECs in a background of escalating inflammation.

LSECs are unique antigen-presenting cells and interact with immune cells including the resident KCs and transient immune cells via the programmed death 1 receptor (PD-1) and its ligands (PD-L1/2; Fig. 1.1) (Keir et al. 2008). LSECs express

PD-L1 which is the ligand for PD-1 expressed by cells within the sinusoids, KCs and CD8+ T cells. This interaction promotes immune tolerance and dampens liver inflammation which is important as the liver sinusoids are directly downstream from gut circulation, containing bacterial and food antigens, via the portal vein (Schurich et al. 2010). Furthermore, PD-1/PD-L1 interactions between LSECs and T_H1 and T_H17 suppress cytokine release from these activated T cells when comparing these activated T cells with dendritic cells, strengthening the hypothesis that LSECs are critical for immune tolerance within the liver (Carambia et al. 2013). PD-1/PD-L1 increases in expression to keep inflammation in check under a variety of simulants including viral infection (Xie et al. 2009; Larrubia et al. 2009), hepatitis, and hepatocellular carcinoma (Wang et al. 2011). Immune tolerance does have its limits in that an undefined threshold may be breached. In mice with sepsis-induced injury (punctured cecum), LSECs and KCs had increased PD-L1 and PD-1 expression, respectively. This resulted in LSEC cell death, with lower cell numbers overall when isolated from the liver, and the few harvested LSECs had lower VEGFR2 expression (a marker for angiogenesis and maintenance of healthy tissue) (Hutchins et al. 2013b). These animals succumbed to septic shock along the same pathways as those exhibiting acute liver failure. The characteristics of PD-1−/PD-L1-induced injury and cell death are very similar to Fas−/FasL-mediated injury model mentioned previously. These pathways are not connected but run in parallel and may synergize each other under the right conditions. The differences between the Fas/FasL and PD-1/PD-L1 are that the former utilizes death receptor-associated kinases and caspases (Wajant 2002) in contrast to the latter which is CD28 dependent and utilizes SHP-1 and SHP-2 phosphatases (Sheppard et al. 2004; Okazaki et al. 2001).

LSECs contain several scavenger receptors that "sense" the extrahepatic environment and respond accordingly (Weigel et al. 2012). These receptors include Stabilin-1, Stabilin-2, mannose receptor, and FC-gamma/CD32b receptors (Sorensen et al. 2015). One of the ligands for Stabilin-1 is SPARC or secreted protein acidic and rich in cysteine which may also serve as a ligand for VCAM-1 (Kzhyshkowska et al. 2006; Kelly et al. 2007). SPARC (also known as osteonectin and BM-40) is a secreted extracellular matrix-associated protein with a major role in the wound healing response, tissue remodeling (Bradshaw et al. 2001), and liver fibrosis (Atorrasagasti et al. 2013). SPARC expression is increased in acute-on-chronic alcohol-induced hepatitis. Using a SPARC knockout mouse, the investigators demonstrated decreased CD4+ T-cell infiltration in the liver and decreased apoptosis of the hepatic endothelium induced by TNF-alpha and IL-6 (Peixoto et al. 2015). This suggests that increased signaling through the Stabilin-1 receptor and, possibly VCAM-1, via SPARC activates LSECs, promotes leukocyte infiltration, and further stimulates death receptors such as Fas/TNFR.

Accumulation of free fatty acids and neutral fats arising from metabolic disorders promotes cellular damage and cell death in many tissues including the liver (Malhi and Gores 2008). It is well known that fat-storing hepatocytes suffer from lipotoxic conditions leading to ballooning and cell death; but how do these metabolic conditions affect LSECs? In most cell-based studies, free fatty acids activate several pathways that lead to cell death including BAX-dependent apoptosis and lysosomal permeabilization, over-generation of reactive oxygen species, activation of p53, and hyperstimulation of the JNK pathway (Malhi and Gores 2008; Brenner et al. 2013). In line with this, rodents fed with a high-fat diet had LSECs with decreased numbers and porosity of fenestrae, also known as capillarization, compared with rodents fed with a balanced chow control diet (Cogger et al. 2016). In contrast, rats that were fasted for 48 h had an increase in LSEC fenestration by 10% (O'Reilly et al. 2010), and rats that were calorically restricted had an increase of 60% of LSEC fenestration and porosity (Jamieson et al. 2007) suggesting that deprivation of calories increases LSEC porosity for nutrient absorption by the hepatocytes, whereas overnutrition decreases LSEC porosity. However, LSECs do not store fatty acids in NAFLD and NASH conditions (author's unpublished observations), and high levels of certain free fatty acids may be neutral or beneficial for LSEC survival that is contrary for hepatocytes and cholangiocytes (Hang et al. 2012). Recent evidence supports this idea in that in vitro cultures of LSECs supplemented with oleic acid promote maintenance of fenestrae and modulate the Akt/PKB signaling followed by MAPK/ERK signaling to promote survival (Hang et al. 2012). Notably, oleic acid is largely nontoxic to hepatocytes and cholangiocytes except on very high concentrations. Overall, these results show that fenestrae of LSECs are a good marker for cell health and survival, and the amount and type of free fatty acids in vivo or in vitro will affect survival and health of LSECs.

Loss of fenestrae or capillarization is not the only marker for LSECs under stressful conditions. In both fatty liver rodent models and human patients, the size of the liver increases by up to ~20%. The increase in size is primarily attributed to the fat stored in the hepatocytes. Excessive fat storage compresses the sinusoidal space within the capillary bed of the liver and increases portal blood pressure (DeLeve et al. 2008b). Increased portal pressure and other associated factors of lipotoxicity will involve KCs and HSCs to recruit inflammatory cells into the liver and foster conditions for inflammation and deposition of extracellular matrix within the space of Disse (fibrosis) (Farrell et al. 2008; McCuskey et al. 2004). Physical parameters within the sinusoids are just as important in cellular health and maintenance as the biochemical signaling molecules. More research needs to be performed to determine the influence of geometric/physical conditions on LSEC autocrine/paracrine signaling.

Capillarization of LSECs is a hallmark of LSEC injury and a precursor for LSEC death. The mechanism for capillarization is beginning to be understood in that hedgehog (Hh) signaling is a primary driver of this phenomenon. Generally, Hh signaling regulates vascular development and increases during all types of liver injury. In addition to hepatocytes, LSECs express Hh receptor, ligands, and the ligand antagonist, Hhip. Using primary cells in vitro, capillarization was stimulated with increased Hh receptor signaling and inhibited with Hhip. Capillarization was compared in parallel with Hh target genes, Gli1/2 and Ptch1, LSEC migration, and tube formation (Xie et al. 2013). Since capillarization is the start of LSEC injury, does it occur before or after stellate cell activation which promotes inflammation and the pre-fibrotic phenotype in NASH? The current paradigm suggested that capillarization of LSECs precedes HSC activation, and decapillarization (healthy fenestrae phenotype) promotes quiescence of HSCs (Deleve et al. 2008a). During liver injury, Hh signaling is increased as a normal repair response for organ regeneration. Injury associated with lipotoxicity (or chronic injury) induces aberrant Hh signaling which induces capillarization.

Liver X receptors (LXRs), oxysterol-activated nuclear receptors for lipogenesis, are found in the liver and other tissues with high lipid metabolism and may play a role in LSEC physiology. There is recent evidence indicating that LXR and Hh pathways cross talk in the development of LSEC capillarization. In LXR-deficient mice, Hh signaling was increased (measured by monitoring expression of Gli2 and CD31) in LSECs in parallel with LSEC capillarization in a carbon tetrachloride injury mouse model (Xing et al. 2016). The conclusions of all these experimental data are that LSEC fenestration is promoted by LXR under normal conditions and shifts to capillarization upon induction of the Hh signaling pathways during severe or chronic injury.

What are the outcomes of LSEC cell death and injury? In chronic injury situations, LSECs become capillarized reducing the oxygen content in the liver. Hypoxia stimulates VEGF production in hepatocytes, HSCs, and LSECs in both autocrine and paracrine manners. Overstimulation by VEGF increases the number of sinusoidal vessels, but these vessels have varying diameter, shape, and flow pattern for blood. While increased blood vessels may serve to decrease portal pressure by increasing avenues of blood flow, the full outcome of angiogenesis in the liver is correlated to portal hypertension and promotion of fibrogenesis (Thabut and Shah 2010). Activated HSCs are in intimate contact with capillarized LSECs which increase the number of HSCs around the blood vessel and also restricts the expansion of the blood vessel when expansion is needed to relieve fluid pressure. As HSCs tighten around the LSECs, portal pressure is increased which exacerbates chronic inflammation (Lee et al. 2007). All of these conditions are forward feeding in that a tenuous chronic inflammation may become more acute with the onset of fibrosis and cirrhosis from either physical or molecular injuries (Gentile and Pagliassotti 2008). Rats fed with a cafeteria diet high in saturated fat and sucrose developed hypertension, obesity, insulin resistance, and elevated triglycerides (i.e., features of metabolic syndrome). Plasma free fatty acids increased from 1600 to 3200 μM. Portal blood flow was decreased and hepatic vascular resistance more than doubled. The dietary intervention was continued for 30 days and resulted in steatosis, but fibrosis did not develop in this short time. In this early stage of disease, LSEC function was impaired with reduced vasodilation response and decreased endothelial nitric oxide synthase (eNOS) phosphorylation and enzymatic function (Pasarin et al. 2012).

In conclusion, LSECs are dynamic gatekeepers of the liver allowing select solutes of the blood to come in contact with the hepatocytes. As a regulatory function, they continually filter out macromolecules from the plasma. LSECs, in coordination with other non-parenchymal cells, keep a check on the immune system to prevent runaway inflammation and destruction of the tissue and promote immune tolerance. The loss of these cells is a precursor for liver failure.

1.6 Hepatic Stellate Cells

Hepatic stellate cells (HSCs) are the major site of storage for fat-soluble retinol esters and retinol in the body (Hendriks et al. 1985). HSCs reside in the space of Disse between LSECs and hepatocytes. HSCs have a transcription factor profile that is similar to adipocytes, consistent with their fat-storing function (She et al. 2005). Their name, stellate, comes from their shape with a triangular cell body containing the nucleus and long, delicate cytoplasmic processes extending into the space of Disse. Transdifferentiation of HSCs into a myofibroblastic phenotype results in loss of lipid droplets, expression of alpha-smooth muscle actin (α-SMA), development of contractile properties, and production of type I and III collagen. This process is referred to as activation of HSCs and is an integral part of hepatic fibrogenesis.

Activation of HSCs is due to cell-cell communication and changes in the extracellular matrix in NASH. For example, a major fibrogenic stimulus that activates HSCs is transforming growth factor-β (TGF-β) produced by HSCs and Kupffer cells (De Bleser et al. 1997; Friedman 2000; Bataller and Brenner 2005). Additionally, HSCs can be activated by conditioned medium from steatotic cultured hepatoma cells. A sub-toxic concentration of palmitic acid (200 μM) was added to primary human hepatocytes or immortalized hepatoma cell lines for 24 h and the conditioned culture medium then transferred to primary HSCs or to the HSC cell line hTERT-HSC, respectively. Conditioned medium increased HSC production of profibrogenic TGF-β, remodeling proteins matrix metalloproteinase-2, tissue inhibitor of metalloproteinase-1 and metalloproteinase-2 and type I collagen, and the contractile protein α-SMA. Importantly, conditioned medium also promoted resistance of HSCs to apoptosis, suggesting that steatotic hepatocytes may promote survival of fibrogenic HSCs in the liver. Lack of toxicity in the palmitic acid-treated hepatocytes was confirmed by the absence of nuclear changes (Hoechst) and lack of release of lactate dehydrogenase or aminotransferases (Wobser et al. 2009). This study demonstrated that hepatocytes release an activating factor upon steatogenesis. The authors used size-exclusion chromatography with a 5-kD cutoff spin column and observed that the removal of proteins over 5 kDa abolished the activating effect. While free fatty acids are smaller than this cutoff, they are also generally bound to proteins in solution, so it is possible that some residual free fatty acids were in the conditioned medium.

Hepatocyte apoptosis is a hallmark of NASH and may provoke HSC activation. The HSC cell line LX-1 was found to have phagocytic capacity to take up apoptotic bodies from dead hepatocytes. This uptake was microtubule dependent and was specific to HSCs as hepatocytes themselves did not engulf apoptotic bodies. Uptake then resulted in HSC activation, measured by α-SMA and type I collagen expression as well as production of TGF-β (Canbay et al. 2003). In this study, apoptosis was induced in hepatoma cells (HepG2 cells) by exposure to UV light, eliminating the possible carry-over of the toxic stimulus in the crude preparation of apoptotic bodies. Apoptotic bodies were prepared by low-speed centrifugation of the collected medium (2000 *g*), so this preparation is not expected to contain soluble cell-free signaling mediators and probably represents a distinct profibrogenic signal.

Activation of HSCs by conditioned medium from steatotic hepatocytes and by hepatocyte apoptotic bodies is relevant in the liver where these two cell types are in close proximity. The predictable result is increased production of extracellular matrix proteins, especially types I and III collagen, resulting in fibrosis and liver dysfunction. Biopsy samples of liver samples from patients with NASH support a role for activated HSCs in fibrosis, with 97% of NASH patients in one study showing an increase in activated HSCs (Washington et al. 2000). Activated HSCs in that study were correlated with the severity of fibrosis but not the degree of steatosis or iron deposition. Resolution of fibrosis requires either deactivation of HSCs or apoptosis of activated HSCs. Evidence supports apoptosis of these cells during fibrosis resolution. Increased HSC apoptosis was observed, for example, in the first 2 days after surgical reversal of bile duct obstruction in the initial stages of fibrosis resolution (Issa et al. 2001). Similarly, non-parenchymal apoptosis was increased during days 3–7 after the last carbon tetrachloride injection in a chemical model of fibrosis and resolution (Iredale et al. 1998). Thus, induction of HSC apoptosis is expected to have a beneficial effect on fibrosis. Indeed, inducing HSC apoptosis in bile duct-ligated mice reduced markers of activated HSCs (Anan et al. 2006a) and reduced the severity of injury (Anan et al. 2006b).

Toxic lipid mediators in NASH appear to indirectly activate HSCs and indirectly promote the survival and fibrogenic potential of these cells. HSC apoptosis would presumably have a beneficial effect on NASH fibrosis, but regrettably lipid mediators seem to act indirectly on HSCs to promote their activation. Apoptosis in this cell compartment may be a therapeutic goal.

1.7 Conclusions

Lipotoxicity in the liver involves induction of cell death (lipoapoptosis) in hepatocytes and potentially cholangiocytes in NASH (Fig. 1.2). Cell death in these epithelial cells may lead to slight elevations of serum markers of hepatobiliary injury but more importantly promote activation of fibrogenic HSCs and inflammatory Kupffer cells. Additionally, lipotoxicity to LSECs leads to capillarization, impaired LSEC function, increased vascular resistance, and ultimately cell death. Increased LSEC and epithelial cell death plus activation and increased survival of Kupffer and stellate cells mean that global attempts to prevent cytotoxicity in the liver may not yield therapeutic benefits. Instead, increasing death of Kupffer cells and HSCs while protecting hepatocytes and cholangiocytes may be a better, if more difficult, goal.

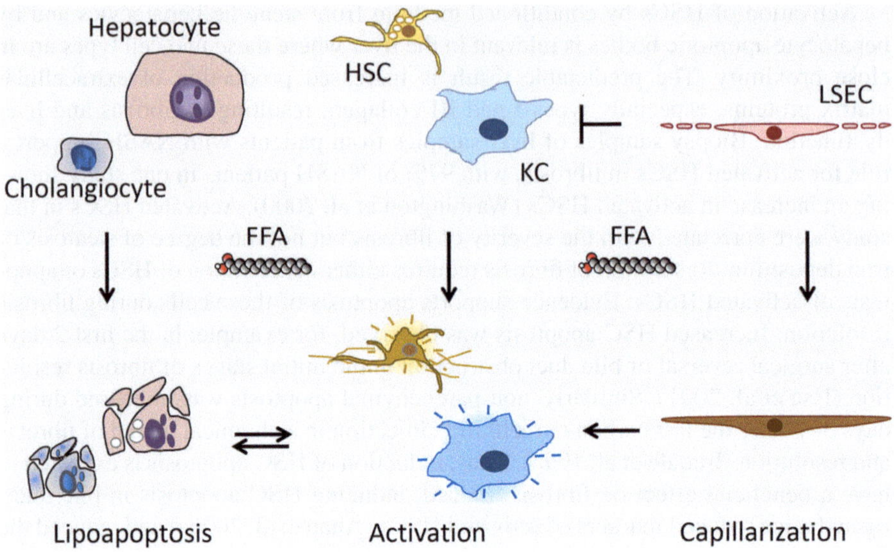

Fig. 1.2 Lipotoxicity in the liver. Healthy cells, hepatocyte, cholangiocyte, hepatic stellate cell (*HSC*), Kupffer cell (*KC*), and liver sinusoidal endothelial cell (*LSEC*) are shown at the top of the figure. Upon pathologic signaling, including by free fatty acids (*FFA*), the normal cell functions and interactions are disrupted. LSEC capillarization and injury as well as lipoapoptosis of hepatocytes and possibly cholangiocytes promote activation of KCs and HSCs. The combined effects include injury, inflammation, and fibrogenesis representing key features of nonalcoholic steatohepatitis

References

Adams LA, Lymp JF, St Sauver J, Sanderson SO, Lindor KD, Feldstein A et al (2005) The natural history of nonalcoholic fatty liver disease: a population-based cohort study. Gastroenterology 129(1):113–121

Ahn JH, Kim MH, Kwon HJ, Choi SY, Kwon HY (2013) Protective effects of oleic acid against Palmitic acid-induced apoptosis in pancreatic AR42J cells and its mechanisms. Korean J Physiol Pharmacol 17(1):43–50. doi:10.4196/kjpp.2013.17.1.43

Akazawa Y, Cazanave S, Mott JL, Elmi N, Bronk SF, Kohno S et al (2010) Palmitoleate attenuates palmitate-induced Bim and PUMA up-regulation and hepatocyte lipoapoptosis. J Hepatol 52(4):586–593. doi:10.1016/j.jhep.2010.01.003

Anan A, Baskin-Bey ES, Bronk SF, Werneburg NW, Shah VH, Gores GJ (2006a) Proteasome inhibition induces hepatic stellate cell apoptosis. Hepatology 43(2):335–344. doi:10.1002/hep.21036

Anan A, Baskin-Bey ES, Isomoto H, Mott JL, Bronk SF, Albrecht JH et al (2006b) Proteasome inhibition attenuates hepatic injury in the bile duct-ligated mouse. Am J Physiol Gastrointest Liver Physiol 291(4):G709–G716. doi:10.1152/ajpgi.00126.2006

Angulo P (2002) Nonalcoholic fatty liver disease. N Engl J Med 346(16):1221–1231. doi:10.1056/NEJMra011775

Angulo P, Kleiner DE, Dam-Larsen S, Adams LA, Bjornsson ES, Charatcharoenwitthaya P et al (2015) Liver fibrosis, but no other histologic features, is associated with long-term outcomes of patients with Nonalcoholic fatty liver disease. Gastroenterology 149(2):389–397.e310. doi:10.1053/j.gastro.2015.04.043

Atorrasagasti C, Peixoto E, Aquino JB, Kippes N, Malvicini M, Alaniz L et al (2013) Lack of the matricellular protein SPARC (secreted protein, acidic and rich in cysteine) attenuates liver fibrogenesis in mice. PLoS One 8(2):e54962. doi:10.1371/journal.pone.0054962

Bataller R, Brenner DA (2005) Liver fibrosis. J Clin Invest 115(2):209–218. doi:10.1172/JCI24282

Bradshaw AD, Reed MJ, Carbon JG, Pinney E, Brekken RA, Sage EH (2001) Increased fibrovascular invasion of subcutaneous polyvinyl alcohol sponges in SPARC-null mice. Wound Repair Regen 9(6):522–530

Braet F, Wisse E (2002) Structural and functional aspects of liver sinusoidal endothelial cell fenestrae: a review. Comp Hepatol 1(1):1

Brenner C, Galluzzi L, Kepp O, Kroemer G (2013) Decoding cell death signals in liver inflammation. J Hepatol 59(3):583–594. doi:10.1016/j.jhep.2013.03.033

Brunt EM (2005) Pathology of nonalcoholic steatohepatitis. Hepatol Res 33(2):68–71. doi:10.1016/j.hepres.2005.09.006

Brunt EM, Kleiner DE, Wilson LA, Belt P, Neuschwander-Tetri BA, Network NCR (2011) Nonalcoholic fatty liver disease (NAFLD) activity score and the histopathologic diagnosis in NAFLD: distinct clinicopathologic meanings. Hepatology 53(3):810–820. doi:10.1002/hep.24127

Canbay A, Taimr P, Torok N, Higuchi H, Friedman S, Gores GJ (2003) Apoptotic body engulfment by a human stellate cell line is profibrogenic. Lab Investig 83(5):655–663

Carambia A, Frenzel C, Bruns OT, Schwinge D, Reimer R, Hohenberg H et al (2013) Inhibition of inflammatory CD4 T cell activity by murine liver sinusoidal endothelial cells. J Hepatol 58(1):112–118. doi:10.1016/j.jhep.2012.09.008

Cardier JE, Schulte T, Kammer H, Kwak J, Cardier M (1999) Fas (CD95, APO-1) antigen expression and function in murine liver endothelial cells: implications for the regulation of apoptosis in liver endothelial cells. FASEB J 13(14):1950–1960

Cazanave SC, Mott JL, Bronk SF, Werneburg NW, Fingas CD, Meng XW et al (2011) Death receptor 5 signaling promotes hepatocyte lipoapoptosis. J Biol Chem 286(45):39336–39348. doi:10.1074/jbc.M111.280420

Cazanave SC, Wang X, Zhou H, Rahmani M, Grant S, Durrant DE et al (2014) Degradation of Keap1 activates BH3-only proteins Bim and PUMA during hepatocyte lipoapoptosis. Cell Death Differ 21(8):1303–1312. doi:10.1038/cdd.2014.49

Charlton MR, Burns JM, Pedersen RA, Watt KD, Heimbach JK, Dierkhising RA (2011) Frequency and outcomes of liver transplantation for nonalcoholic steatohepatitis in the United States. Gastroenterology 141(4):1249–1253. doi:10.1053/j.gastro.2011.06.061

Clouston AD, Powell EE, Walsh MJ, Richardson MM, Demetris AJ, Jonsson JR (2005) Fibrosis correlates with a ductular reaction in hepatitis C: roles of impaired replication, progenitor cells and steatosis. Hepatology 41(4):809–818. doi:10.1002/hep.20650

Cogger VC, Mohamad M, Solon-Biet S, Senior AM, Warren A, O'Reilly JN et al (2016) Dietary macronutrients and the aging liver sinusoidal endothelial cell. Am J Physiol Heart Circ Physiol. doi:10.1152/ajpheart.00949.2015

De Bleser PJ, Niki T, Rogiers V, Geerts A (1997) Transforming growth factor-beta gene expression in normal and fibrotic rat liver. J Hepatol 26(4):886–893

Deleve LD, Wang X, Guo Y (2008a) Sinusoidal endothelial cells prevent rat stellate cell activation and promote reversion to quiescence. Hepatology 48(3):920–930. doi:10.1002/hep.22351

DeLeve LD, Wang X, Kanel GC, Atkinson RD, McCuskey RS (2008b) Prevention of hepatic fibrosis in a murine model of metabolic syndrome with nonalcoholic steatohepatitis. Am J Pathol 173(4):993–1001. doi:10.2353/ajpath.2008.070720. ajpath.2008.070720 [pii]

Demir M, Lang S, Steffen HM (2015) Nonalcoholic fatty liver disease – current status and future directions. J Dig Dis 16(10):541–557. doi:10.1111/1751-2980.12291

Depraetere V, Golstein P (1997) Fas and other cell death signaling pathways. Semin Immunol 9(2):93–107. doi:10.1006/smim.1997.0062

Ekstedt M, Hagstrom H, Nasr P, Fredrikson M, Stal P, Kechagias S et al (2015) Fibrosis stage is the strongest predictor for disease-specific mortality in NAFLD after up to 33 years of follow-up. Hepatology 61(5):1547–1554. doi:10.1002/hep.27368

Falkowska-Hansen B, Falkowski M, Metharom P, Krunic D, Goerdt S (2007) Clathrin-coated vesicles form a unique net-like structure in liver sinusoidal endothelial cells by assembling along undisrupted microtubules. Exp Cell Res 313(9):1745–1757. doi:10.1016/j.yexcr.2007.02.026

Falkowski O, An HJ, Ianus IA, Chiriboga L, Yee H, West AB et al (2003) Regeneration of hepatocyte 'buds' in cirrhosis from intrabiliary stem cells. J Hepatol 39(3):357–364

Farrell GC, Teoh NC, McCuskey RS (2008) Hepatic microcirculation in fatty liver disease. Anat Rec (Hoboken) 291(6):684–692. doi:10.1002/ar.20715

Feldstein AE, Canbay A, Angulo P, Taniai M, Burgart LJ, Lindor KD et al (2003) Hepatocyte apoptosis and fas expression are prominent features of human nonalcoholic steatohepatitis. Gastroenterology 125(2):437–443

Fraser R, Bowler LM, Day WA, Dobbs B, Johnson HD, Lee D (1980) High perfusion pressure damages the sieving ability of sinusoidal endothelium in rat livers. Br J Exp Pathol 61(2):222–228

Fraser R, Clark SA, Day WA, Murray FE (1988) Nicotine decreases the porosity of the rat liver sieve: a possible mechanism for hypercholesterolaemia. Br J Exp Pathol 69(3):345–350

Frenzel H, Kremer B, Richter IE, Hucker H (1976) On the influence of hypoxia on the sinus endothelial cells of rat liver. A scanning and transmission electron microscopic investigation (author's transl). Virchows Arch B Cell Pathol 21(1):79–90

Friedman SL (2000) Molecular regulation of hepatic fibrosis, an integrated cellular response to tissue injury. J Biol Chem 275(4):2247–2250

Friedrich K, Rupp C, Hov JR, Steinebrunner N, Weiss KH, Stiehl A et al (2013) A frequent PNPLA3 variant is a sex specific disease modifier in PSC patients with bile duct stenosis. PLoS One 8(3):e58734. doi:10.1371/journal.pone.0058734

Gadd VL, Skoien R, Powell EE, Fagan KJ, Winterford C, Horsfall L et al (2014) The portal inflammatory infiltrate and ductular reaction in human nonalcoholic fatty liver disease. Hepatology 59(4):1393–1405. doi:10.1002/hep.26937

Gentile CL, Pagliassotti MJ (2008) The role of fatty acids in the development and progression of nonalcoholic fatty liver disease. J Nutr Biochem 19(9):567–576. doi:10.1016/j.jnutbio.2007.10.001

Goh GB, McCullough AJ (2016) Natural history of nonalcoholic fatty liver disease. Dig Dis Sci 61(5):1226–1233. doi:10.1007/s10620-016-4095-4

Hang TC, Lauffenburger DA, Griffith LG, Stolz DB (2012) Lipids promote survival, proliferation, and maintenance of differentiation of rat liver sinusoidal endothelial cells in vitro. Am J Physiol Gastrointest Liver Physiol 302(3):G375–G388. doi:10.1152/ajpgi.00288.2011

Hanigan MH, Frierson HF Jr (1996) Immunohistochemical detection of gamma-glutamyl transpeptidase in normal human tissue. J Histochem Cytochem 44(10):1101–1108

Hardonk MJ, Dijkhuis FW, Hulstaert CE, Koudstaal J (1992) Heterogeneity of rat liver and spleen macrophages in gadolinium chloride-induced elimination and repopulation. J Leukoc Biol 52(3):296–302

Hatting M, Zhao G, Schumacher F, Sellge G, Al Masaoudi M, Gabetaler N et al (2013) Hepatocyte caspase-8 is an essential modulator of steatohepatitis in rodents. Hepatology 57(6):2189–2201. doi:10.1002/hep.26271

Hendriks HF, Verhoofstad WA, Brouwer A, de Leeuw AM, Knook DL (1985) Perisinusoidal fat-storing cells are the main vitamin A storage sites in rat liver. Exp Cell Res 160(1):138–149

Hindi M, Levy C, Couto CA, Bejarano P, Mendes F (2013) Primary biliary cirrhosis is more severe in overweight patients. J Clin Gastroenterol 47(3):e28–e32. doi:10.1097/MCG.0b013e318261e659

Holt MP, Yin H, Ju C (2010) Exacerbation of acetaminophen-induced disturbances of liver sinusoidal endothelial cells in the absence of Kupffer cells in mice. Toxicol Lett 194(1–2):34–41. doi:10.1016/j.toxlet.2010.01.020

Huang W, Metlakunta A, Dedousis N, Zhang P, Sipula I, Dube JJ et al (2010) Depletion of liver Kupffer cells prevents the development of diet-induced hepatic steatosis and insulin resistance. Diabetes 59(2):347–357. doi:10.2337/db09-0016

Hutchins NA, Chung CS, Borgerding JN, Ayala CA, Ayala A (2013a) Kupffer cells protect liver sinusoidal endothelial cells from Fas-dependent apoptosis in sepsis by down-regulating gp130. Am J Pathol 182(3):742–754. doi:10.1016/j.ajpath.2012.11.023. S0002-9440(12)00889-9 [pii]

Hutchins NA, Wang F, Wang Y, Chung CS, Ayala A (2013b) Kupffer cells potentiate liver sinusoidal endothelial cell injury in sepsis by ligating programmed cell death ligand-1. J Leukoc Biol 94(5):963–970. doi:10.1189/jlb.0113051

Idrissova L, Malhi H, Werneburg NW, LeBrasseur NK, Bronk SF, Fingas C et al (2015) TRAIL receptor deletion in mice suppresses the inflammation of nutrient excess. J Hepatol 62(5):1156–1163. doi:10.1016/j.jhep.2014.11.033

Iredale JP, Benyon RC, Pickering J, McCullen M, Northrop M, Pawley S et al (1998) Mechanisms of spontaneous resolution of rat liver fibrosis. Hepatic stellate cell apoptosis and reduced hepatic expression of metalloproteinase inhibitors. J Clin Invest 102(3):538–549. doi:10.1172/JCI1018

Irie M, Suzuki N, Sohda T, Anan A, Iwata K, Takeyama Y et al (2007) Hepatic expression of gamma-glutamyltranspeptidase in the human liver of patients with alcoholic liver disease. Hepatol Res 37(11):966–973. doi:10.1111/j.1872-034X.2007.00151.x

Issa R, Williams E, Trim N, Kendall T, Arthur MJ, Reichen J et al (2001) Apoptosis of hepatic stellate cells: involvement in resolution of biliary fibrosis and regulation by soluble growth factors. Gut 48(4):548–557

Jacobs F, Wisse E, De Geest B (2009) Early effect of a single intravenous injection of ethanol on hepatic sinusoidal endothelial fenestrae in rabbits. Comp Hepatol 8:4. doi:10.1186/1476-5926-8-4

Jamieson HA, Hilmer SN, Cogger VC, Warren A, Cheluvappa R, Abernethy DR et al (2007) Caloric restriction reduces age-related pseudocapillarization of the hepatic sinusoid. Exp Gerontol 42(4):374–378. doi:10.1016/j.exger.2006.11.004. S0531-5565(06)00340-8 [pii]

Ji J, Zhang L, Wang P, Mu YM, Zhu XY, Wu YY et al (2005) Saturated free fatty acid, palmitic acid, induces apoptosis in fetal hepatocytes in culture. Exp Toxicol Pathol 56(6):369–376. doi:10.1016/j.etp.2005.02.003

Keir ME, Butte MJ, Freeman GJ, Sharpe AH (2008) PD-1 and its ligands in tolerance and immunity. Annu Rev Immunol 26:677–704. doi:10.1146/annurev.immunol.26.021607.090331

Kellendonk C, Opherk C, Anlag K, Schutz G, Tronche F (2000) Hepatocyte-specific expression of Cre recombinase. Genesis 26(2):151–153

Kelly KA, Allport JR, Yu AM, Sinh S, Sage EH, Gerszten RE et al (2007) SPARC is a VCAM-1 counter-ligand that mediates leukocyte transmigration. J Leukoc Biol 81(3):748–756. doi:10.1189/jlb.1105664

Kzhyshkowska J, Workman G, Cardo-Vila M, Arap W, Pasqualini R, Gratchev A et al (2006) Novel function of alternatively activated macrophages: stabilin-1-mediated clearance of SPARC. J Immunol 176(10):5825–5832. doi:176/10/5825 [pii]

Lanthier N, Molendi-Coste O, Cani PD, van Rooijen N, Horsmans Y, Leclercq IA (2011) Kupffer cell depletion prevents but has no therapeutic effect on metabolic and inflammatory changes induced by a high-fat diet. FASEB J 25(12):4301–4311. doi:10.1096/fj.11-189472

Lanthier N, Molendi-Coste O, Horsmans Y, van Rooijen N, Cani PD, Leclercq IA (2010) Kupffer cell activation is a causal factor for hepatic insulin resistance. Am J Physiol Gastrointest Liver Physiol 298(1):G107–G116. doi:10.1152/ajpgi.00391.2009

Larrubia JR, Benito-Martinez S, Miquel J, Calvino M, Sanz-de-Villalobos E, Parra-Cid T (2009) Costimulatory molecule programmed death-1 in the cytotoxic response during chronic hepatitis C. World J Gastroenterol 15(41):5129–5140

Le Couteur DG, Cogger VC, Markus AM, Harvey PJ, Yin ZL, Ansselin AD et al (2001) Pseudocapillarization and associated energy limitation in the aged rat liver. Hepatology 33(3):537–543. doi:10.1053/jhep.2001.22754. S0270-9139(01)85119-3 [pii]

Lee JS, Semela D, Iredale J, Shah VH (2007) Sinusoidal remodeling and angiogenesis: a new function for the liver-specific pericyte? Hepatology 45(3):817–825. doi:10.1002/hep.21564

Lefkowitch JH, Haythe JH, Regent N (2002) Kupffer cell aggregation and perivenular distribution in steatohepatitis. Mod Pathol 15(7):699–704. doi:10.1097/01.MP.0000019579.30842.96

Mak KM, Leo MA, Lieber CS (1984) Alcoholic liver injury in baboons: transformation of lipocytes to transitional cells. Gastroenterology 87(1):188–200

Malhi H, Bronk SF, Werneburg NW, Gores GJ (2006a) Free fatty acids induce JNK-dependent hepatocyte lipoapoptosis. J Biol Chem 281(17):12093–12101. doi:10.1074/jbc.M510660200

Malhi H, Gores GJ (2008) Molecular mechanisms of lipotoxicity in nonalcoholic fatty liver disease. Semin Liver Dis 28(4):360–369. doi:10.1055/s-0028-1091980

Malhi H, Gores GJ, Lemasters JJ (2006b) Apoptosis and necrosis in the liver: a tale of two deaths? Hepatology 43(2 Suppl 1):S31–S44. doi:10.1002/hep.21062

McCuskey RS, Ito Y, Robertson GR, McCuskey MK, Perry M, Farrell GC (2004) Hepatic microvascular dysfunction during evolution of dietary steatohepatitis in mice. Hepatology 40(2):386–393. doi:10.1002/hep.20302

McLean AJ, Cogger VC, Chong GC, Warren A, Markus AM, Dahlstrom JE et al (2003) Age-related pseudocapillarization of the human liver. J Pathol 200(1):112–117. doi:10.1002/path.1328

Mei S, Ni HM, Manley S, Bockus A, Kassel KM, Luyendyk JP et al (2011) Differential roles of unsaturated and saturated fatty acids on autophagy and apoptosis in hepatocytes. J Pharmacol Exp Ther 339(2):487–498. doi:10.1124/jpet.111.184341

Munshi MK, Priester S, Gaudio E, Yang F, Alpini G, Mancinelli R et al (2011) Regulation of biliary proliferation by neuroendocrine factors: implications for the pathogenesis of cholestatic liver diseases. Am J Pathol 178(2):472–484. doi:10.1016/j.ajpath.2010.09.043

Muschen M, Warskulat U, Douillard P, Gilbert E, Haussinger D (1998) Regulation of CD95 (APO-1/Fas) receptor and ligand expression by lipopolysaccharide and dexamethasone in parenchymal and nonparenchymal rat liver cells. Hepatology 27(1):200–208. doi:10.1002/hep.510270131

Natarajan SK, Ingham SA, Mohr AM, Wehrkamp CJ, Ray A, Roy S et al (2014) Saturated free fatty acids induce cholangiocyte lipoapoptosis. Hepatology 60(6):1942–1956. doi:10.1002/hep.27175

Nehra V, Angulo P, Buchman AL, Lindor KD (2001) Nutritional and metabolic considerations in the etiology of nonalcoholic steatohepatitis. Dig Dis Sci 46(11):2347–2352

Neuschwander-Tetri BA, Clark JM, Bass NM, Van Natta ML, Unalp-Arida A, Tonascia J et al (2010) Clinical, laboratory and histological associations in adults with nonalcoholic fatty liver disease. Hepatology 52(3):913–924. doi:10.1002/hep.23784

Neyrinck AM, Cani PD, Dewulf EM, De Backer F, Bindels LB, Delzenne NM (2009) Critical role of Kupffer cells in the management of diet-induced diabetes and obesity. Biochem Biophys Res Commun 385(3):351–356. doi:10.1016/j.bbrc.2009.05.070

Nopanitaya W, Lamb JC, Grisham JW, Carson JL (1976) Effect of hepatic venous outflow obstruction on pores and fenestration in sinusoidal endothelium. Br J Exp Pathol 57(5):604–609

O'Reilly JN, Cogger VC, Fraser R, Le Couteur DG (2010) The effect of feeding and fasting on fenestrations in the liver sinusoidal endothelial cell. Pathology 42(3):255–258. doi:10.3109/00313021003636469

Okazaki T, Maeda A, Nishimura H, Kurosaki T, Honjo T (2001) PD-1 immunoreceptor inhibits B cell receptor-mediated signaling by recruiting src homology 2-domain-containing tyrosine phosphatase 2 to phosphotyrosine. Proc Natl Acad Sci U S A 98(24):13866–13871. doi:10.1073/pnas.231486598

Park SY, Kim SY, Jung MY, Bae DJ, Kim IS (2008) Epidermal growth factor-like domain repeat of stabilin-2 recognizes phosphatidylserine during cell corpse clearance. Mol Cell Biol 28(17):5288–5298. doi:10.1128/MCB.01993-07. MCB.01993-07 [pii]

Pasarin M, La Mura V, Gracia-Sancho J, Garcia-Caldero H, Rodriguez-Vilarrupla A, Garcia-Pagan JC et al (2012) Sinusoidal endothelial dysfunction precedes inflammation and fibrosis in a model of NAFLD. PLoS One 7(4):e32785. doi:10.1371/journal.pone.0032785

Peixoto E, Atorrasagasti C, Aquino JB, Militello R, Bayo J, Fiore E et al (2015) SPARC (secreted protein acidic and rich in cysteine) knockdown protects mice from acute liver injury by reducing vascular endothelial cell damage. Gene Ther 22(1):9–19. doi:10.1038/gt.2014.102

Perry IJ, Wannamethee SG, Shaper AG (1998) Prospective study of serum gamma-glutamyltransferase and risk of NIDDM. Diabetes Care 21(5):732–737

Postic C, Shiota M, Niswender KD, Jetton TL, Chen Y, Moates JM et al (1999) Dual roles for glucokinase in glucose homeostasis as determined by liver and pancreatic beta cell-specific gene knock-outs using Cre recombinase. J Biol Chem 274(1):305–315

Rantala AO, Lilja M, Kauma H, Savolainen MJ, Reunanen A, Kesaniemi YA (2000) Gamma-glutamyl transpeptidase and the metabolic syndrome. J Intern Med 248(3):230–238

Richardson MM, Jonsson JR, Powell EE, Brunt EM, Neuschwander-Tetri BA, Bhathal PS et al (2007) Progressive fibrosis in nonalcoholic steatohepatitis: association with altered regeneration and a ductular reaction. Gastroenterology 133(1):80–90. doi:10.1053/j.gastro.2007.05.012

Rivera CA, Adegboyega P, van Rooijen N, Tagalicud A, Allman M, Wallace M (2007) Toll-like receptor-4 signaling and Kupffer cells play pivotal roles in the pathogenesis of non-alcoholic steatohepatitis. J Hepatol 47(4):571–579. doi:10.1016/j.jhep.2007.04.019

Romeo S, Kozlitina J, Xing C, Pertsemlidis A, Cox D, Pennacchio LA et al (2008) Genetic variation in PNPLA3 confers susceptibility to nonalcoholic fatty liver disease. Nat Genet 40(12):1461–1465. doi:10.1038/ng.257

Roskams T, Yang SQ, Koteish A, Durnez A, DeVos R, Huang X et al (2003) Oxidative stress and oval cell accumulation in mice and humans with alcoholic and nonalcoholic fatty liver disease. Am J Pathol 163(4):1301–1311. doi:10.1016/S0002-9440(10)63489-X

Sarphie G, D'Souza NB, Van Thiel DH, Hill D, McClain CJ, Deaciuc IV (1997) Dose- and time-dependent effects of ethanol on functional and structural aspects of the liver sinusoid in the mouse. Alcohol Clin Exp Res 21(6):1128–1136

Schurich A, Berg M, Stabenow D, Bottcher J, Kern M, Schild HJ et al (2010) Dynamic regulation of CD8 T cell tolerance induction by liver sinusoidal endothelial cells. J Immunol 184(8):4107–4114. doi:10.4049/jimmunol.0902580

Schwimmer JB, Behling C, Newbury R, Deutsch R, Nievergelt C, Schork NJ et al (2005) Histopathology of pediatric nonalcoholic fatty liver disease. Hepatology 42(3):641–649. doi:10.1002/hep.20842

She H, Xiong S, Hazra S, Tsukamoto H (2005) Adipogenic transcriptional regulation of hepatic stellate cells. J Biol Chem 280(6):4959–4967. doi:10.1074/jbc.M410078200

Sheppard KA, Fitz LJ, Lee JM, Benander C, George JA, Wooters J et al (2004) PD-1 inhibits T-cell receptor induced phosphorylation of the ZAP70/CD3zeta signalosome and downstream signaling to PKCtheta. FEBS Lett 574(1–3):37–41. doi:10.1016/j.febslet.2004.07.083

Shiratori Y, Tananka M, Kawase T, Shiina S, Komatsu Y, Omata M (1993) Quantification of sinusoidal cell function in vivo. Semin Liver Dis 13(1):39–49. doi:10.1055/s-2007-1007336

Sorensen KK, Simon-Santamaria J, McCuskey RS, Smedsrod B (2015) Liver sinusoidal endothelial cells. Compr Physiol 5(4):1751–1774. doi:10.1002/cphy.c140078

Sorrentino P, Tarantino G, Perrella A, Micheli P, Perrella O, Conca P (2005) A clinical-morphological study on cholestatic presentation of nonalcoholic fatty liver disease. Dig Dis Sci 50(6):1130–1135

Stienstra R, Saudale F, Duval C, Keshtkar S, Groener JE, van Rooijen N et al (2010) Kupffer cells promote hepatic steatosis via interleukin-1beta-dependent suppression of peroxisome proliferator-activated receptor alpha activity. Hepatology 51(2):511–522. doi:10.1002/hep.23337

Tahan V, Canbakan B, Balci H, Dane F, Akin H, Can G et al (2008) Serum gamma-glutamyltranspeptidase distinguishes non-alcoholic fatty liver disease at high risk. Hepato-Gastroenterology 55(85):1433–1438

Taira K (1994) Trabecular meshworks in the sinusoidal endothelial cells of the golden hamster liver: a freeze-fracture study. J Submicrosc Cytol Pathol 26(2):271–277

Takahashi Y, Inui A, Fujisawa T, Takikawa H, Fukusato T (2011) Histopathological characteristics of non-alcoholic fatty liver disease in children: comparison with adult cases. Hepatol Res 41(11):1066–1074. doi:10.1111/j.1872-034X.2011.00855.x

Tchorz JS, Kinter J, Muller M, Tornillo L, Heim MH, Bettler B (2009) Notch2 signaling promotes biliary epithelial cell fate specification and tubulogenesis during bile duct development in mice. Hepatology 50(3):871–879. doi:10.1002/hep.23048

Thabut D, Shah V (2010) Intrahepatic angiogenesis and sinusoidal remodeling in chronic liver disease: new targets for the treatment of portal hypertension? J Hepatol 53(5):976–980. doi:10.1016/j.jhep.2010.07.004

Van Rooijen N, Sanders A (1994) Liposome mediated depletion of macrophages: mechanism of action, preparation of liposomes and applications. J Immunol Methods 174(1–2): 83–93

Volpes R, van den Oord JJ, De Vos R, Desmet VJ (1992) Hepatic expression of type a and type B receptors for tumor necrosis factor. J Hepatol 14(2–3):361–369

Wajant H (2002) The Fas signaling pathway: more than a paradigm. Science 296(5573):1635–1636. doi:10.1126/science.1071553

Wang BJ, Bao JJ, Wang JZ, Wang Y, Jiang M, Xing MY et al (2011) Immunostaining of PD-1/PD-Ls in liver tissues of patients with hepatitis and hepatocellular carcinoma. World J Gastroenterol 17(28):3322–3329. doi:10.3748/wjg.v17.i28.3322

Warren A, Bertolino P, Cogger VC, McLean AJ, Fraser R, Le Couteur DG (2005) Hepatic pseudo-capillarization in aged mice. Exp Gerontol 40(10):807–812. doi:10.1016/j.exger.2005.06.012. S0531-5565(05)00147-6 [pii]

Washington K, Wright K, Shyr Y, Hunter EB, Olson S, Raiford DS (2000) Hepatic stellate cell activation in nonalcoholic steatohepatitis and fatty liver. Hum Pathol 31(7):822–828. doi:10.1053/hupa.2000.8440

Weigel PH, Pandey SM, Harris EN (eds) (2012) A HARE/Stab2 mediated sensing system may monitor tissue biomatrix homeostasis and stress, Structure and function of biomatrix: control of cell behavior and Gene expression, vol 5, 1st edn. Matrix Biology Institute, Edgewater

Welters HJ, Tadayyon M, Scarpello JH, Smith SA, Morgan NG (2004) Mono-unsaturated fatty acids protect against beta-cell apoptosis induced by saturated fatty acids, serum withdrawal or cytokine exposure. FEBS Lett 560(1–3):103–108. doi:10.1016/S0014-5793(04)00079-1

Wieckowska A, Zein NN, Yerian LM, Lopez AR, McCullough AJ, Feldstein AE (2006) In vivo assessment of liver cell apoptosis as a novel biomarker of disease severity in nonalcoholic fatty liver disease. Hepatology 44(1):27–33. doi:10.1002/hep.21223

Wisse E (1970) An electron microscopic study of the fenestrated endothelial lining of rat liver sinusoids. J Ultrastruct Res 31(1):125–150

Wisse E, De Zanger RB, Charels K, Van Der Smissen P, McCuskey RS (1985) The liver sieve: considerations concerning the structure and function of endothelial fenestrae, the sinusoidal wall and the space of Disse. Hepatology 5(4):683–692

Wobser H, Dorn C, Weiss TS, Amann T, Bollheimer C, Buttner R et al (2009) Lipid accumulation in hepatocytes induces fibrogenic activation of hepatic stellate cells. Cell Res 19(8):996–1005. doi:10.1038/cr.2009.73

Wong RJ, Aguilar M, Cheung R, Perumpail RB, Harrison SA, Younossi ZM et al (2015) Nonalcoholic steatohepatitis is the second leading etiology of liver disease among adults awaiting liver transplantation in the United States. Gastroenterology 148(3):547–555. doi:10.1053/j.gastro.2014.11.039

Xie G, Choi SS, Syn WK, Michelotti GA, Swiderska M, Karaca G et al (2013) Hedgehog signalling regulates liver sinusoidal endothelial cell capillarisation. Gut 62(2):299–309. doi:10.1136/gutjnl-2011-301494. gutjnl-2011-301494 [pii]

Xie G, Wang X, Wang L, Atkinson RD, Kanel GC, Gaarde WA et al (2012) Role of differentiation of liver sinusoidal endothelial cells in progression and regression of hepatic fibrosis in rats. Gastroenterology 142(4):918–927.e916. doi:10.1053/j.gastro.2011.12.017. S0016-5085(11)01699-4 [pii]

Xie Z, Chen Y, Zhao S, Yang Z, Yao X, Guo S et al (2009) Intrahepatic PD-1/PD-L1 up-regulation closely correlates with inflammation and virus replication in patients with chronic HBV infection. Immunol Investig 38(7):624–638
Xing Y, Zhao T, Gao X, Wu Y (2016) Liver X receptor alpha is essential for the capillarization of liver sinusoidal endothelial cells in liver injury. Sci Rep 6:21309. doi:10.1038/srep21309
Xu WH, Ye QF, Xia SS (2004) Apoptosis and proliferation of intrahepatic bile duct after ischemia-reperfusion injury. Hepatobiliary Pancreat Dis Int 3(3):428–432

Chapter 2
The Role of Cholangiocyte Cell Death in the Development of Biliary Diseases

April O'Brien, Chad Hall, Laurent Ehrlich, Tianhao Zhou, Fanyin Meng, Gianfranco Alpini, and Shannon S. Glaser

Abbreviations

ADPKD	Autosomal dominant polycystic kidney disease
AEA	Anandamide
AKT	Protein kinase B
AMA	Antimitochondrial antibodies
BA	Biliary atresia
BDL	Bile duct ligation
CBDL	Common bile duct ligation
CCA	Cholangiocarcinoma
CD	Cluster of differentiation
DEN	Diethylnitrosamine
FADD	Fas-associated death domain
FFA	Free fatty acid
HSC	Hepatic stellate cells
IFN-γ	Interferon- γ
IL	Interleukin
JNK	c-Jun N-terminal kinase
LMBDL	Left median bile duct ligation

This material is the result of work supported by resources at the Central Texas Veterans Health Care System. The content is the responsibility of the author(s) alone and does not necessarily reflect the views or policies of the Department of Veterans Affairs or the United States Government.

A. O'Brien • C. Hall • L. Ehrlich • T. Zhou • F. Meng • G. Alpini • S.S. Glaser (✉)
Department of Internal Medicine, College of Medicine | Texas A&M Health Science Center, Central Texas Veterans Health Care System, VA Bld 205 | 1901 S. 1st Street, Temple, TX 76504, USA
e-mail: SGlaser@medicine.tamhsc.edu

W.-X. Ding, X.-M. Yin (eds.), *Molecules, Systems and Signaling in Liver Injury*, Cell Death in Biology and Diseases, DOI 10.1007/978-3-319-58106-4_2

X.-M. Yin and Z. Dong (Series eds.), *Cell Death in Biology and Diseases*

NAFLD	Nonalcoholic fatty liver disease
NF-kB	Nuclear factor kappa-light-chain-enhancer of activated B cells
NK	Natural killer
NOD	Nucleotide-binding oligomerization domain
PAMPs	Pathogen-associated molecular patterns
pANCA	Peri-neutrophil cytoplasmic antibody
PBC	Primary biliary cholangitis
PKD	Polycystic kidney disease
PNPLA3	Patatin-like phospholipase domain-containing protein 3
PRR	Pattern-recognition receptors
PSC	Primary sclerosing cholangitis
PUMA	p53 upregulated modulator of apoptosis
STAT3	Signal transducer and activator of transcription 3
TLRs	Toll-like receptors
TNF	Tumor necrosis factor
TRAIL	Tumor necrosis factor-related apoptosis-inducing ligand
UDCA	Ursodeoxycholic acid

2.1 Introduction

The liver is comprised primarily of two types of epithelia: hepatocytes and cholangiocytes. Hepatocytes make up approximately 70% of the endogenous population, whereas cholangiocytes constitute only 3–5% (Glaser et al. 2009). Cholangiocytes are heterogeneous, varying in size, shape, and function, and are generally categorized as either small or large based upon both their morphology and functionality. While both kinds of cholangiocytes line the biliary tree, large cholangiocytes are more columnar in shape and line larger bile ducts, whereas small cholangiocytes tend to be cuboidal and are found in smaller bile ducts (Maroni et al. 2015). In addition, small cholangiocytes are less specialized with a higher nucleus/cytoplasm ratio. Large cholangiocytes have a smaller nucleus/cytoplasm ratio and are ciliated, allowing them to act as both chemo- and mechanosensors (Maroni et al. 2015). Together, their primary role is to secrete, modify, and transport the bile into the duodenum. Cholangiocytes also facilitate repair following hepatic injury, sense, and respond to inflammatory signals and are the target of cholangiopathies. Effective targeted therapies for this subset of biliary diseases are lacking; therefore, it is critical to establish a thorough understanding of the biological pathways and mechanisms of cholangiocyte death that may play a role in the pathogenesis of cholestatic liver diseases.

2.2 Cholangiopathies

Cholangiopathies are diseases of the biliary epithelium and commonly include primary sclerosing cholangitis (PSC), primary biliary cholangitis (PBC), biliary atresia, polycystic liver disease, and cholangiocarcinoma (Lazaridis and LaRusso 2015).

Cholangiopathies can be categorized, but are not limited to the following groups: autoimmune, drug- or toxin-induced, ischemic, idiopathic, infectious, and genetic (O'Hara et al. 2013). The five major cholangiopathies – PSC, PBC, autosomal dominant polycystic kidney disease (ADPKD), biliary atresia, and cholangiocarcinoma – affect both genders and all ages worldwide.

2.3 Primary Sclerosing Cholangitis

PSC is a chronic, progressive disease that results in periductular inflammation and fibrosis that progresses to end-stage liver diseases (Primary Biliary Cirrhosis 2016). Unlike other cholestatic liver diseases, PSC affects both the small and large bile ducts with 20% of small-duct PSC presenting patients progressing to large-duct PSC (Primary Biliary Cirrhosis 2016). Additionally, the majority of PSC patients will also have inflammatory bowel disease (Trivedi and Chapman 2012). PSC predominantly strikes males (66%) more than females and ranges from adolescence to advanced ages with the average age of a PSC patient being 41 years at diagnosis (Primary Biliary Cirrhosis 2016). Currently there are no effective treatments that halt disease progression, and liver transplantation is often necessary (Hirschfield et al. 2013; Karlsen et al. 2010). Complicating the need for liver transplantation is the reoccurrence of the disease, the predominance toward hepatobiliary cancers and secondary autoimmune disorders.

2.3.1 Primary Biliary Cholangitis

PBC is a chronic inflammatory cholestatic disease of autoimmune origins that affects the small interlobular and septal bile ducts. Without treatment, PBC will progress to cirrhosis and end-stage liver failure (Corrigan and Hirschfield; Czul and Levy 2016). It presents more frequently in women than men, and individuals with a diagnosis before 50 years of age have a 50% chance that treatment options will fail (Corrigan and Hirschfield). Approximately 30% of PBC patients will also be diagnosed with additional autoimmune disorders, the most common being Sjögren's syndrome (17.4%) and Raynaud's syndrome (12.5%) (Primary Biliary Cirrhosis 2016). Currently the treatment of choice is ursodeoxycholic acid (UDCA) which aids in slowing the disease progression and delays transplantation necessity or death (Qian and Dinghang 2014). Due to the pruritus that accompanies PBC, antihistamines will often be prescribed to improve the patient's quality of life. In addition, anti-inflammatories may help reduce the ductal inflammation allowing for better bile flow (Primary Biliary Cirrhosis 2016).

2.3.2 Autosomal Dominant Polycystic Kidney Disease

ADPKD is an inherited renal disorder characterized by the development and growth of cysts in the kidneys, liver, pancreas, and spleen (Ali et al. 2015; Srivastava and Patel 2014; Tong et al. 2015). ADPKD is the most common form of genetic renal failure in adults. ADPKD can result from two types of genetic mutations: ADPKD-1 sufferers carry a genetic mutation on the gene-encoding protein polycystin-1 found on chromosome 16, while ADPKD-2 is caused by a gene mutation on chromosome 4 that encodes the protein polycystin-2 (Erickson et al. 2008; Tong et al. 2015). ADPKD-1 mutations tend to be more severe with patients entering end-stage renal disease at a younger median age. It also accounts for 85% of clinical cases overall (Ali et al. 2015; Srivastava and Patel 2014). Both mutations result in the development of cysts, resembling grape clusters and primarily filled with serous fluid, blood, and urine. These cysts eventually compromise normal renal function through crowding and compression of the renal tubes (Srivastava and Patel 2014). Within the liver, cysts arise from cholangiocytes and ultimately replace healthy liver tissue to a point where they may compress adjacent organs, leading to discomfort and anorexia (Gordon 2015). Consequently many patients remain asymptomatic until the sheer size and number of cysts necessitate medical intervention, generally not until adulthood. Treatment options include cyst fenestration, partial hepatectomy, or liver transplantation (Drenth 2010).

2.3.3 Biliary Atresia

Biliary atresia is a congenital disorder that affects more females than males and predominantly children of Asian or African American descent. There are two types of biliary atresia: fetal and perinatal. Of the two, fetal biliary atresia poses a greater threat to the infant as it is often accompanied by defects of the heart, spleen, or intestines. (Hartley et al. 2009). Models of biliary atresia and livers from affected children express increased levels of proapoptosis molecules such as caspase 1 and 4 and TNFα (Petersen and Davenport 2013). While there are no known hereditary components, several theories have been proposed to explain this phenomenon, including viral or bacterial infections, autoimmunity, or dysregulated prenatal liver and bile duct development. A Kasai procedure is often required to restore enteric drainage of bile; however, many of these patients will go on to require hepatic transplantation or die from complications of the disease (Hartley et al. 2009).

2.3.4 *Cholangiocarcinoma*

Cholangiocarcinoma is a rare malignancy of the biliary epithelium that commonly presents late in the disease progression. The malignancy is difficult to treat due to the advanced stage at presentation and is often fatal with an overall 5-year survival rate of less than 15% (Ciombor and Goff 2013). Cholangiocarcinomas are categorized by the tumor's origins, either arising from the bile ducts lying within the liver (intrahepatic CCA) or outside the liver (extrahepatic CCA) (Keller and Schub 2016). Difficulties in the treatment of cholangiocarcinoma stem from the fact that it is most often diagnosed at an advanced stage limiting available treatment options. To date, the most palliative option is liver resection, which carries additional risks of bile leakage, abdominal abscesses, and liver failure (Ciombor and Goff 2013).

2.4 Animal Models

Currently there are many commercially available live animal models of cholestatic liver disease that include mechanical, chemical, genetically modified, and parasitic etiologies (Primary Biliary Cirrhosis 2016). Bile duct ligation (BDL) (mechanical), Mdr2$^{-/-}$ mice (genetic modification), IL-12p35$^{-/-}$ (genetic modification), and dnTGFβRII (genetic modification) will be discussed in further detail. BDL is a well-studied mechanical representation of liver fibrosis used to model cholestatic liver disease. This is achieved through surgical ligation of the common bile duct, which impedes the flow of bile, inducing a strong fibrotic response after 21–28 days (Tag et al. 2015). Recent years have seen the technical advancements of partial BDL or complete BDL with surgical reanastomosis. Bile is a necessary component of the digestive system to emulsify fats and digest lipids while excreting bilirubin and additional wastes. Blockage of the common bile duct causes toxic buildup of waste products resulting in inflammatory cascades and upregulation of fibrotic gene expression. This model has the added benefit of being highly reproducible.

Another reliable animal model of fibrosis is the generation of Mdr2 (Abcb4)$^{-/-}$ mice, which, as a result of a canalicular phospholipid flippase deficiency, spontaneously develop liver injury and present with features of PSC. The resultant cascade of events includes bile leakage into the portal tract due to a disruption of both the tight junctions and basement membranes of the biliary epithelium (Popov et al. 2005). It is also thought that the presence of these phospholipids emulsifies certain molecules such as bile acids, reducing their overall toxicity.

PBC, like biliary atresia, is notoriously difficult to mimic in its pathogenesis. Previous models demonstrate the development of the hallmark antimitochondrial antibodies (AMA) yet lack in their replication of the trademark fibrosis or cirrhosis, the underlying cause for liver transplantation in PBC patients (Popov 2013). Recently, a double-mutant mouse generated by crossing two PBC mouse models, dnTGFβRII and IL-12p35$^{-/-}$, to generate an IL-12p35$^{-/-}$; dnTGFβRII strain, features both circulating AMA and periportal fibrosis (Popov 2013). This advancement represents an outstanding opportunity for researchers to test novel therapies and targets in order to gain insight into this devastating disease.

Two rat models with ADPKD are utilized: one where the animals exhibit spontaneous manifestations of polycystic kidney disease (PKD) through heritable traits and the other where genetic modifications mimic the mutation of human orthologous genes (Erickson et al. 2008). In the spontaneous hereditary model Han:SPRD-Cy, rats form renal cysts as a result of a missense mutation. This mutation shares the human PKD phenotype through non-orthologous genes. Homozygous rats will have initial cysts discoverable in the neonate stage, whereas the heterozygotes develop the disease at a much slower rate. PKD rats carry a mutation orthologous to the ADPKD gene in humans, and these animals express this gene mutation in the kidney, liver, and pancreas (Erickson et al. 2008). Both the PKD and Cy/+ strains have an approximate life span of 1.5 years.

Historically, several animal models have been developed to mimic biliary atresia, from lampreys and zebra fish to rhesus monkeys and mice. Lampreys have been found to naturally develop degeneration of the biliary system. Although it has proven to be useful for the study of bile duct transport, it lacks the disease pathology of human biliary atresia. Zebra fish are advantageous for studying biliary atresia because of their rapidly developing gastrointestinal system and clear elucidation of biliary development. Rhesus monkeys provided the first evidence that viral infections play a role in the etiology of biliary atresia when infant monkeys with biliary pathology tested positive for reovirus type 3. Hepatic viral-induced mice have provided further details into the pathogenic pathways, but to date no individual animal model has been discovered that truly mimics the disease pathology of biliary atresia in human infants (Petersen 2012).

Cholangiocarcinomas arise from chronic cholestasis, which ultimately contributes to the overall hepatocellular injury. One way to effectively model this pathogenesis is through a combination of left median bile duct ligation (LMBDL) and diethylnitrosamine (DEN). Due to the catastrophic insult suffered by the liver after common bile duct ligation (CBDL), many animals display high mortality with severe liver damage or greater survival rates associated with a significant decrease in liver damage (Yang et al. 2011). Intraperitoneally injected diethylnitrosamine, a known carcinogen, initiates liver cancer formation, and when coupled with the LMBDL procedure, a model of cholestasis results in a working model of human cholangiocarcinoma which effectively mimics the disease progression and chronic cholestasis and fibrosis (Yang et al. 2011).

2.5 Cell Death

2.5.1 *Innate Immunity in Cholangiocytes*

The liver's unique histological structure provides the basis for its role as the first line of defense against xenobiotics and infection (Ishibashi et al. 2009). The functional unit of the liver is called the hepatic lobule consisting of a hexagonal-like sheet of cells surrounding a central vein with portal triads at the periphery. Blood rich in metabolites, toxins, hormones, and immune cells returning from splanchnic and mesenteric circulation enters the liver from the portal vein and flows into the sinusoids lined by fenestrated endothelial cells. Here hepatocytes, along with Kupffer cells, hepatic stellate cells (HSCs), endothelial cells, and various lymphocytes, coordinate immune surveillance and detoxify as the blood flows to the central vein. Additionally, hepatocytes and cholangiocytes secrete and modify the bile into bile canaliculi, which flows the opposite direction from the central vein to the portal triad and is drained by the biliary tree into the gallbladder and duodenum (Gao et al. 2008; Turcotte and Jarnagin 2015).

Cholangiocytes play a major role in hepatobiliary innate immunity. As opposed to the fenestrated endothelial cells lining the sinusoid, cholangiocytes maintain an epithelial barrier through regulation of tight junction proteins. Cholangiocytes secrete antimicrobial peptides such as defensins and secretory IgA as well as mucosal layer that protects the epithelial surface (Chen et al. 2008). Cholangiocytes also express primary cilia that extend into the biliary lumen whose function is both mechanical and sensory. Defects in primary cilia are associated with ADPKD and BA (Chu et al. 2012). Furthermore, cholangiocytes express pattern-recognition receptors (PRR) that recognize pathogen-associated molecular patterns (PAMPs) to initiate a localized, nonspecific inflammatory response (Fig. 2.1). Examples of PRR include surface Toll-like receptors (TLRs) and cytoplasmic NOD and RIG receptors (Gao et al. 2008; Turcotte and Jarnagin 2015).

PAMP recognition by these receptors can lead to programmed gene expression changes and coordinated defense against pathogens through secretion of inflammatory cytokines (TNF-α, IFN-γ, IL-1, IL-6, and TGF-β) and recruitment of immune cells (Kupffer cells, neutrophils, and various lymphocytes) (Gao et al. 2008; Turcotte and Jarnagin 2015). Cholangiocytes also respond in various ways to cytokines secreted by other cells including, but not limited to, changes in proliferation, apoptosis, cytotoxicity, and expression of surface adhesion ligands (Chen et al. 2008). It is widely believed that adhesion ligands can facilitate direct cell-to-cell cytotoxicity between cholangiocytes and immune cells.

Natural killer (NK) cells are a major player in innate immunity and are implicated in cholangiopathies (Gao et al. 2008). Recent evidence revealed that cholangiocytes express surface CD1d ligands that recruit NKT cells (a subtype of NK cells) and initiate cell death (Schrumpf et al. 2015). NK cells are crucial in fighting viruses, intracellular pathogens, and tumors because a) they do not express antigen receptors like T and B cells and b) they target cells with decreased expression of

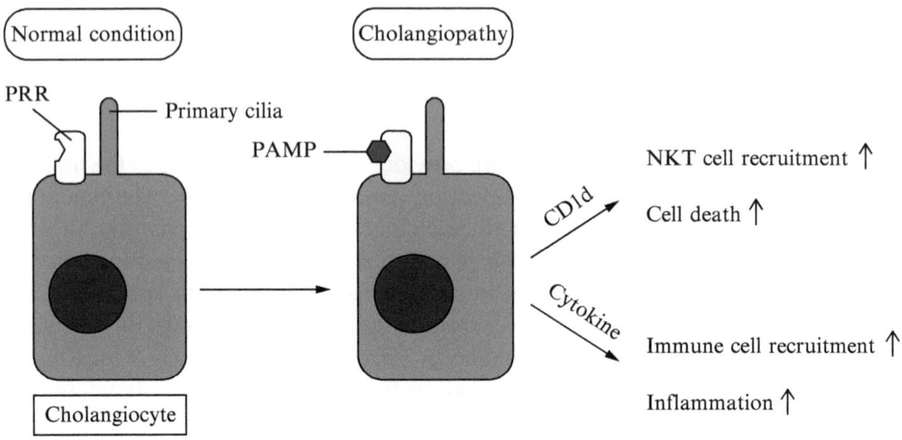

Fig. 2.1 Cholangiocyte innate immunity in cholangiopathies. Cholangiocytes express primary cilia and pattern-recognition receptors (*PRR*) as part of the innate immune system. Defects in primary cilia or uncontrolled activation of PRR can lead to cholangiopathies

MHC class I molecules (presumably due to evasive actions of the disease state) (Vivier et al. 2008). NK cells can also have a protective effect through clearing of profibrotic, proinflammatory cells. NK cells target activated hepatic stellate cells (HSCs) for cell death, but not quiescent HSCs (Gao et al. 2008), and are responsible for clearing self-recognizing T cells and dendritic cells (Tian et al. 2013).

Cholangiocytes also play a role in maintaining homeostasis between innate and adaptive immunity. Both cytokine secretion and cognate interaction via MHC class II molecules can drive activation and differentiation of various T and B lymphocytes. Dysregulation of this balance can lead to cholangiopathies. PBC is characterized by the presence of antimitochondrial autoantibodies (AMA) that drive sustained TLR signaling, increased proinflammatory Th1 and Th17 and decreased Treg phenotypes, and increased cholangiocyte cytotoxicity and HSC activation (Syal et al. 2012; Wang et al. 2016). PSC has a similar pathophysiology however with a few key differences. For instance, p-ANCA autoantibodies predominate instead of AMA, and cholangiocyte senescence is more prolonged (Nakanuma et al. 2015).

2.5.2 Apoptosis in Cholangiocytes

Cholangiopathies are rare disorders with no common genetic defects, yet chronic inflammation seems to be a universal contributing factor. Apoptosis is an important regulatory mechanism to keep cellular transformation in check through the removal of damaged cells and reduction in malignant progression (Humphreys et al. 2010). Apoptosis also serves to maintain homeostasis and promote liver tissue regeneration as apoptotic bodies release growth signals that in turn stimulate the progenitor cells to initiate proliferation (Guicciardi et al. 2013). Apoptosis is the process by which a cell exposes a phosphatidylserine on the outer leaflet while condensing its chromatin and fragmenting its DNA. As a result, the cell shrinks and blebs and is reduced to smaller membrane-enclosed organelles called apoptotic bodies which are then marked for phagocytosis (Guicciardi et al. 2013). This fragmentation and signaling cascade, through a family of caspases, act upon various molecules to regulate normal cellular turnover and maintenance during times of stress.

Cholangiocytes undergo apoptosis through a variety of molecular pathways. Common pathways include Fas (CD95), TRAIL-R1/2, TNFR1/2, and CD40 (Fig. 2.2). In addition to the aforementioned receptors, there is evidence that autophagy and lipoapoptosis are also key players in the apoptotic regulatory system. Fas (CD95) receptors, a member of the TNF receptor superfamily, are widely expressed in the liver at low levels, which dramatically increase during inflammation, via its ligand FasL (CD178). In PBC, there is cross-linking of the ligand, resulting a progressive loss of the bile ducts (Guicciardi et al. 2013). Furthermore, there is evidence that both hepatocytes and cholangiocytes form regenerative clusters that overexpress death receptor Fas and mito-inhibitory protein glypican 3 (Hattoum et al. 2013). Additionally, previous research has shown that anandamide (AEA), an endocannabinoid, is both antiproliferative and proapoptotic. Cannabinoid receptors, Cb1 and Cb2, are found in cells of the central nervous system, various peripheral tissues, the gastrointestinal tract, and the immune system. However, blocking these G-protein-coupled receptors does not alter the effects of AEA on cholangiocarcinomas. This is due, in part, to the reaction of AEA with other receptor-independent microdomains that recruits lipid rafts, along with Fas and FasL, in the outer leaflet of the plasma membrane. Disruption of these rafts with methyl-β-cyclodextrin mitigated the AEA suppression of cellular proliferation (DeMorrow et al. 2007).

Another indispensable member of the TNF receptor superfamily is the TNF-related apoptosis-inducing ligand, or TRAIL, additionally noted as death receptor 4/5. In healthy cells, proapoptotic pathways are blocked, and TRAIL activates NF-kB and JNK to promote cellular differentiation and proliferation (Ishimura et al. 2006). However, in cholestatic liver disease, such as human PSC and PBC, TRAIL expression and apoptosis are significantly increased causing cholangiocyte cell death and cholangitis (Takeda et al. 2008). In this pathway, the ligand binds to the death receptors TRAIL-R1/2 and forms a trimeric structure. It is this structure that recruits the Fas-associated death domain (FADD) and procaspases 8 and 10. Through self-activation apoptosis is initiated (Ishimura et al. 2006).

Tumor necrosis factor alpha (TNF-α) is a necessary immunological cytokine that plays a critical role in the inflammatory response of the innate immune system. It is produced in response to viruses, parasites, bacterial endotoxins, and various cytokines. Its course of action is determined by its cell surface receptors, TNF-α receptor 1 (TNF-αR1) and TNF-α receptor 2 (TNF-αR2). TNF-αR1 is widely distributed

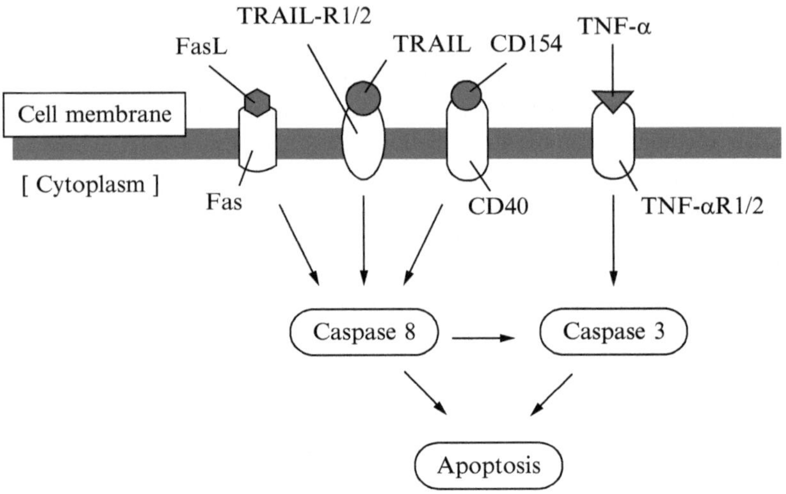

Fig. 2.2 Extrinsic apoptosis signaling in cholangiocytes. Various external signaling pathways that lead to caspase cascade and apoptosis

and responsible for the majority of the TNF-α cellular responses, including apoptosis (Idriss et al. 2014). Previous research has found that this pathway operates through caspase 3, a common end point caspase in the apoptosis cascade. Generally, cholangiocytes are markedly resistant to TNF-α and proliferate in response to ductal secretions of secretin in an autocrine and paracrine manner following injury. However, cholestatic injury, a primary feature of cholangiopathies, sensitizes cholangiocytes to TNF-α signaling in a way that alters ductular secretions and increases apoptosis (Alpini et al. 2003).

CD40 is a major player in cholangiocyte apoptosis through a variety of signaling pathways such as Fas-mediated signaling, STAT3 upregulation, and CD154-bearing Kupffer cells. Fas, or CD95, is a typical death receptor of the TNF family that, in response to cross-linking with its ligand FasL, results in cholangiocyte apoptosis and subsequent bile duct loss. However, research has demonstrated that Fas-FasL cross-linking on its own is not enough to induce apoptosis in either normal or malignant cells. Additionally, Fas has been shown to be downregulated as the tumor moves toward genetic instability. Thus, it appears that CD40, a cell surface glycoprotein, is the main route for apoptosis signaling. In addition to autocrine and paracrine death induction, cholangiocytes are also susceptible to immune-related death through the CD154-expressing Kupffer cells. In chronic liver diseases, circulating macrophages may be activated by bacterial endotoxins or proinflammatory cytokines. These activated macrophages then overexpress CD154 directly inducing apoptosis through binding its cognate receptor, CD40 (Alabraba et al. 2008). Other key players also regulate CD40 signaling in apoptotic cell death. For example, inhibition of STAT3 yields a 50% reduction in overall cholangiocyte cell death (Ahmed-Choudhury et al. 2006). While this reduction is significant and offers a promising therapeutic target, it also indicates there are other, equally important, players to be explored along the CD40 axis.

In addition to cholangiocyte innate death receptors, nonconventional pharmaceutical treatments have also been found to increase apoptosis in cholangiocarcinoma cell lines. For example, Simvastatin, a cholesterol-lowering drug readily used in the prevention and treatment of atherosclerosis, has been shown to increase apoptosis in cholangiocarcinoma cell lines while simply inhibiting proliferation in normal cholangiocytes with no effect on apoptosis (Miller et al. 2011). Several cholesterol pathway intermediates facilitate proper membrane localization and activity of the Rho family of GTPases, including Rac1. Simvastatin interferes with cholangiocyte's ability to localize Rac1 to cholesterol rafts, presumably through inhibition of these crucial intermediates, subsequently decreasing Rac1 activity and increasing apoptosis. Specifically, simvastatin was shown to increase caspase 3 and caspase 7 activities. These results were completely reversible when the cells were pretreated with cholesterol (Miller et al. 2011). As cholangiocytes are typically quiescent unless under duress, the treatment poses no threat to the healthy epithelial cells (Maroni et al. 2015).

2.5.3 *Senescence and Autophagy*

Recent studies in biliary physiology have shown that cholangiocytes use senescence and autophagy in response to biliary injury, including fibrosing cholangiopathies, such as PSC and PBC (Nakanuma et al. 2015; Wang 2015). Cellular senescence is a cellular process by which proliferation halts and cells become arrested in the G1 phase of the cell cycle. Senescent cells remain metabolically active; however, they do not respond to external stimuli (Kuilman et al. 2010). Cellular senescence can occur following numerous insults, such as oxidative stress and oncogene activation, and is thought to be a protective mechanism against additional insults (Nakanuma et al. 2015). Senescent cells have also been shown to take on a secretory phenotype, which persistently secrete cytokines, chemokines, growth factors, and profibrotic factors (Kuilman et al. 2010; Lawless et al. 2010).

Autophagy is a regulated program of self-degradation that allows cells to remove long-lived or damaged proteins and/or organelles. When autophagy is activated during cellular stress, organelles and proteins become sequestered in autophagosomes and digested through fusion with lysosomes (Mizushima 2007). These mechanisms can act as a protective response to cellular stress and enable cells to maintain cellular viability and homeostasis (Kroemer et al. 2010). These cellular mechanisms, along with apoptosis, play a critical role in cholangiocyte response to injury. Regulation of these cellular processes occurs through a balance between the tumor suppressor p53 and expression of anti-apoptotic signaling, specifically the AKT/mTOR pathway (Wang 2015). The cross talk between p53 and AKT/mTOR determines whether or not cells survive injury (Nakanuma et al. 2015).

Previous studies have shown that these mechanisms are implicated in PBC and may be associated with the progressive bile duct loss that is characteristic of the disease. For example, cholangiocytes from PBC patients exhibit shortened telomeres and increased expression of SA-β-Gal, $p16^{INK4a}$, and $p21^{WAF1/Cip}$, which are characteristics of senescent cells (Sasaki et al. 2005, 2008b). It is hypothesized that the senescent cholangiocytes in PBC maintain a secretory phenotype, which promotes the release of cytokines and chemokines into the periductular environment, creating an inflammatory reaction resulting in biliary injury (Sasaki et al. 2008a). Additionally, studies have shown that the autophagy markers are upregulated in bile ducts from PBC patients (Sasaki et al. 2012). This includes microtubule-associated protein light chain 3β (LC3) and p62, a protein involved in the ubiquitin-bound cargo to autophagosomes (Wang 2015). Accumulation of p62 is suggestive of deregulated autophagy and insufficient processing of damaged proteins, which could be a contributing factor in the pathogenesis of PBC (Nakanuma et al. 2015).

Additionally, studies have shown that senescence and autophagy may also contribute to the pathogenesis of PSC. Cultured cholangiocytes isolated from PSC patients demonstrate a senescent phenotype, such as overexpression of $p16^{INK4a}$ and SA-β-Gal. These cholangiocytes also express a secretory phenotype, such as in PBC, which may contribute to the recruitment of inflammatory cells and modulation of the periductal environment (Tabibian et al. 2014a, b). Less is known about the role of autophagy in PSC; however, it has been suggested that the deregulated autophagy in respect to processing microbial pathogens from the intestinal flora may contribute to the development of peri-neutrophil cytoplasmic antibody (pANCA), an autoantibody commonly found in PSC patients (Nakanuma et al. 2015; Schwarze et al. 2003). These findings suggest that senescence and autophagy may play a role in the progression of PBC and PSCs, but more research is needed to elucidate the differences between the two etiologies of biliary injury.

2.5.4 Lipoapoptosis

Nonalcoholic fatty liver disease (NAFLD) is not recognized in the cholangiopathy family, but it is the most common chronic liver disease in Western countries (Gautheron et al. 2014). The involvement of cholangiocyte damage in the pathogenesis of NAFLD has not been thoroughly investigated. A recent study has demonstrated that the PNPLA3 rs738409 variant (I148M) that is associated with an increased risk for the development of NAFLD and alcoholic liver disease due to hepatocyte lipotoxicity is also a risk factor for reduced survival in male PSC patients with dominant bile duct stenosis presumably due to lipotoxicity at the cholangiocyte level (Friedrich et al. 2013; Romeo et al. 2008; Tian et al. 2010). A recent in vitro study has demonstrated that saturated FFAs palmitate and stearate induced cholangiocyte lipoapoptosis via caspase activation, nuclear translocation of FoxO3, and increased proapoptotic PUMA expression (Natarajan et al. 2014). These findings may help explain why a subset of NAFLD patients with symptoms of cholestasis have biliary injury not consistent with what is normally observed in the majority of patients with NAFLD. Further in vivo model studies are necessary to clearly elucidate the role of cholangiocyte lipoapoptosis in the pathogenesis of NAFLD.

2.6 Conclusions and Future Directions

Cholangiocytes play an integral role in the overall function of the liver and its ability to clear toxins, aid in the digestion and absorption of fats, and expel waste through the bile. Cholangiopathies continue to carry a large disease burden due to the lack of effective treatment strategies despite extensive research over the past two decades. The use of transgenic mice to mimic these diseases will hopefully provide more insight into the mechanisms of liver injury. Studies suggest that cholangiocytes use numerous mechanisms to respond to hepatic injury including apoptosis, autophagy, and senescence. A better understanding of the biological functions and cell death pathways of cholangiocytes in both normal and disease states will aid in the development of novel targeted therapies for cholangiopathies.

References

Ahmed-Choudhury J, Williams KT, Young LS, Adams DH, Afford SC (2006) CD40 mediated human cholangiocyte apoptosis requires JAK2 dependent activation of STAT3 in addition to activation of JNK1/2 and ERK1/2. Cell Signal 18(4):456–468. doi:http://dx.doi.org/10.1016/j.cellsig.2005.05.015

Alabraba EB, Lai V, Boon L, Wigmore SJ, Adams DH, Afford SC (2008) Coculture of human liver macrophages and cholangiocytes leads to CD40-dependent apoptosis and cytokine secretion. Hepatology 47(2):552–562. doi:10.1002/hep.22011

Ali H, Hussain N, Naim M, Zayed M, Al-Mulla F, Kehinde EO, Seaburg LM, Sundsbak JL, Harris PC (2015) A novel PKD1 variant demonstrates a disease-modifying role in trans with a truncating PKD1 mutation in patients with autosomal dominant polycystic kidney disease. BMC Nephrol 16(1):26. doi:10.1186/s12882-015-0015-7

Alpini G, Ueno Y, Tadlock L, Glaser S, LeSage G, Francis H, Taffetani S, Marzioni M, Alvaro D, Patel T (2003) Increased susceptibility of cholangiocytes to tumor necrosis factor-α cytotoxicity after bile duct ligation. Am J Physiol Cell Physiol 285(1):C183–C194. Retrieved from http://ajpcell.physiology.org/content/285/1/C183.abstract

Chen X-M, O'Hara SP, LaRusso NF (2008) The immunobiology of cholangiocytes. Immunol Cell Biol 86(6):497–505. doi:10.1038/icb.2008.37

Chu AS, Russo PA, Wells RG (2012) Cholangiocyte cilia are abnormal in syndromic and non-syndromic biliary atresia. Mod Pathol 25(5):751–757. doi:http://www.nature.com/modpathol/journal/v25/n5/suppinfo/modpathol2011212s1.html

Ciombor KK, Goff LW (2013) Advances in the management of biliary tract cancers. Clin Adv Hematol Oncol 11(1):28–34. Retrieved from http://www.ncbi.nlm.nih.gov/pmc/articles/PMC3713630/

Corrigan M, Hirschfield GM (2015) Primary biliary cirrhosis. Medicine 43(11):645–647. doi:10.1016/j.mpmed.2015.08.012

Czul F, Levy C (2016) Novel therapies on primary biliary cirrhosis. Clinic Liver Dis 20(1):113–130. doi:10.1016/j.cld.2015.08.006
DeMorrow S, Glaser S, Francis H, Venter J, Vaculin B, Vaculin S, Alpini G (2007) Opposing actions of Endocannabinoids on cholangiocarcinoma growth: recruitment of Fas and Fas ligand to lipid rafts. J Biol Chem 282(17):13098–13113. doi:10.1074/jbc.M608238200
Drenth JPH (2010) 4: congenital fibrocystic liver diseases. Best Pract Res Clin Gastroenterol 24(5):573–584. Retrieved from http://linkresolver.tamu.edu:9003/tamu?genre=article&issn=15216918&isbn=&title=BestPractice&Research Clinical Gastroenterology=&volume=24&issue=5&date=20100101&atitle=4%3A Congenital fibrocystic liver diseases&aulast=Drenth%2C Joost P.H.&spage=573&pages=573-584&sid=EBSCO%3AScienceDirect&pid=
Erickson N, Mohanty SK, Shivakumar P, Sabla G, Chakraborty R, Bezerra JA (2008) Temporal-spatial activation of apoptosis and epithelial injury in murine experimental biliary atresia. Hepatology 47(5):1567–1577. doi:10.1002/hep.22229
Friedrich K, Rupp C, Hov JR, Steinebrunner N, Weiss K-H, Stiehl A, Brune M, Schaefer PK, Schemmer P, Sauer P, Schirmacher P, Runz H, Karlsen TH, Stremmel W, Gotthardt DN (2013) A frequent PNPLA3 variant is a sex specific disease modifier in PSC patients with bile duct stenosis. PLoS One 8(3):e58734. doi:10.1371/journal.pone.0058734
Gao B, Jeong WI, Tian Z (2008) Liver: an organ with predominant innate immunity. Hepatology 47(2):729–736. doi:10.1002/hep.22034
Gautheron J, Vucur M, Reisinger F, Cardenas DV, Roderburg C, Koppe C, Kreggenwinkel K, Schneider AT, Bartneck M, Neumann UP, Canbay A, Reeves HL, Luedde M, Tacke F, Trautwein C, Heikenwalder M, Luedde T (2014) A positive feedback loop between RIP3 and JNK controls non-alcoholic steatohepatitis. EMBO Mol Med 6(8):1062–1074. doi:10.15252/emmm.201403856
Glaser S, Gaudio E, Miller T, Alvaro D, Alpini G (2009) Cholangiocyte proliferation and liver fibrosis. Expert Rev Mol Med 11:e7–e7. doi:10.1017/S1462399409000994
Gordon SC (2015) Polycystic liver disease. Gastroenterol Hepatol 11(8):542–544. 543p. Retrieved from http://lib-ezproxy.tamu.edu:2048/login?url=http://search.ebscohost.com/login.aspx?direct=true&db=ccm&AN=109839455&site=eds-live
Guicciardi ME, Malhi H, Mott JL, Gores GJ (2013) Apoptosis and necrosis in the liver. Compr Physiol 3:977–1010. John Wiley & Sons Inc
Hartley JL, Davenport M, Kelly DA (2009) Biliary atresia. Lancet 374(9702):1704–1713. doi:10.1016/s0140-6736(09)60946-6
Hattoum A, Rubin E, Orr A, Michalopoulos GK (2013) Expression of hepatocyte epidermal growth factor receptor, FAS and glypican 3 in EpCAM-positive regenerative clusters of hepatocytes, cholangiocytes, and progenitor cells in human liver failure. Hum Pathol 44(5):743–749. doi:10.1016/j.humpath.2012.07.018
Hirschfield GM, Karlsen TH, Lindor KD, Adams DH (2013) Primary sclerosing cholangitis. Lancet 382(9904):1587–1599. doi:10.1016/S0140-6736(13)60096-3
Humphreys EH, Williams KT, Adams DH, Afford SC (2010) Primary and malignant cholangiocytes undergo cd40 mediated fas dependent apoptosis, but are insensitive to direct activation with exogenous fas ligand. PLoS One 5(11):e14037. doi:10.1371/journal.pone.0014037
Idriss NK, Sayyed HG, Zakhary MM, Sayed S (2014) Implication of tumor necrosis factor alpha receptor 1 and hexosaminidase: relationship to pathogenesis of liver diseases. Comp Clin Pathol 23(4):1095–1102. doi:10.1007/s00580-013-1747-z
Ishibashi H, Nakamura M, Komori A, Migita K, Shimoda S (2009) Liver architecture, cell function, and disease. Semin Immunopathol 31(3):399–409. doi:10.1007/s00281-009-0155-6
Ishimura N, Isomoto H, Bronk SF, Gores GJ (2006) Trail induces cell migration and invasion in apoptosis-resistant cholangiocarcinoma cells. Am J Physiol Gastrointest Liver Physiol 290(1):G129–G136. doi:10.1152/ajpgi.00242.2005
Kalt C (2016) Primary biliary cirrhosis. MEDSURG Nursing 25(1):4–7. issn.10920811
Karlsen TH, Schrumpf E, Boberg KM (2010) Primary sclerosing cholangitis. Best Pract Res Clin Gastroenterol 24(5):655–666. doi:10.1016/j.bpg.2010.07.005
Keller BRA, Schub TB (2016) Cholangiocarcinoma. EBSCO Publishing, Ipswich

Kroemer G, Marino G, Levine B (2010) Autophagy and the integrated stress response. Mol Cell 40(2):280–293. doi:10.1016/j.molcel.2010.09.023
Kuilman T, Michaloglou C, Mooi WJ, Peeper DS (2010) The essence of senescence. Genes Dev 24(22):2463–2479. doi:10.1101/gad.1971610
Lawless C, Wang C, Jurk D, Merz A, Zglinicki T, Passos JF (2010) Quantitative assessment of markers for cell senescence. Exp Gerontol 45(10):772–778. doi:10.1016/j.exger.2010.01.018
Lazaridis KN, LaRusso NF (2015) The cholangiopathies. Mayo Clin Proc 90(6):791–800. doi:10.1016/j.mayocp.2015.03.017
Maroni L, Haibo B, Ray D, Zhou T, Wan Y, Meng F, Marzioni M, Alpini G (2015) Functional and structural features of cholangiocytes in health and disease. Cell Mol Gastroenterol and Hepatol 1(4):368–380. doi:10.1016/j.jcmgh.2015.05.005
Miller T, Yang F, Wise CE, Meng F, Priester S, Munshi MK, Guerrier M, Dostal DE, Glaser SS (2011) Simvastatin stimulates apoptosis in cholangiocarcinoma by inhibition of Rac1 activity. Dig Liver Dis 43(5):395–403. doi:10.1016/j.dld.2011.01.010
Mizushima N (2007) Autophagy: process and function. Genes Dev 21(22):2861–2873. doi:10.1101/gad.1599207
Nakanuma Y, Sasaki M, Harada K (2015) Autophagy and senescence in fibrosing cholangiopathies. J Hepatol 62(4):934–945. doi:10.1016/j.jhep.2014.11.027. http://dx.doi.org/10.1016/j.jhep.2014.11.027
Natarajan SK, Ingham SA, Mohr AM, Wehrkamp CJ, Ray A, Roy S, Cazanave SC, Phillipp MA, Mott JL (2014) Saturated free fatty acids induce cholangiocyte lipoapoptosis. Hepatology 60(6):1942–1956. doi:10.1002/hep.27175
O'Hara SP, Tabibian JH, Splinter PL, LaRusso NF (2013) The dynamic biliary epithelia: molecules, pathways, and disease. J Hepatol 58(3) doi:10.1016/j.jhep.2012.10.011
Petersen C (2012) Biliary atresia: the animal models. Semin Pediatr Surg 21(3):185–191. doi:http://dx.doi.org/10.1053/j.sempedsurg.2012.05.002
Petersen C, Davenport M (2013) Aetiology of biliary atresia: what is actually known? Orphanet J Rare Dis 8(1):1–13. doi:10.1186/1750-1172-8-128
Popov Y (2013) Mouse model of primary biliary cirrhosis with progressive fibrosis: are we there yet? Hepatology 57(2):429–431. doi:10.1002/hep.25969
Popov Y, Patsenker E, Fickert P, Trauner M, Schuppan D (2005) Mdr2 (Abcb4)−/− mice spontaneously develop severe biliary fibrosis via massive dysregulation of pro- and antifibrogenic genes. J Hepatol 43(6):1045–1054. doi:10.1016/j.jhep.2005.06.025
Qian W, Dinghang YAO (2014) Research progress in medical treatment of primary biliary cirrhosis. J Clin Hepatol 30(5):462–465. doi:10.3969/j.issn.1001-5256.2014.05.022
Romeo S, Kozlitina J, Xing C, Pertsemlidis A, Cox D, Pennacchio LA, Boerwinkle E, Cohen JC, Hobbs HH (2008) Genetic variation in PNPLA3 confers susceptibility to nonalcoholic fatty liver disease. Nat Genet 40(12):1461–1465. doi:10.1038/ng.257
Sasaki M, Ikeda H, Haga H, Manabe T, Nakanuma Y (2005) Frequent cellular senescence in small bile ducts in primary biliary cirrhosis: a possible role in bile duct loss. J Pathol 205(4):451–459. doi:10.1002/path.1729
Sasaki M, Ikeda H, Sato Y, Nakanuma Y (2008a) Proinflammatory cytokine-induced cellular senescence of biliary epithelial cells is mediated via oxidative stress and activation of ATM pathway: a culture study. Free Radic Res 42(7):625–632. doi:10.1080/10715760802244768
Sasaki M, Ikeda H, Yamaguchi J, Nakada S, Nakanuma Y (2008b) Telomere shortening in the damaged small bile ducts in primary biliary cirrhosis reflects ongoing cellular senescence. Hepatology 48(1):186–195. doi:10.1002/hep.22348
Sasaki M, Miyakoshi M, Sato Y, Nakanuma Y (2012) Autophagy may precede cellular senescence of bile ductular cells in ductular reaction in primary biliary cirrhosis. Dig Dis Sci 57(3):660–666. doi:10.1007/s10620-011-1929-y
Schrumpf E, Tan C, Karlsen TH, Sponheim J, Bjorkstrom NK, Sundnes O, Alfsnes K, Kaser A, Jefferson DM, Ueno Y, Eide TJ, Haraldsen G, Zeissig S, Exley MA, Blumberg RS, Melum E (2015) The biliary epithelium presents antigens to and activates natural killer T cells. Hepatology 62(4):1249–1259. doi:10.1002/hep.27840

Schwarze C, Terjung B, Lilienweiss P, Beuers U, Herzog V, Sauerbruch T, Spengler U (2003) IgA class antineutrophil cytoplasmic antibodies in primary sclerosing cholangitis and autoimmune hepatitis. Clin Exp Immunol 133(2):283–289. doi:10.1046/j.1365-2249.2003.02195.x

Srivastava A, Patel N (2014) Autosomal dominant polycystic kidney disease. Am Fam Physician 90(5):303–307. Retrieved from http://lib-ezproxy.tamu.edu:2048/login?url=http://search.ebscohost.com/login.aspx?direct=true&db=edswsc&AN=000341285900006&site=eds-live

Syal G, Fausther M, Dranoff JA (2012) Advances in cholangiocyte immunobiology. Am J Physiol Gastrointest Liver Physiol 303(10):G1077–G1086. doi:10.1152/ajpgi.00227.2012

Tabibian JH, O'Hara SP, Splinter PL, Trussoni CE, LaRusso NF (2014a) Cholangiocyte senescence by way of N-ras activation is a characteristic of primary sclerosing cholangitis. Hepatology 59(6):2263–2275. doi:10.1002/hep.26993

Tabibian JH, Trussoni CE, O'Hara SP, Splinter PL, Heimbach JK, LaRusso NF (2014b) Characterization of cultured cholangiocytes isolated from livers of patients with primary sclerosing cholangitis. Lab Investig 94(10):1126–1133. doi:10.1038/labinvest.2014.94

Tag CG, Sauer-Lehnen S, Weiskirchen S, Borkham-Kamphorst E, Tolba RH, Tacke F, Weiskirchen R (2015) Bile duct ligation in mice: induction of inflammatory liver injury and fibrosis by obstructive cholestasis. J Vis Exp 96:52438. doi:10.3791/52438

Takeda K, Kojima Y, Ikejima K, Harada K, Yamashina S, Okumura K, Aoyama T, Frese S, Ikeda H, Haynes NM, Cretney E, Yagita H, Sueyoshi N, Sato N, Nakanuma Y, Smyth MJ, Okumura K (2008) Death receptor 5 mediated-apoptosis contributes to cholestatic liver disease. Proc Natl Acad Sci U S A 105(31):10895–10900. doi:10.1073/pnas.0802702105

Tian C, Stokowski RP, Kershenobich D, Ballinger DG, Hinds DA (2010) Variant in PNPLA3 is associated with alcoholic liver disease. Nat Genet 42(1):21–23. doi:10.1038/ng.488

Tian Z, Chen Y, Gao B (2013) Natural killer cells in liver disease. Hepatology 57(4):1654–1662. doi:10.1002/hep.26115

Tong A, Rangan GK, Ruospo M, Saglimbene V, Strippoli GFM, Palmer SC, Tunnicliffe DJ, Craig JC (2015) A painful inheritance--patient perspectives on living with polycystic kidney disease: thematic synthesis of qualitative research. Nephrol Dial Transplant 30(5):790–800. doi:10.1093/ndt/gfv010

Trivedi PJ, Chapman RW (2012) Mini review: PSC, AIH and overlap syndrome in inflammatory bowel disease. Clin Res Hepatol Gastroenterol 36:420–436. doi:10.1016/j.clinre.2011.10.007

Turcotte S, Jarnagin WR (2015) Liver & portal venous system. In: Doherty GM (ed) Current diagnosis & treatment: surgery. McGraw-Hill Education, New York, p 14e

Vivier E, Tomasello E, Baratin M, Walzer T, Ugolini S (2008) Functions of natural killer cells. Nat Immunol 9(5):503–510. doi:http://www.nature.com/ni/journal/v9/n5/suppinfo/ni1582_S1.html

Wang J, Yang G, Dubrovsky AM, Choi J, Leung PSC (2016) Xenobiotics and loss of tolerance in primary biliary cholangitis. World J Gastroenterol 22(1):338–348. doi:10.3748/wjg.v22.i1.338

Wang K (2015) Autophagy and apoptosis in liver injury. Cell Cycle 14(11):1631–1642. doi:10.1080/15384101.2015.1038685

Yang H, Li TWH, Peng J, Tang X, Ko KS, Xia M, Aller MA (2011) A mouse model of cholestasis-associated cholangiocarcinoma and transcription factors involved in progression. Gastroenterology 141(1):378–388.e374. doi:http://dx.doi.org/10.1053/j.gastro.2011.03.044

Chapter 3
Cell Death and Autophagy in Hepatic Stellate Cell Activation and Function

Fatemeh P. Parvin-Nejad and Scott L. Friedman

Abbreviations

aHSC	Activated hepatic stellate cell
AMPK	Adenosine monophosphate-activated protein kinase
ASK1	Apoptosis signal-regulating kinase 1
ATF6	Activating transcription factor 6
BiP	Binding immunoglobulin protein
ECM	Extracellular matrix
ER	Endoplasmic reticulum
ERK	Extracellular signal-regulated kinase
FasL	Fas ligand
HIF-1α	Hypoxia-inducible factor 1 alpha
HSC	Hepatic stellate cell
IFNγ-	Interferon gamma
IP3	Inositol 1,4,5-trisphosphate
IRE1	Inositol-requiring kinase 1
JNK	c-Jun-N-terminal kinase
MAPK	Mitogen-activated protein kinase

Grant support Fatemeh Parvin-Nejad is supported by a Howard Hughes Medical Institute Medical Research Fellows grant. Scott L. Friedman is supported by grant NIDDK RO156621.

Author contributions FPN– wrote the initial draft and designed figures. SLF –reviewed and finalized text, modified figures.

F.P. Parvin-Nejad • S.L. Friedman, MD (✉)
Division of Liver Diseases, Department of Medicine, Icahn School of Medicine at Mount Sinai, New York, NY 10029, USA
e-mail: Scott.Friedman@mssm.edu

W.-X. Ding, X.-M. Yin (eds.), *Molecules, Systems and Signaling in Liver Injury*, Cell Death in Biology and Diseases, DOI 10.1007/978-3-319-58106-4_3

X.-M. Yin and Z. Dong (Series eds.), *Cell Death in Biology and Diseases*

MMP	Matrix metalloproteinase
mTOR	Mammalian target of rapamycin
NF-κB	Nuclear factor-kappa B
NGF	Nerve growth factor
NK cell	Natural killer cell
Nrf2	NF-E2-related factor 2
PDGF	Platelet-derived growth factor
PERK	Protein kinase-like endoplasmic reticulum kinase
PI3K	Phosphatidylinositol 3-kinase
PKCθ	Protein kinase C theta
PUMA	p53-upregulated modulator of apoptosis
qHSC	Quiescent hepatic stellate cell
TGF-β	Transforming growth factor beta
TIMP	Tissue inhibitor of metalloproteinases
TLR	Toll-like receptor
TNF	Tumor necrosis factor
TRAIL	TNF-related apoptosis-inducing ligand
UPR	Unfolded protein response
VEGF	Vascular endothelial growth factor
XBP1	X-box binding protein 1

3.1 Introduction

Hepatic fibrosis contributes to a significant burden of end-stage liver disease worldwide. Tremendous advances in uncovering mechanisms of fibrosis have begun to yield significant progress in defining therapeutic targets and developing new treatments. Some of this success is based on the observations that hepatic fibrosis is often reversible (Iredale et al. 1998), but the underlying mechanisms have not yet been fully elucidated. Understanding these mechanisms may unearth additional therapeutic targets to prevent or even reverse fibrosis in patients with hepatic disease, since despite the remarkable progress, there are still no approved antifibrotic drugs.

Central to the pathogenesis of hepatic fibrosis is the hepatic stellate cell. This resident perisinusoidal mesenchymal cell is the primary storage site for vitamin A in normal liver, contained within cytoplasmic droplets as retinyl esters. In liver injury of any cause, stellate cells undergo a characteristic "activation" in which they transdifferentiate from a resting nonproliferative cell to a highly fibrogenic, proliferative, and contractile cell type that orchestrates hepatic inflammation and fibrogenesis (Lee et al. 2015). This article summarizes emerging knowledge about the contributions of cell death and autophagy to the pathways driving conversion of quiescent (qHSC) to activated (aHSC) hepatic stellate cells. This summary is especially timely, since clarification of cell death and autophagy in other tissues and biological contexts has begun to illuminate our understanding of stellate cell biology.

3.2 Apoptosis and Autophagy in aHSCs

Two mechanisms of cellular recycling and energy homeostasis in aHSCs—apoptosis and autophagy—have been implicated in the regulation of aHSC function and survival and may prove to be central in the balance of fibrogenesis and fibrosis resolution. Apoptosis, or programmed cell death, has a clear role in the resolution of fibrosis; in fact, the fraction of apoptotic cells undergoing apoptosis increases in resolving hepatic injury (Iredale et al. 1998). Importantly, aHSCs are at increased susceptibility to apoptosis mediated by the soluble Fas ligand, possibly related to decreased expression of anti-apoptotic proteins Bcl-2 and Bcl-xl (Gong et al. 1998). On the other hand, autophagy, which displays increased rates of flux during liver injury (Hernández-Gea et al. 2012), drives activation of HSCs (Thoen et al. 2011), ultimately contributing to increased fibrogenesis.

Autophagy is a mechanism of intracellular energy homeostasis, degrading intracellular organelles and proteins to provide energy substrates for cellular processes (Czaja 2011). In activated stellate cells, autophagy degrades lipid stored within cytoplasmic droplets to provide energy supporting their activation (Hernández-Gea et al. 2012). Here we examine the signaling pathways driving cell death and autophagy that contribute to stellate cell activation and fibrogenesis.

3.3 HSC Activation and Apoptosis

Apoptosis of aHSCs contributes to the spontaneous resolution of hepatic fibrosis, as best characterized in carbon tetrachloride-induced liver injury in rodents (discussed in detail below) (Iredale et al. 1998). Apoptosis is induced in HSCs through several mechanisms, setting in motion a negative feedback loop that can attenuate fibrogenesis. Moreover, an apoptotic fate in activated HSCs is controlled by the balance between a complex array of pro-apoptotic and anti-apoptotic factors that include:

3.3.1 Natural Killer Cell-Mediated Apoptosis

One mechanism by which aHSCs undergo apoptosis is through natural killer (NK) cell-mediated killing. NK cells, a component of the innate immune system, can induce apoptosis of target cells through a variety of mechanisms, including the transfer of cytoplasmic granules from NK cells to their cellular targets as well as the

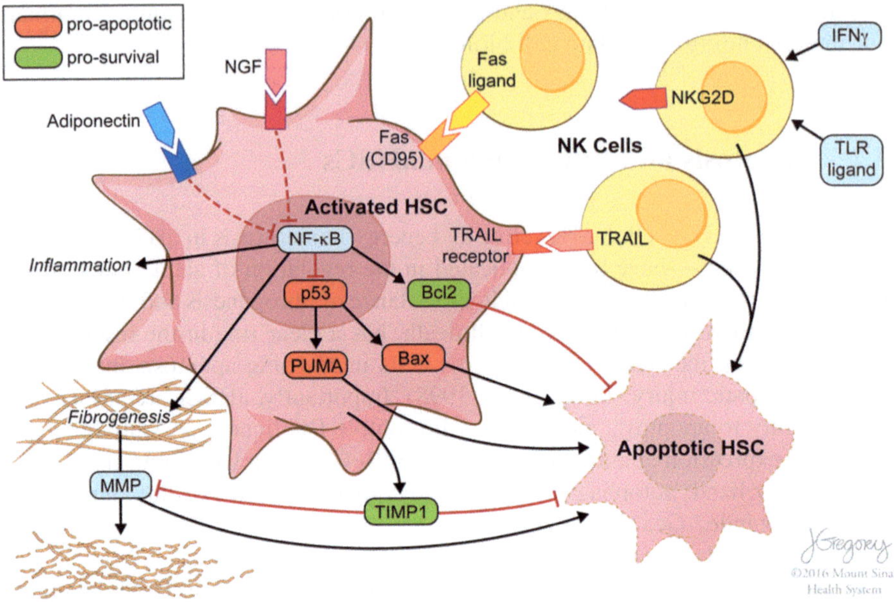

Fig. 3.1 Apoptotic pathways in activated HSCs. Activated HSCs (aHSCs) negatively regulate their own fibrogenic function through several pathways of apoptosis (factors colored green are pro-survival; those colored red are pro-apoptotic). Natural killer (NK) cells, activated by interferon gamma and TLR ligands, recognize aHSCs through NKG2D, Fas ligand, and TRAIL, acting either through the release of cytoplasmic granules or the caspase 8/caspase 3 pathway to provoke aHSC apoptosis. MMPs, which can drive fibrinolysis, also lead to apoptosis, whereas their inhibitors, TIMPs, are anti-apoptotic, contributing to persistent fibrogenesis from surviving aHSCs. NF-κB, which is inhibited by downstream effectors of adiponectin and NGF, modulates transcription in the nucleus, upregulating the anti-apoptotic Bcl2 while downregulating pro-apoptotic proteins Bax and PUMA by reducing expression of p53. NF-κB also contributes to inflammation and fibrogenesis (Printed with permission from ©Mount Sinai Health System)

engagement of NK cell-expressed Fas ligand (FasL) or tumor necrosis factor (TNF) with their respective effectors on target cells, Fas (also known as CD95) and TNF receptor (TNFR) (Warren and Smyth 1999). In the liver, NK cells can target and kill aHSCs, thus reducing hepatic fibrosis (Radaeva et al. 2006). Activated by interferon gamma (IFNγ) and the Toll-like receptor (TLR) ligand polyinosinic:polycytidylic acid, NK cells induce apoptosis of aHSCs through cell receptors NKG2D, TNF-related apoptosis-inducing ligand (TRAIL) (Radaeva et al. 2006), and FasL (Fig. 3.1) (Saile et al. 1997). Fas and TRAIL receptor can also induce apoptosis mediated by the caspase-8/caspase-3 pathway (Iredale 2001). Interestingly, while Fas is expressed ubiquitously on both hepatocytes and HSCs, TRAIL receptor is upregulated only on aHSCs, thereby presenting a more specific target for future therapeutics that could potentially attenuate fibrosis without damaging the hepatic parenchyma (Taimr et al. 2003; Fasbender et al. 2016; Oh et al. 2016). NK cells can also mediate apoptosis of senescent HSCs, leading to reduced fibrogenesis (Krizhanovsky et al. 2008).

3.3.2 *Matrix Metalloproteinases and Their Inhibitors: Linking Fibrinolysis and Apoptosis*

Besides NK cell-mediated killing, HSC apoptosis is also regulated by the pathways that control fibrosis resolution, or fibrinolysis, which in turn depends on the balance between matrix metalloproteinases (MMPs), which cleave collagen and ECM materials, and their inhibitors, tissue inhibitors of metalloproteinases (TIMPs). TIMP-1, which is produced by aHSCs (Ramachandran et al. 2015), has direct anti-apoptotic activity in addition to its inhibitory function on MMPs, some of which are in fact pro-apoptotic (Fig. 3.1) (Iredale 2001). However, TIMP-1 and TIMP-2 levels decrease in association with the resolution of liver injury (Iredale et al. 1998). This reduction in TIMP-1 levels relieves the inhibition on both MMP function and aHSC apoptosis, not only increasing the degradation of extracellular matrix (ECM) already deposited but also decreasing further fibrogenesis by reducing the number of aHSCs.

3.3.3 *Nuclear Factor-Kappa B: An Anti-Apoptotic Factor*

Nuclear factor-kappa B (NF-κB) regulates aHSC survival by inhibiting apoptotic factors in the nucleus. Levels of transcriptionally active NF-κB rise sharply during HSC activation, leading to the inhibition of apoptosis as well as the expression of several pro-inflammatory genes leading to increased fibrogenesis. NF-κB's anti-apoptotic function is achieved through the continued expression of the anti-apoptotic protein Bcl-2 and the inhibition of pro-apoptotic factors Bax and PUMA via the suppression of cell cycle regulator p53 (Fig. 3.1). NF-κB is inhibited by nerve growth factor (NGF), a neurotrophin, as well as by adiponectin through downstream AMPK pathway activation (Hernandez-Gea and Friedman 2011).

3.4 Autophagy and HSC Activation

Autophagy plays a key role in the activation of HSCs. Autophagic flux increases during liver injury and HSC activation, whereas inhibiting autophagy partly inhibits HSC activation (Thoen et al. 2011). Increased autophagic activity is thought to liberate energy from lipid droplets contained within HSCs, thereby providing the required energy for HSC activation; in fact, in HSCs from an autophagy-knockout mouse strain, activation and resulting fibrogenesis can be restored by adding oleic acid as an energy substrate (Hernández-Gea et al. 2012). The engagement of autophagy to provide fuel for extracellular matrix (ECM) production occurs in fibrogenic cells from the kidney and lung as well, suggesting that autophagy is part of a core pathway of fibrogenesis across several tissue types (Hilscher et al. 2013). Several pathways and processes mediate the activation of HSCs via autophagy, including an mTOR-dependent pathway and the hypoxia-inducible factor 1 alpha (HIF-1α) pathway following hypoxia, among other pathways.

3.4.1 mTOR-Dependent Autophagy

The mTOR pathway induces autophagy leading to HSC activation in periods of metabolic (Hilscher et al. 2013) as well as hypoxic stress (Jin et al. 2016). Under nutrient-depleted conditions, the inhibitory function of mTOR complexes (mTORC) 1 and 2 on their usual target, the ULKs-Atg13-FIP200 complex (a mammalian autophagy complex), is relieved, leading to activation of the target complex and induction of autophagy (Fig. 3.2) (Hosokawa et al. 2009). mTORC1 is regulated by several factors, including the phosphatidylinositol 3-kinase (PI3K) complex, components of which have opposing effects: class I PI3K inhibits autophagy through Akt signaling, while class III PI3K (Vps34) stimulates autophagy by increasing concentrations of phosphatidylinositol 3-phosphate. Other kinases influencing the effect of mTOR on autophagy include adenosine monophosphate-activated protein kinase (AMPK), IκB kinase, and mitogen-activated protein kinase (MAPK) pathway components, including extracellular signal-regulated kinase (ERK), p38, and c-Jun-N-terminal kinase (JNK) (Hilscher et al. 2013); the functions of p38 and JNK with respect to endoplasmic reticulum (ER) stress are discussed below. Hypoxic stress also provokes autophagy and HSC activation in rats through the calcium-dependent AMPK-mTOR and protein kinase C-theta pathways (Fig. 3.2) (Jin et al. 2016). Interestingly, calcium release mediated by a different factor, inositol 1,4,5-trisphosphate (IP3), leads to the *inhibition* of autophagy through a Ca^{2+}-calpain-Gsα pathway (Hilscher et al. 2013).

3.4.2 HIF-1α: The Role of Hypoxia in Autophagy and HSC Activation

HIF-1α, a protective factor in periods of hypoxic stress, regulates autophagy-mediated activation of HSCs. HIF-1α can be induced by several pathways, including hypoxia, growth factors (such as platelet-derived growth factor (PDGF), transforming growth factor beta (TGF-β), and TNF-α), oncogenic signaling pathways (p53 and Ras), and vascular endothelial growth factor (VEGF) (through a positive feedback loop involving PI3K and MAPK), as well as epigenetic mechanisms (Fig. 3.2) (Zhan et al. 2015). In conditions of hypoxic stress, HIF-1α stimulates the activation of HSCs (Wang et al. 2013), which is attributable to HIF-1α-induced autophagic activity (Deng et al. 2014). HIF-1α is also implicated in the self-perpetuating cycle of HSC activation and fibrogenesis. Hypoxia, oncogenic

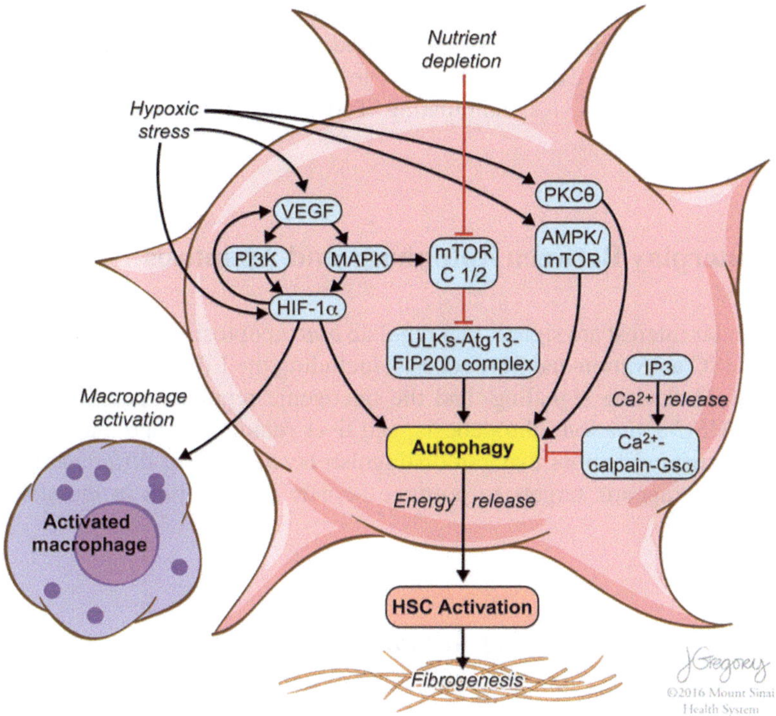

Fig. 3.2 Autophagy-inducing pathways leading to HSC activation. In HSCs, autophagy generates energy through cleavage of retinyl esters to yield fatty acids, which support HSC activation and fibrogenesis. Under conditions of nutrient depletion, mTOR complex activity is inhibited, relieving its inhibition of the ULKs-Atg13-FIP200 complex, which leads to autophagy. Conditions of hypoxic stress lead to activation of several other pathways triggering autophagy, including PKCθ, the calcium-dependent AMPK/mTOR pathway, and HIF-1α. Conversely, calcium release also allows IP3 to activate the *anti-autophagic* Ca^{2+}-calpain-Gsα complex. HIF-1α is regulated by a positive feedback loop: hypoxia stimulates VEGF, which upregulates HIF-1α through the actions of PI3K and MAPK. The upregulated HIF-1α in turn further induces VEGF. MAPK also activates the anti-autophagic mTORC, providing an internal balance mechanism. HIF-1α also has an anti-fibrogenic function, activating macrophages which clear TGF-β-secreting necrotic hepatocytes during liver injury, thus removing a stimulus for fibrogenesis (Printed with permission from ©Mount Sinai Health System)

stimuli, and growth factors can lead to increased VEGF expression, which triggers PI3K- and MAPK-dependent stimulation of HIF-1α, which in addition to provoking autophagic-mediated HSC activation feeds back to further stimulate VEGF production and continue the cycle (Zhan et al. 2015). Furthermore, the VEGF feedback loop may comprise part of a common pathway between HIF-1α and mTOR (which is also regulated by PI3K and MAPK as described above) in the induction of autophagy. Notably, HIF-1α may have an anti-fibrogenic role as well (Fig. 3.2). In knockout mice lacking HIF-1α only in HSCs, loss of HIF-1α increases collagen I production and reduces conversion of macrophages to a more active phenotype (which clear necrotic hepatocytes) following carbon tetrachloride-induced hepatic injury. These findings suggest that increased fibrogenesis following liver injury could be mediated by exposure of HSCs to necrotic products from inadequately cleared, dying hepatocytes, stimulating TGF-β-dependent collagen production (Mochizuki et al. 2014).

3.5 Interplay Between Autophagy and Apoptosis

Of particular interest are signals that regulate *both* apoptosis and autophagy in their roles in HSC activation and fibrogenesis, including the ER stress response, TGF-β signaling, Hedgehog signaling, and the oncogenic regulator p62. These signals integrate HSC activation, fibrogenesis, and liver remodeling and may prove to be key in identifying factors that can distinguish between these different fates, thus providing therapeutic targets to promote hepatic regeneration in the absence of fibrogenesis.

3.5.1 ER Stress: Linking Autophagy and Apoptosis

One well-described pathway that moderates both apoptosis and autophagy is the unfolded protein response (UPR) following ER stress. Oxidative stress has already been shown to activate HSCs (Svegliati-Baroni et al. 2001). Conditions of oxidative stress and increased protein synthesis may overwhelm ER folding capacity, leading to activation of the UPR, a compensatory mechanism which functions to restore ER homeostasis (Hernández-Gea et al. 2013). In this mechanism, unfolded proteins sequester a regulatory factor known as binding immunoglobulin protein (BiP), releasing and thus activating its three targets, all of which are ER transmembrane-signaling proteins. These proteins—activating transcription factor 6 (ATF6), inositol-requiring kinase 1 (IRE1), and protein kinase-like endoplasmic reticulum kinase (PERK)—provide signals to upregulate genes encoding proteins which contribute to protein folding and otherwise help reduce the burden on the overloaded ER (Fig. 3.3) (Schröder and Kaufman 2005). Of these, ATF6 and IRE1α can upregulate autophagy and resulting collagen 1 expression (Hernández-Gea et al. 2013; de Galarreta et al. 2016), whereas the PERK pathway has been implicated in HSC activation and fibrogenesis via overexpression of SMAD2 (an effector of the TGF-β pathway) (Koo et al. 2016). Conversely, IRE1 mediates ER stress-induced aHSC apoptosis after treatment with cannabidiol (Lim et al. 2011) as well as after exposure to the senescence-inducing factor CCN1 (Borkham-Kamphorst et al. 2016).

Given the multitude of downstream effects of IRE1α signaling, its role must be further clarified. IRE1α acts through apoptosis signal-regulating kinase 1 (ASK1) to activate downstream stress kinases JNK and p38 MAPK (Borkham-Kamphorst et al. 2016).

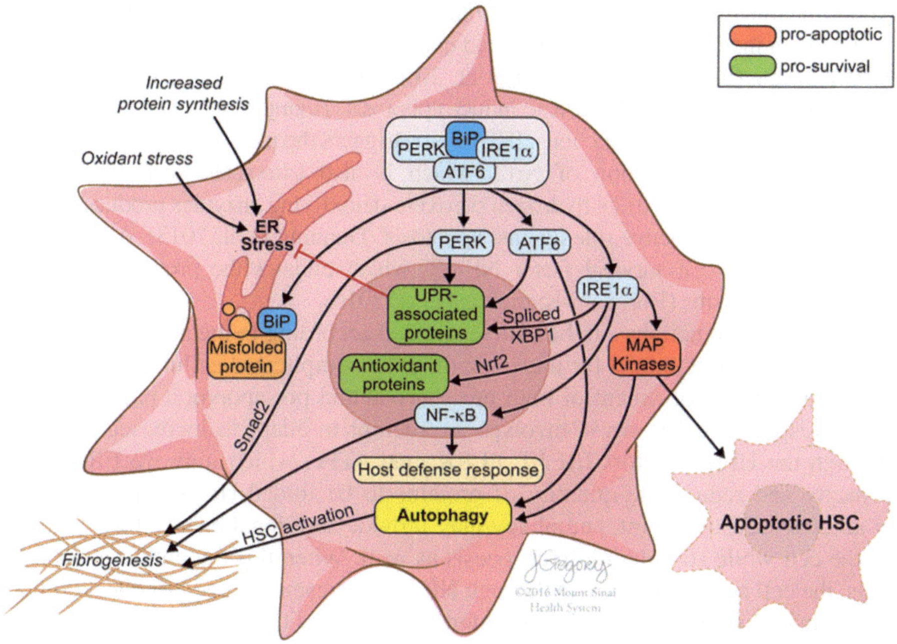

Fig. 3.3 ER stress and the unfolded protein response in HSCs. Oxidants and conditions requiring increased protein synthesis increase stress on the ER, leading to the misfolding of proteins (factors colored green are pro-survival; those colored red are pro-apoptotic). Misfolded proteins trigger the compensatory unfolded protein response (UPR) by sequestering the inhibitory BiP away from the three UPR effectors, PERK, ATF6, and IRE1α. These three arms of the UPR increase the transcription and expression of UPR-associated proteins such as molecular chaperones, which improve protein folding and reduce protein expression, reducing the burden on the ER in an effort to adapt to changing demands on the cell. The effectors also modulate autophagy, apoptosis, and fibrosis. PERK induces the overexpression of Smad2 by destabilizing downregulatory transcription factors, leading to Smad2-mediated fibrogenesis. ATF6 has also been implicated in stimulating autophagy. IRE1α is the most complex effector arm, leading to *adaptive* responses such as increasing expression of UPR-associated proteins (by splicing XBP1) and antioxidant proteins (by upregulating Nrf2), *alarm* responses such as activating NF-κB (which initiates a host defense response as well as contributing to fibrogenesis), and, when these strategies prove unsuccessful in maintaining cellular viability, the *apoptotic* response through various MAP kinases (some of which also contribute to autophagy) (Printed with permission from ©Mount Sinai Health System)

Yet the functions of these effectors remain to be fully distinguished. While the autophagic and fibrogenic properties of IRE1α in activated HSCs are dependent on p38 MAPK (Hernández-Gea et al. 2013), some studies have found increased levels of activated JNK as well (though no causality in the autophagy pathway) (de Galarreta et al. 2016), and JNK has been independently implicated in autophagy induction in other cell types. In starvation models using HeLa and breast cancer cell lines, JNK1 phosphorylates Bcl-2, releasing Beclin 1 to stimulate autophagy (Wei et al. 2008). Both factors appear to have a role in HSC apoptosis as well. The overexpression of CCN1 (which induces senescence in cutaneous myofibroblasts during wound healing) in HSC and myofibroblast cell lines leads to HSC apoptosis associated with an increase in JNK and reduced collagen 1 expression, but no change in p38 MAPK levels (Borkham-Kamphorst et al. 2016). Yet thapsigargin-induced ER stress acts through JNK *and* p38 MAPK to induce apoptosis (Huang et al. 2014). Parallel to its autophagic and apoptotic functions, IRE1α also triggers the transcription of antioxidant proteins through the action of Nrf2, which is also associated with fibrogenesis (Hernández-Gea et al. 2013); hydrogen peroxide-detoxifying enzymes produced in this response play a protective role against apoptosis (Dunning et al. 2013). Additionally, IRE1α activates cell survival factor NF-κB, which initiates the production of host defense mechanisms (Borkham-Kamphorst et al. 2016).

These many functions can be conceptualized according to the adaptation-alarm-apoptosis ER stress model for understanding the complicated cellular response via IRE1α and the involvement of both pro-survival and pro-apoptotic pathways. The initial response to ER stress involves an attempt to adapt to increased demands through the UPR effectors IRE1α, PERK, and ATF6, which improve ER folding capacity and reduce the quantity of proteins to be folded (Borkham-Kamphorst et al. 2016). At this point, autophagy may provide additional energy for activation of the cell while stimulating fibrogenesis as well. When this attempt at adaptation fails, the cell next signals alarm through NF-κB, representing one last checkpoint to maintain cell viability through host defenses before the balance ultimately shifts to apoptosis, which reduces the population of fibrogenic cells and contributes to the resolution of fibrosis (Borkham-Kamphorst et al. 2016). This adaptation-alarm-apoptosis model may provide a framework for designing future studies, which must better delineate a temporal sequence of function for the various components of the ER stress response in order to reconcile the conflicting conclusions of the current literature.

3.5.2 *TGF-β: A Fibrotic Stimulus Regulating Autophagy and Apoptosis*

Another possible common pathway is that mediated by TGF-β, which reduces apoptosis and increases autophagy, leading to HSC activation and increased fibrogenesis. TGF-β is well-described as an important fibrotic stimulus, triggering the production of ECM in HSCs (Friedman 2000). In the liver, TGF-β1 is primarily produced by macrophages and can activate HSCs (Karlmark et al. 2009), causing enhanced expression of TIMPs (Wynn and Barron 2010) and thereby contributing to the persistence of ECM. TGF-β has both pro-apoptotic and anti-apoptotic functions, mediated by Smad family proteins as well as other mediators, as seen in a number of cell types and developmental contexts (Heldin et al. 2009). One study reports the downregulation of apoptosis in HSCs treated with TGF-β, which was reversed upon the application of bafilomycin, an autophagy blocker, or anti-LC3 siRNA, an inhibitor of the autophagy pathway (Fu et al. 2014). However, further studies are needed to confirm this effect.

3.5.3 *miR-148a and Hedgehog Signaling: Incorporating Autophagy and Apoptosis into Liver Regeneration*

The Hedgehog signaling pathway is yet another node linked to both autophagy and apoptosis in HSC regulation. Hedgehog signaling has a central role in embryonic development, but is also significantly activated during chronic liver injury to mediate hepatic repair (Machado and Diehl 2015). Hedgehog ligands contribute to the activation of HSCs, which also produce these ligands once activated (Machado and Diehl 2015). The Hedgehog pathway also inhibits autophagy, a function which is relieved by miR-148a, a micro-RNA which itself inhibits Hedgehog signaling. Treatment of HSC cell lines with miR-148a also increases apoptosis, suggesting another common pathway of regulation (Liu et al. 2015). Interestingly, Hedgehog also activates HIF-1α, which stimulates the epithelial-to-mesenchymal transition in HSCs, facilitating their function in remodeling the injured liver (Chen et al. 2012). A better understanding of the contributions of this signaling pathway may provide insight into the molecular switches regulating regeneration versus fibrosis in the liver.

3.5.4 *p62: A Linchpin of Autophagic and Apoptotic Balance*

p62 has recently emerged as a key factor integrating autophagy, apoptosis, and HSC activation and function. Through its scaffolded protein structure, it activates NF-κB by interacting with TRAF6, Nrf2 by sequestering the inhibitory Keap1, and mTORC1 by interacting with mTORC1 regulators at its PB1 domain (Moscat et al. 2016). Through these multiple interactions, p62 is connected to the inflammatory, fibrogenic, and anti-apoptotic roles of NF-κB, the antioxidant activity of Nrf2, and the anti-autophagic function of mTORC1. Whereas the global overexpression of p62 leads to the development of hepatocellular carcinoma (Umemura et al. 2016), the function of p62 specifically in HSCs, acting through the vitamin D receptor, corresponds to the inhibition of HSC activation and produces anti-fibrogenic and anti-tumorigenic effects (Duran et al. 2016). Thus, p62, along with the ER stress response, TGF-β, and Hedgehog signaling, may belong to a panel of multi-regulatory factors preserving the delicate balance of autophagy and apoptosis in HSCs and contributing to liver injury and fibrosis when dysregulated.

3.6 Conclusion

The activation of HSCs and their resulting fibrogenesis are regulated by a number of pathways involving autophagy and apoptosis. Here we have examined the contributions of NK cells, MMPs, and NF-κB on apoptosis and of mTOR signaling and HIF-1α on autophagy, as well as the integrated actions of the ER stress response, TGF-β, Hedgehog signaling, and p62. Much remains to be learned about the function of HSCs in fibrosis and healing of the liver, and carefully detailing the influences of these pathways may be key to our understanding of liver function and the future development of therapeutic agents aimed at healing liver injury without causing fibrosis.

Bibliography

Borkham-Kamphorst E, Steffen BT, de Leur EV, Haas U, Tihaa L, Friedman SL, Weiskirchen R (2016) CCN1/CYR61 overexpression in hepatic stellate cells induces ER stress-related apoptosis. Cell Signal 28(1):34–42. doi:10.1016/j.cellsig.2015.10.013. The Authors

Chen Y, Choi SS, Michelotti GA, Chan IS, Swiderska-Syn M, Karaca GF, Xie G et al (2012) Hedgehog controls hepatic stellate cell fate by regulating metabolism. Gastroenterology 143(5):1319–1329.e11. doi:10.1053/j.gastro.2012.07.115. Elsevier Inc.

Czaja MJ (2011) Functions of autophagy in hepatic and pancreatic physiology and disease. Gastroenterology 140(7):1895–1908. doi:10.1053/j.gastro.2011.04.038. Elsevier Inc.

Deng J, Huang Q, Wang Y, Shen P, Guan F, Li J, Huang H, Shi C (2014) Hypoxia-inducible factor-1alpha regulates autophagy to activate hepatic stellate cells. Biochem Biophys Res Commun 454(2):328–334. doi:10.1016/j.bbrc.2014.10.076. Elsevier Inc.

Dunning S, Atta ur Rehman, Tiebosch MH, Hannivoort RA, Haijer FW, Woudenberg J, van den Heuvel FAJ, Buist-Homan M, Faber KN, Moshage H (2013) Glutathione and antioxidant enzymes serve complementary roles in protecting activated hepatic stellate cells against hydrogen peroxide-induced cell death. Biochim Biophys Acta Mol basis Dis 1832(12):2027–2034. doi:10.1016/j.bbadis.2013.07.008. Elsevier B.V.

Duran A, Hernandez ED, Reina-Campos M, Castilla EA, Subramaniam S, Raghunandan S, Roberts LR et al (2016) p62/SQSTM1 by binding to vitamin D receptor inhibits hepatic stellate cell activity, fibrosis, and liver cancer. Cancer Cell 30(4):595–609. doi:10.1016/j.ccell.2016.09.004. Elsevier Inc.

Fasbender F, Widera A, Hengstler JG, Watzl C (2016) Natural killer cells and liver fibrosis. Front Immunol 7(JAN):1–7. doi:10.3389/fimmu.2016.00019

Friedman SL (2000) Molecular regulation of hepatic fibrosis, an integrated cellular response to tissue injury. Am Soc Fot Biochem Mol Biol 275(10):2247–2250

Fu M-Y, He Y-J, Lv X, Liu Z-H, Shen Y, Ye G-R, Deng Y-M, Shu J-C (2014) Transforming growth factor-β1 reduces apoptosis via autophagy activation in hepatic stellate cells. Mol Med Rep 10(3):1282–1288. doi:10.3892/mmr.2014.2383

Gong W, Pecci A, Roth S, Lahme B, Beato M, Gressner AM (1998) Transformation-dependent susceptibility of rat hepatic stellate cells to apoptosis induced by soluble Fas Ligand. Hepatology (Baltimore, Md) 28(2):492–502. doi:10.1002/hep.510280229

Heldin CH, Landstrom M, Moustakas A (2009) Mechanism of TGF-Beta signaling to growth arrest, apoptosis, and epithelial-mesenchymal transition. Curr Opin Cell Biol 21(2):166–176. doi:10.1016/j.ceb.2009.01.021

Hernandez-Gea V, Friedman SL (2011) Pathogenesis of liver fibrosis. Annu Rev Pathol 6:425–456. doi:10.1146/annurev-pathol-011110-130246

Hernández-Gea V, Ghiassinejad Z, Rozenfeld R, Gordon R, Fiel MI, Yue Z, Czaja MJ, Friedman SL (2012) Autophagy releases lipid that promotes fibrogenesis by activated hepatic stellate cells in mice and in human tissues. Gastroenterology 142(4):938–946. doi:10.1053/j.gastro.2011.12.044. Elsevier Inc.

Hernández-Gea V, Hilscher M, Rozenfeld R, Lim MP, Nieto N, Werner S, Devi LA, Friedman SL (2013) Endoplasmic reticulum stress induces fibrogenic activity in hepatic stellate cells through autophagy. J Hepatol 59(1):98–104. doi:10.1016/j.jhep.2013.02.016

Hilscher M, Hernandez-Gea V, Friedman SL (2013) Autophagy and mesenchymal cell fibrogenesis. Biochim Biophys Acta Mol basis Dis 1832(7):972–978. doi:10.1016/j.bbadis.2012.11.001

Hosokawa N, Hara T, Kaizuka T, Kishi C, Takamura A, Miura Y, Iemura S-i et al (2009) Nutrient-dependent mTORC1 association with the ULK1–Atg13–FIP200 complex required for autophagy. Mol Biol Cell 20(7):1981–1991. doi:10.1091/mbc.E08-12-1248

Huang Y, Li X, Wang Y, Wang H, Cheng H, Li J (2014) Endoplasmic reticulum stress-induced hepatic stellate cell apoptosis through calcium-mediated JNK/P38 MAPK and Calpain/Caspase-12 pathways. Mol Cell Biochem 394(1–2):1–12. doi:10.1007/s11010-014-2073-8

Iredale JP, Benyon RC, Pickering J, McCullen M, Northrop M, Pawley S, Hovell C, Arthur MJ (1998) Mechanisms of spontaneous resolution of rat liver fibrosis. Hepatic stellate cell apoptosis and reduced hepatic expression of metalloproteinase inhibitors. J Clin Invest 102(3):538–549. doi:10.1172/JCI1018

Iredale JP (2001) Hepatic stellate cell behavior during resolution of liver injury. Inflammation 21(3):427–436

Jin Y, Bai Y, Ni H, Qiang L, Ye L, Shan Y, Zhou M (2016) Activation of autophagy through calcium-dependent AMPK/mTOR and PKCθ pathway causes activation of rat hepatic stellate cells under hypoxic stress. FEBS Lett 590(5):672–682. doi:10.1002/1873-3468.12090

Karlmark KR, Weiskirchen R, Zimmermann HW, Gassler N, Ginhoux F, Weber C, Merad M, Luedde T, Trautwein C, Tacke F (2009) Hepatic recruitment of the inflammatory Gr1+ Monocyte subset upon liver injury promotes hepatic fibrosis. Hepatology 50(1):261–274. doi:10.1002/hep.22950

Koo JH, Lee HJ, Kim W, Kim SG (2016) Endoplasmic reticulum stress in hepatic stellate cells promotes liver fibrosis via PERK-mediated degradation of HNRNPA1 and up-regulation of SMAD2. Gastroenterology 150(1):181–193.e8. doi:10.1053/j.gastro.2015.09.039. Elsevier, Inc.

Krizhanovsky V, Yon M, Dickins RA, Hearn S, Simon J, Miething C, Yee H, Zender L, Lowe SW (2008) Senescence of activated stellate cells limits liver fibrosis. Cell 134(4):657–667. doi:10.1016/j.cell.2008.06.049

Lee YA, Wallace MC, Friedman SL (2015) Pathobiology of liver fibrosis: a translational success story. Gut 64(5):830–841. doi:10.1136/gutjnl-2014-306842

Lim MP, Devi LA, Rozenfeld R (2011) Cannabidiol causes activated hepatic stellate cell death through a mechanism of endoplasmic reticulum stress-induced apoptosis. Cell Death Dis 2(6):e170. doi:10.1038/cddis.2011.52. Nature Publishing Group

Liu X-Y, He Y-J, Yang Q-H, Huang W, Liu Z-H, Ye G-R, Tang S-H, Shu J-C (2015) Induction of autophagy and apoptosis by miR-148a through the sonic Hedgehog signaling pathway in hepatic Stellate cells. Am J Cancer Res 5(9):2569–2589. http://www.pubmedcentral.nih.gov/articlerender.fcgi?artid=4633891&tool=pmcentrez&rendertype=abstract

Machado MV, Diehl AM (2015) Review Hedgehog signaling in the liver. Signaling Pathways in Liver Diseases 54(2):262–274. doi:10.1002/9781118663387.ch19

Mochizuki A, Pace A, Rockwell CE, Roth KJ, Chow A, O'Brien KM, Albee R et al (2014) Hepatic stellate cells orchestrate clearance of necrotic cells in a hypoxia-inducible factor-1 -dependent manner by modulating macrophage phenotype in mice. J Immunol 192(8):3847–3857. doi:10.4049/jimmunol.1303195

Moscat J, Karin M, Diaz-Meco MT (2016) p62 in cancer: signaling adaptor beyond Autophagy. Cell 167(3):606–609. doi:10.1016/j.cell.2016.09.030. Elsevier Inc.

Oh Y, Park O, Swierczewska M, Hamilton JP, Park JS, Kim TH, Lim SM et al (2016) Systemic PEGylated TRAIL treatment ameliorates liver cirrhosis in rats by eliminating activated hepatic stellate cells. Hepatology 64(1):209–223. doi:10.1002/hep.28432

Radaeva S, Sun R, Jaruga B, Nguyen VT, Tian Z, Gao B (2006) Natural killer cells ameliorate liver fibrosis by killing activated stellate cells in NKG2D-dependent and tumor necrosis factor-related apoptosis-inducing Ligand-dependent manners. Gastroenterology 130(2):435–452. doi:10.1053/j.gastro.2005.10.055

Ramachandran P, Iredale J, Fallowfield J (2015) Resolution of liver fibrosis: basic mechanisms and clinical relevance. Semin Liver Dis 35(2):119–131. doi:10.1055/s-0035-1550057

de Galarreta R, Marina AN, Ansorena E, Garzón AG, Mòdol T, López-Zabalza MJ, Martínez-Irujo JJ, Iraburu MJ (2016) Unfolded protein response induced by Brefeldin A increases collagen type I levels in hepatic stellate cells through an IRE1α, p38 MAPK and Smad-dependent pathway. Biochimica et Biophysica Acta (BBA) – Molecular Cell Research 1863(8):2115–2123. doi:10.1016/j.bbamcr.2016.05.002. Elsevier B.V.

Saile B, Knittel T, Matthes N, Schott P, Ramadori G (1997) CD95/CD95L-mediated apoptosis of the hepatic Stellate cell. A mechanism terminating uncontrolled hepatic Stellate cell proliferation during hepatic tissue repair. Am J Pathol 151(5):1265–1272. http://www.pubmedcentral.nih.gov/articlerender.fcgi?artid=1858076&tool=pmcentrez&rendertype=abstract

Schröder M, Kaufman RJ (2005) The mammalian unfolded protein response. Annu Rev Biochem 74(1):739–789. doi:10.1146/annurev.biochem.73.011303.074134

Svegliati-Baroni G, Saccomanno S, Van Goor H, Jansen P, Benedetti A, Moshage H (2001) Involvement of reactive oxygen species and nitric oxide radicals in activation and proliferation of rat hepatic stellate cells. Liver 21(1):1–12. doi:10.1034/j.1600-0676.2001.210101.x

Taimr P, Higuchi H, Kocova E, Rippe RA, Friedman S, Gores GJ (2003) Activated stellate cells express the TRAIL receptor-2/death receptor-5 and undergo TRAIL-mediated apoptosis. Hepatology 37(1):87–95. doi:10.1053/jhep.2003.50002

Thoen LFR, Guimarães ELM, Dollé L, Mannaerts I, Najimi M, Sokal E, Van Grunsven LA (2011) A role for autophagy during hepatic stellate cell activation. J Hepatol 55(6):1353–1360. doi:10.1016/j.jhep.2011.07.010. European Association for the Study of the Liver

Umemura A, He F, Taniguchi K, Nakagawa H, Yamachika S, Font-Burgada J, Zhong Z et al (2016) p62, upregulated during preneoplasia, induces hepatocellular carcinogenesis by maintaining survival of stressed HCC-initiating cells. Cancer Cell 29(6):935–948. doi:10.1016/j.ccell.2016.04.006. Elsevier Inc.

Wang Y, Huang Y, Guan F, Xiao Y, Deng J, Chen H, Chen X, Li J, Huang H, Shi C (2013) Hypoxia-inducible factor-1alpha and MAPK co-regulate activation of hepatic stellate cells upon hypoxia stimulation. PLoS One 8(9):1–8. doi:10.1371/journal.pone.0074051

Warren H, Smyth M (1999) NK cells and apoptosis. Immunol Cell Biol 77(October 1998):64–75. doi:10.1046/j.1440-1711.1999.00790.x

Wei Y, Pattingre S, Sinha S, Bassik M, Levine B (2008) JNK1-mediated phosphorylation of Bcl-2 regulates starvation-induced autophagy. Mol Cell 30(6):678–688. doi:10.1016/j.molcel.2008.06.001

Wynn TA, Barron L (2010) Macrophages: master regulators of inflammation and fibrosis. Semin Liver Dis 30(3):245–257. doi:10.1055/s-0030-1255354

Zhan L, Cheng H, Meng X-M, Yang S, Wu XQ, Yang Y, Li J (2015) Hypoxia-inducible factor-1alpha in hepatic fibrosis: a promising therapeutic target. Biochimie 108:1–7. doi:10.1016/j.biochi.2014.10.013. Elsevier Ltd.

Chapter 4
The Switch: Mechanisms Governing Macrophage Phenotypic Variability in Liver Disease

John Marentette and Cynthia Ju

Abbreviations

AILI	Acetaminophen-induced liver injury
ALA	Amebic liver abscess
ALD	Alcoholic liver disease
AMI	Acute myocardial infarction
APAP	Acetaminophen
Arg1	Arginase 1
CCL2	Chemokine (C-C motif) ligand 2
CCR2	C-C chemokine receptor type 2
CGD	Chronic granulomatous disease
COX-2	Cyclooxygenase-2
CSF	Colony-stimulating factor
CTGF	Connective tissue growth factor
CX_3CR1	CX3C chemokine receptor 1
DHA	Docosahexaenoic acid
EGFR	Epidermal growth factor receptor
EPA	Eicosapentaenoic acid
GFP	Green fluorescent protein
GM-CSF	Granulocyte/macrophage colony-stimulating factor
HBV	Hepatitis B virus
HCC	Hepatocellular carcinoma

J. Marentette • C. Ju (✉)
Skaggs School of Pharmacy and Pharmaceutical Sciences, University of Colorado Anschutz Medical Campus, 12850 East Montview Blvd, Aurora, CO 80045, USA
e-mail: Cynthia.Ju@UCDenver.edu

W.-X. Ding, X.-M. Yin (eds.), *Molecules, Systems and Signaling in Liver Injury*, Cell Death in Biology and Diseases, DOI 10.1007/978-3-319-58106-4_4

X.-M. Yin and Z. Dong (Series eds.), *Cell Death in Biology and Diseases*

HCV	Hepatitis C virus
HIF	Hypoxia-inducible factor
IFNγ	Interferon γ
IGF	Insulin-like growth factor
IL	Interleukin
iNOS	Inducible nitric oxide synthase
LTB_4	Leukotriene B_4
LPS	Lipopolysaccharide
Ly6C	Lymphocyte antigen 6 complex
SIRS	Systemic inflammatory response syndrome
SREBP-1c	Sterol regulatory element-binding protein 1c
TLR	Toll-like receptor
TNFα	Tumor necrosis factor α
NALD	Non-alcoholic liver disease
NGF	Nerve growth factor
NO	Nitric oxide
Nr4a1	Nuclear receptor subfamily 4 group A member 1
PDGF	Platelet-derived growth factor
PGE_2	Prostaglandin E_2
PHD	Prolyl hydroxylase
PPARα	Peroxisome proliferator-activated receptor α
RNS	Reactive nitrogen species
ROS	Reactive oxygen species
SPM	Specialized pro-resolving mediators
TGFβ	Transforming growth factor β
Treg	Regulatory T cell
VEGF	Vascular endothelial growth factor
VHL	Von Hippel-Lindau

4.1 Introduction

Macrophages are key effectors of innate immune response to pathogens and the initiation of inflammation, and they contribute to maintaining tissue homeostasis, tissue repair, and development (Wynn et al. 2013b). Macrophages also help shape the adaptive immune response through the presentation of foreign antigens and production of chemokines which facilitate the trafficking of adaptive immune cells to sights of inflammation and tissue damage (Newson et al. 2014). Macrophages are classified as either tissue resident (i.e., liver Kupffer cells, brain microglia, and alveolar macrophages in the lung) or recruited (monocyte−/bone marrow-derived macrophages). In mice, monocytes are subdivided based on their expression of Ly6C and CCR2. $Ly6C^{hi}$ $CCR2^{+}$ monocytes rapidly infiltrate tissue upon injury (Karlmark et al. 2009). During acute or chronic liver injury, $Ly6C^{hi}$ monocyte-derived macrophages are the predominate macrophage subtype (Tacke and Randolph 2006;

Karlmark et al. 2009). Ly6C^{low}CCR2^{-} monocytes have been shown to exhibit a "patrolling" behavior and have been found crawling along the hepatic endothelium (Carlin et al. 2013). While monocyte-derived Ly6C^{hi} initially exert tissue-destructive properties, they can differentiate into Ly6C^{low} monocyte-derived macrophages which promote tissue repair and resolution of inflammation (Dal-Secco et al. 2015). The human monocyte counterparts are classified as classical (CD14^{hi}CD16^{-}), intermediate (CD14^{+}CD16^{+}), and nonclassical (CD14^{dim}CD16^{+}) (Ingersoll et al. 2010). Mouse and human monocytes share much similarity but differ in several aspects that must be taken into consideration when translating mouse and human studies. For instance, based on receptor expression CD16^{-} and Ly6C^{hi}, monocytes highly express CCR2, whereas CD16^{hi} and Ly6C^{lo} monocytes express high levels of CX$_3$CR1 (Geissmann et al. 2003; Tacke et al. 2007). However, in regard to FcγR1 expression, mRNA and protein expression is conserved between CD16^{-} and Ly6C^{hi} monocytes, whereas human CD16^{+} monocytes lose FcγR1 surface protein expression which is retained on Ly6C^{low} monocytes (Ingersoll et al. 2010). Furthermore, Ly6C^{low} monocytes are more instrumental in phagocytosis, whereas in humans, CD16^{-} monocytes are more phagocytic (Wildgruber et al. 2009; Tacke et al. 2006). Macrophages display a remarkable ability to alter their phenotype based on tissue microenvironmental cues such as hypoxia, cytokines, lipids, and exposure to dead cells. In acute and chronic liver diseases, various subsets of hepatic macrophages have been found to be involved in disease pathogenesis. In this review, we will summarize the development, heterogeneity, and plasticity of macrophages in response to microenvironmental stimuli in physiological and pathophysiological conditions of the liver.

4.2 Types and Origins of Hepatic Macrophages

In normal healthy liver, tissue-resident Kupffer cells comprise 90% of the macrophage population (Ju and Tacke 2016). The remaining population of liver macrophages under homeostatic conditions are comprised of monocyte-derived, Ly6C^{low}, blood vessel patrolling macrophages (Fig. 4.1). The origin of Kupffer cells has been a long-standing debate, and recent fate mapping studies have provided some insights. The developmental fate of macrophages has been deduced using time course analysis of *Csf1r*-, *Flt3*-, and *Tie2*-expressing cells. It was shown that adult tissue-resident macrophages in the brain, liver, lung, skin, and spleen arise almost exclusively from a *Tie2* progenitor pathway (Gomez Perdiguero et al. 2014). Furthermore, the data indicate that liver-resident Kupffer cells develop from fetal yolk-sac progenitor cells during embryonic day 6.5 (E6.5)–E8.0 and are only marginally replaced by hematopoietic stem cells during development (Gomez Perdiguero et al. 2014). Additionally, analysis of mice utilizing green fluorescent protein (GFP) reporter or Cre recombinase in their CX$_3$CR1 gene (Yona et al. 2013) and parabiosis experiments (Hashimoto et al. 2013) have shown that during homeostatic scenarios, monocytes do not significantly contribute to most tissue-resident

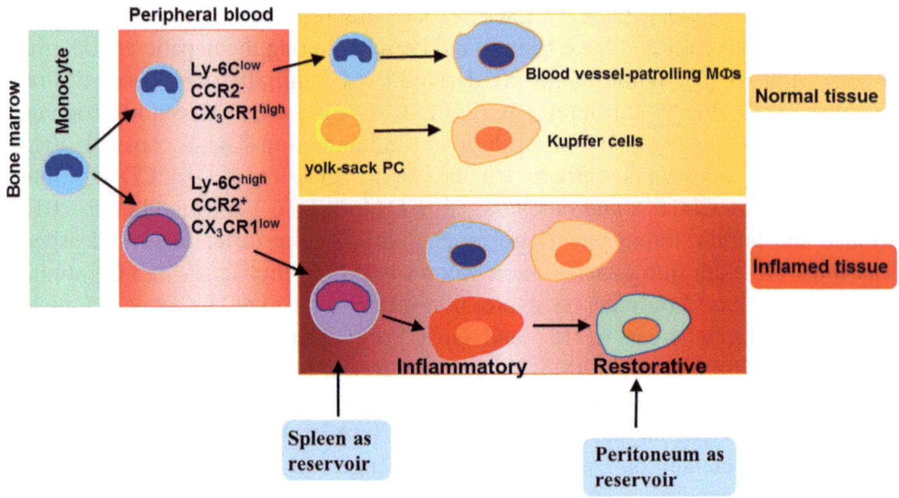

Fig. 4.1 Macrophages in normal and inflamed tissues. Monocyte development in the bone marrow and reside in the blood as Ly6C^{high} monocyte and Ly6C^{low} monocytes. In normal liver tissue, the two predominant types of macrophages are tissue yolk-sac progenitor-derived tissue-resident Kupffer cells and blood vessel patrolling Ly6C^{low} monocyte-derived macrophages. Following injury and subsequent inflammation, the liver is rapidly infiltrated by Ly6C^{hi} monocytes that differentiate into pro-inflammatory Ly6C^{hi} macrophages. Additional Ly6C^{hi} monocytes reside in the spleen and can be mobilized following injury. As inflammation progresses, the Ly6C^{hi} pro-inflammatory macrophages convert to the tissue-restorative Ly6C^{low} macrophage phenotype to promote resolution of the inflammatory response. Additionally, a population of Ly6C^{low} tissue-restorative macrophages in the peritoneum can be rapidly recruited into the liver during injury

macrophage populations. Several studies have indicated that tissue-resident macrophages develop from different lineage depending on the cells tissue fate. For instance, microglial cells arise predominately from yolk-sac-derived macrophages (Ginhoux et al. 2010), while Langerhans cells mostly derive from fetal liver progenitors but retain yolk-sac-derived components (Hoeffel et al. 2012). Additionally, it has been proposed that macrophages of the lung, heart, kidney, and liver derive from fetal liver-derived progenitors (Guilliams et al. 2013; Epelman et al. 2014; Ginhoux and Jung 2014). Finally it has been shown that tissue macrophages can replenish themselves (Hashimoto et al. 2013).

Kupffer cells are principally involved in maintaining tissue homeostasis by serving as an immune sentinel. They sense antigens, pathogens, or damaged cells through interaction with a number of cell surface receptors (such as complement receptor, toll-like receptors (TLR), scavenger receptors, and Fcγ receptor) to initiate and potentiate the inflammatory response (Sica et al. 2015). The roles of macrophages are elicited through the production of trophic factors [i.e., macrophage colony-stimulating factor (CSF), granulocyte/macrophage CSF (GM-CSF), insulin-like growth factor (IGF), nerve growth factor (NGF)], immune mediators [i.e., interferon γ (IFNγ), tumor necrosis factor α (TNFα), chemokine (C-C motif) ligand 2 (CCL2), interleukin-4 (IL-4)], and effector signaling molecules [inducible nitric oxide synthase (iNOS)] (Gautier et al. 2012; Ju and Tacke 2016). In addition, Kupffer cells are involved in the development of immune tolerance, acting as an antigen-presenting cell. Studies have shown that Kupffer cells are involved in liver-mediated tolerance to allografts (Kamei et al. 1990; Sato et al. 1996), to induce tolerance to soluble antigens (Ju et al. 2003) and suppress dendritic cell-mediated T cell proliferation (You et al. 2008).

During liver injury, resident Kupffer cells mount initial responses by rapid production of inflammatory cytokines, such as IL-1β and TNFα, and chemokines (CCL2) which facilitate the recruitment of additional inflammatory cells, such as monocytes. $Ly6C^{hi}$ monocytes are recruited to inflamed tissue and differentiate into $Ly6C^{hi}$ macrophages which promote inflammatory progression (Serbina et al. 2012) (Fig. 4.1). The trafficking of monocytes to inflamed tissue is facilitated through the interaction of monocyte CCR2 with CCL2. Several studies using either infection or sterile tissue damage have demonstrated the influx of $Ly6C^{hi}$ monocytes (Lin et al. 2009; Morias et al. 2015; Helk et al. 2013; Karlmark et al. 2009; You et al. 2013; Serbina et al. 2012). The $Ly6C^{+}$ cells switch to a tissue-restorative $Ly6C^{low}$ macrophage as injury resolves (Wang et al. 2014). The $Ly6C^{low}$ macrophages are instrumental in the resolution of the inflammatory and wound healing response. Ample evidence supports the concept that $Ly6C^{low}$ macrophages are derived from $Ly6C^{hi}$ macrophages in injured tissues rather than recruited from circulation (Nahrendorf et al. 2007; Ramachandran et al. 2012; Ludovic et al. 2007; Lin et al. 2009; Hilgendorf et al. 2014; Varga et al. 2013). In an animal model of kidney injury, it was shown that the generation of $Ly6C^{low}$ monocytes during disease progression is from differentiating $Ly6C^{high}$ monocytes and that this phenotypic conversion counteracts inflammation and facilitates tissue repair (Lin et al. 2009). This concept has been validated in studies utilizing $Nr4a1^{-/-}$ mice, in which the absence of Nr4a1 inhibits $Ly6C^{low}$ monocyte production in the bone marrow (Hanna et al. 2011). In these studies focused on myocardial infarction and muscle injury, it was observed that wild-type and $Nr4a1^{-/-}$ mice display similar accumulation of $Ly6C^{hi}$ and $Ly6C^{low}$ macrophages (Hilgendorf et al. 2014; Varga et al. 2013). These observations suggest that $Ly6C^{hi}$ macrophages give rise to $Ly6C^{low}$ macrophages.

Aside from the bone marrow, the spleen serves as an immediate reservoir for the mobilization of large quantities of $Ly6C^{hi}$ monocytes following injury (Swirski et al. 2009). It was shown that following myocardial ischemia, the spleen harbors fewer monocytes in comparison with spleens in non-ischemic animals, and the number of monocytes in the blood significantly increased indicating a flux of splenic monocytes into circulation. Furthermore, $GATA6^{+}$ peritoneal macrophages, which are $Ly6C^{low}$ and are important in facilitating tissue repair, have recently been shown to infiltrate directly into the liver within 1 h of liver injury in response to signals from necrotic cells and that depletion of peritoneal macrophages increases mortality in models of acute liver injury (Wang and Kubes 2016) (Fig. 4.1).

4.3 Tissue Microenvironmental Cues Governing Macrophage Phenotypes and Functions

Besides the heterogeneity, the phenotype and functions of macrophages change based on the local tissue environment. The plasticity of macrophages is coordinated through microenvironmental cues and the subsequent response of macrophages. Several inciting factors (hypoxia, cytokines, lipids, and dead cells) have been elucidated that stimulate the diverse array of macrophage functions.

4.3.1 Impact of Tissue Hypoxia on Macrophages

Inflammatory signaling typically results in a severely oxygen-deprived tissue environment due to increased oxygen demand, as well as increased consumption by infiltrated inflammatory cells. For example, neutrophils consume copious amounts of oxygen for respiratory burst (Eltzschig and Carmeliet 2011). Macrophages often accumulate in hypoxic areas and macrophage functions are strongly influenced by hypoxia (Murdoc et al. 2005). The expression of hypoxia-inducible factor (HIF) allows for organismal adaptation to hypoxia by increasing oxygen delivery or allowing for the adaptation of the organism to decreased oxygen availability (Wenge 2002). HIFs consist of a constitutively expressed β-subunit and an oxygen-sensitive α-subunit, whose function is controlled via prolyl hydroxylase (PHD) and von Hippel-Lindau (VHL)-mediated ubiquitination and degradation (Ohh et al. 2000; Maxwell et al. 1999). The two predominant α-subunit isoforms of HIF are HIF1α and HIF2α. Increased HIF1α and HIF2α protein expressions have been reported in several liver diseases including non-alcoholic liver disease (NALD), alcoholic liver disease (ALD), and hepatocellular carcinoma (HCC) (Bangoura et al. 2004; Corpechot et al. 2002; Novo et al. 2012). Additionally, HIFs may contribute to fibrosis development as hypoxic areas co-localize with those of fibrosis, and HIF stabilization has been reported in liver fibrosis models (Corpechot et al. 2002; Moon et al. 2009). Critical cellular processes such as glycolysis, oxygen homeostasis, fat metabolism, tissue remodeling, angiogenesis, and cell proliferation are influenced by HIF1α signaling (Semenza 2001; Hirota 2002). Several pathological processes in which macrophage are involved (i.e., inflammation, wound healing, tumor development) are characterized by hypoxia (Semenza 2014). Macrophages rely on HIFs for energy production and activity (Burke et al. 2002). HIF1α and HIF1β have been shown to be increased during the differentiation of monocytes to macrophages in vitro and from human peripheral blood monocytes (Oda et al. 2006). Both HIF1α and HIF2α accumulate in primary human macrophages and mouse bone marrow-derived macrophages under hypoxic conditions in vitro (Griffiths et al. 2000; Talks et al. 2000). HIF1α protein has been shown to be stabilized in pro-inflammatory macrophages and is linked to nitric oxide (NO) production (Takeda et al. 2010). HIF2α has been shown to be stabilized in tissue-restorative macrophages and is linked to arginase 1 (Arg1) expression (Takeda et al. 2010). These studies highlight the importance of HIF expression in regulating macrophage phenotype and function.

4.3.2 Macrophage Responses to Cytokines

Cytokine-induced macrophage activation and cytokines produced by activated macrophages have been used to group macrophages into two distinct groups, the classically activated and alternatively activated macrophages (Murray et al. 2014). LPS either alone or in combination with cytokines (i.e., IFNγ, TNFα, and GM-CSF) can induced classically activated macrophages, which produce immune-stimulatory cytokines (TNFα, IL-1β, IL-6, IL-12, and IL-23) as well as reactive oxygen species (ROS) and reactive nitrogen species (RNS) (Steevels and Meyaard 2011). IL-4, IL-10, IL-13, and transforming growth factor β (TGFβ) have been shown to promote alternatively activated macrophages. In a model of muscle injury and repair, ablation of regulatory T cells (Treg) produced a more severe injury due the inability of the infiltrated, pro-inflammatory Ly6C^{hi} macrophages to switch to the tissue-restorative Ly6C^{low} macrophages. Among transcripts shown to be downregulated due to the ablation of Tregs was IL-10, a known polarizer of the tissue-restorative Ly6C^{low} macrophages (Burzyn et al. 2013). Similarly, in a mouse model of acute myocardial infarction (AMI) depletion of Tregs exacerbated the infarct area and delayed resolution (Weirather et al. 2014). This observation was attributed to the Treg-deficient mice displaying mainly pro-inflammatory macrophages compared to control-treated AMI mice, indicative of a delayed transition to the tissue-restorative macrophage phenotype (Weirather et al. 2014). These observations are further supported in a mouse model of angiotensin II-induced kidney injury, in which Treg depletion resulted in more pro-inflammatory macrophage polarization and increased injury. As in the studies focused on muscle injury and repair, the injurious phenotype was attributed to reduced anti-inflammatory cytokine expression, such as IL-10 (Mian et al. 2016). While cytokine stimulation generally activates macrophages to one phenotype or another, certain pathogens exploit cytokine stimulation as a means of dampening immune responses. Infection with *Salmonella typhimurium* (*S. typhimurium*) resulted in a delayed immune response and increased bacterial burden due to macrophage necroptosis. This observed macrophage necroptosis was mediated through the induction of IFNγ (Robinson et al. 2012).

4.3.3 Effects of Lipid Mediators on Macrophages

Lipid mediators such as resolvins, protectins, and maresins embody a class of molecules termed specialized pro-resolving mediators (SPM) which facilitate the resolution of inflammatory responses (Dalli and Serhan 2016). In the process of resolution, pro-inflammatory macrophages undergo phenotypic changes to that of the tissue-restorative macrophages. SPMs have been shown to influence the ability of conversion to the tissue-restorative macrophages, thereby promoting inflammation resolution (Serhan and Savill 2005; Dalli and Serhan 2016). During the initiation of inflammation, arachidonic acid-derived lipid mediators actively facilitate the egress of leukocytes to a site of inflammation. Prostaglandin E_2 (PGE_2) and leukotrienes B_4 (LTB_4) which are generated via metabolism of arachidonic acid play vital roles in endothelial changes and diapedesis of neutrophils to inflammatory sites (Levy et al. 2001; Serhan and Savill 2005). During inflammatory response progression, lipid-derived mediators switch from the inflammation-inducing leukotrienes to the inflammation-prohibitive lipoxins A_4 and B_4 (Serhan and Savill 2005). Lipoxin A_4 increases TGF-β1 expression during resolution thereby mediating the resolution response (Bannenberg et al. 2005). Serhan et al. have shown that resolution of inflammation is a biochemically active process involving the biosynthesis of a set of SPMs including lipoxins, E-series resolvins derived from eicosapentaenoic acid (EPA), D-series resolvins derived from docosahexaenoic acid (DHA), protectins, and maresins (Bannenberg et al. 2005; Serhan and Savill 2005). Although their actions are specific, each class of SPMs participates in limiting neutrophil recruitment, limiting microbial invasion and enhancing the uptake of cellular debris and apoptotic neutrophils in resolution (Bannenberg et al. 2005).

4.3.4 Phagocytosis of Dead Cells by Macrophages

Cell death helps shape the inflammatory response and the programming of macrophages. The recognition of apoptotic cells by macrophages has a profound impact on altering the macrophages response and promoting tissue restoration (Gordy et al. 2011; Goren et al. 2009). During inflammation resolution, uptake of apoptotic neutrophils by macrophages is an essential step in the reduction of inflammation (Savill 1997). Uptake of apoptotic cells may stimulate macrophages to release anti-inflammatory mediators, such as TGF-β1, which suppress inflammation (Huynh et al. 2002; Lucas et al. 2003). Efferocytosis is the process by which macrophages ingest and remove apoptotic and necrotic cells. Efferocytic receptors on macrophages sense surface changes on apoptotic cells, such as the flipping of phosphatidylserine from the inner to the outer leaflet of the cell membrane, to distinguish apoptotic cells from healthy ones (Fadok et al. 1992). Effective removal of dead cells is paramount for the resolution of the inflammatory response by preventing the release of intracellular contents (such as proteases and inflammatory mediators from neutrophils) of dying cells into the tissue environment perpetuating inflammation (Vandivier et al. 2006). Efferocytosis of dead hepatocytes has been shown to promote the transition from the $Ly6C^{hi}$ macrophages to the tissue-restorative $Ly6C^{low}$ (Wang et al. 2014). Recognition and removal of neutrophils following the initial response to an injury help shape the inflammatory response toward a restorative process. Exposure of macrophages to apoptosing neutrophils suppresses pro-inflammatory cytokine and chemokine production through the production of TGFβ and IL-10 (Bratton and Henson 2011; Korns et al. 2011). Studies using a mouse model of chronic granulomatous disease (CGD), in which autoimmunity and excessive inflammation are known to occur, have shown that macrophage efferocytosis of neutrophils is impaired resulting in hyperinflammation (Fernandez-Boyanapalli et al. 2009). Macrophages from CGD mice displayed characteristics of pro-inflammatory macrophages with impaired efferocytosis compared to macrophages from wild-type (WT) mice which display a tissue-reparative phenotype. In addition, WT macrophages expressed more IL-4 compared to CGD macrophages, and the impaired efferocytosis of CGD macrophages was attenuated by pretreatment with IL-4 (Fernandez-Boyanapalli et al. 2009). Recent studies have demonstrated the impact of Kupffer cell death on mediating infiltrating macrophage phenotype. In this study, infection with *Listeria monocytogenes* resulted in rapid Kupffer cell necroptosis within the first hours of infection. Kupffer cell death triggered the release of IL-33, which triggers production of IL-4 (Pecaric-Petkovic et al. 2009), resulting in recruited monocyte-derived macrophages to proliferate and switch from a pro-inflammatory to an anti-inflammatory phenotype (Blériot et al. 2015).

4.4 Hepatic Macrophages in Acute and Chronic Liver Injury

The heterogeneous phenotype of liver macrophages has been exemplified in numerous studies of both acute and chronic liver injury. As described below, macrophages exert opposing roles in a variety of liver injury models. They can facilitate the perpetuation of the inflammatory response or act to dampen inflammation and promote tissue restoration. However, the functions of both pro-inflammatory and tissue-restorative macrophages must remain in check as excessive responses from either macrophage population can contribute to disease progression in both acute and chronic liver diseases.

4.4.1 Acetaminophen-Induced Liver Injury (AILI)

Acetaminophen (APAP) overdose causes severe damage to the liver which can progress to liver failure. Following hepatocyte damage from APAP, resident Kupffer cells are activated and release a number of pro-inflammatory mediators, TNFα and IL-1β, and chemoattractants, CCL2, which promote inflammation and a massive influx of circulating monocytes (Zigmond et al. 2014; Holt et al. 2008). In addition to activation of Kupffer cells, CCR2-depenent hepatic accumulation of circulating monocytes has been observed (Holt et al. 2008). Recent studies have also shown that both $Ly6C^{hi}$ and $Ly6C^{low}$ monocyte-derived macrophages are present in the liver during AILI (Zigmond et al. 2014). In the early resolution phase, the infiltrating

Ly6C^{hi} monocytes differentiate into Ly6C^{lo} macrophages and become the predominant liver macrophage phenotype (Zigmond et al. 2014). Mice lacking CCR2 or those treated with anti-CCR2 antibodies exhibit delayed recovery following APAP administration suggesting a prominent role of monocyte-derived macrophages in tissue repair following acute injury (Zigmond et al. 2014; Holt et al. 2008). Interestingly, it was shown that the number of resident Kupffer cells significantly reduces after APAP administration but recovers during the resolution phase by self-renewal (Zigmond et al. 2014). When both hepatic macrophage populations, resident and infiltrating, are depleted, the recovery from APAP-induced liver injury is even more delayed further highlighting the influence of hepatic macrophages in recovery after acute liver damage (You et al. 2013). Moreover, it has been shown that enhancing both infiltrating and resident macrophage populations via administration of CSF1 improved AILI (Stutchfield et al. 2015). However, debate still continues as to whether yolk-sac-derived tissue-resident Kupffer cells repopulate the liver or if the reestablishment of macrophage populations is from circulating monocytes.

4.4.2 Parasite-Induced Liver Inflammation

Liver macrophages contribute to the severity and protection against parasite-induced liver inflammation by producing tissue-destructive TNFα as well as being the major source of anti-inflammatory cytokine, IL-10. In a mouse model of *Trypanosoma congolense* infection, three distinct myeloid cell subsets were observed including Ly6C^{hi} "inflammatory" monocytes that produce TNFα and contribute to the development of systemic inflammatory response syndrome (SIRS); Ly6C^{-} "patrolling" monocytes, which produce IL-10 limiting TNFα-mediated tissue damage; and IL-10-producing macrophages (Morias et al. 2015). In these studies, it was shown that 7 days postinfection, C57Bl/6 mice displayed high accumulation of Ly6C^{hi} monocytes which corresponded to a peak in pro-inflammatory immune response. After 21 days postinfection, the number of Ly6C^{hi} monocytes reduced, and the predominant myeloid cell type was that of Ly6C^{low} monocytes and macrophages. By the use of CCR2$^{-/-}$ mice, the investigators determined that following *T. congolense* infection, CCR2$^{-/-}$ mice display less Ly6C^{hi} monocytes, Ly6C^{low} monocytes, and macrophages compared to their wild-type counterparts (Morias et al. 2015). This data shows during parasite infection that the majority of Ly6C^{low} monocytes and macrophages arise from infiltrating Ly6C^{hi} monocytes and that the Ly6C^{low} cells promote this phenotypic switch. In addition to infiltrating monocytes/macrophages, Kupffer cells have been shown to influence the severity of parasite-induced liver inflammation. In a mouse model of *Entamoeba histolytica* infection, amebic liver abscess (ALA) formation was almost completely abolished by depletion of Kupffer cells by gadolinium chloride or clodronate liposomes (Helk et al. 2013). Kupffer cells were shown to substantially contribute to liver damage during ALA formation. In addition to Kupffer cells, Ly6C^{hi} monocytes were shown to be instrumental in abscess formation (Helk et al. 2013). In $CCR2^{-/-}$ mice, it was observed that abscess formation was almost abolished following infection and was accompanied by a significant reduction in $Ly6C^{hi}$ monocytes. Abscess formation was restored following adoptive transfer of $CD115^{+}$ monocytes into $CCR2^{-/-}$ mice (Helk et al. 2013). These studies further show the dual role of tissue destruction and restoration in monocyte/macrophage populations.

4.5 Chronic Liver Disease

Multiple etiological contributors have been shown to result in the development of chronic liver disease such as viral infection, alcohol abuse, metabolic disorders, and autoimmune disease. With the exception of viral hepatitis, chronic liver injury stems from the accumulation of fat in the liver (steatosis) and the development of liver inflammation (hepatitis) which over time progresses to liver fibrosis with continued fibrosis resulting in cirrhosis and finally progressing to hepatocellular carcinoma. Macrophages play major roles in each stage in the progression of chronic liver injury.

4.5.1 Non-alcoholic and Alcoholic Liver Disease (NALD and ALD)

Non-alcoholic and alcoholic liver disease are leading causes of liver-related morbidity and mortality in western countries (Gao and Bataller 2011; Sakaguchi et al. 2011). The pathogenesis is similar between both diseases starting with fatty liver and developing into hepatitis. Kupffer cells have been shown to play a major role in hepatocytes metabolic adaptation by regulating the oxidation of fatty acids and promoting lipid storage in hepatocytes (Olefsky and Glass 2010). Kupffer cells also become activated in response to overload of lipids and cholesterol in the liver (Dixon et al. 2013) and produce high levels of TNFα and IFNγ (Bergheim et al. 2008). The earliest response of the liver to alcohol abuse is the development of steatosis in hepatocytes. Early studies show that alcohol intake disrupts mitochondrial β-oxidation of fatty acids resulting to steatosis development (Baraona and Lieber 1979). Additionally, ethanol increases fatty acid synthesis in hepatocytes by upregulating sterol regulatory element-binding protein 1c (SREBP-1c) (You et al. 2002) and inhibits peroxisome proliferator-activated receptor (PPAR)-α reducing fatty acid oxidation (Yu et al. 2003). Ablation of Kupffer cells has been shown to inhibit the development of fatty liver after alcohol exposure (Adachi et al. 1994) and in response to a high-fat diet (Leroux et al. 2012).

A large number of people develop fatty liver without any other complications; however, others progress from steatosis to hepatitis. Persistent inflammation and injury can lead to scaring in the liver which can progress to cirrhosis (Marra et al. 2008). Accumulation of fat in liver cells has been shown to induce hepatocyte apoptosis through mitochondrial impairment which is associated with oxidative stress and the production of reactive oxygen species (ROS) and reactive nitrogen species (RNS) (Rubio et al. 2007). The cytotoxic effects of oxidative stress on hepatocytes lead to cell death via apoptosis and necrosis further promoting inflammation and fibrosis (Feldstein et al. 2003; Wieckowska et al. 2006). The major source of inflammatory mediators, TNFα, IL-1β, and ROS, during non-alcoholic liver disease pathogenesis is from Kupffer cells and bone marrow-derived macrophages (Ramadori and Armbrust 2001; Rivera et al. 2007). In a mouse model of NALD, using the methionine-choline-deficient (MCD) diet, it was observed that conventional Kupffer cells were lost early in disease development and were replaced by monocyte-derived macrophages. Characterization of the infiltrating macrophages revealed high CCR2 expression, low Ly6C expression, and reduced expression of genes associated with anti-inflammatory macrophages. This is conflicting to the typical view of macrophages with high CCR2 expression also displaying high Ly6C expression. This data further exemplifies the plasticity of macrophage phenotype in a given microenvironment.

In patients with alcoholic hepatitis and fibrosis, clinical observations support the notion that macrophages promote disease progression (Tapia-Abellán et al. 2012). Macrophage inflammatory genes are upregulated in cirrhosis patients (Tapia-Abellán et al. 2012), and several indicators of macrophage activation are elevated in ALD patients. Circulating monocytes from ALD patients are more susceptible to lipopolysaccharide (LPS) stimulation (Gobejishvili et al. 2006; Zhang et al. 2001). Activation of hepatic macrophages via ethanol administration has been shown to enhance the production of TNFα, IL-6, CCL2, and ROS in animal model of ALD (Enomoto et al. 2000; Petrasek et al. 2012). Depletion of hepatic macrophages reduces alcohol-induced inflammation (Koop et al. 1997). Pro-inflammatory Kupffer cells have been shown to perpetuate the disease progression in ALD, and reducing pro-inflammatory Kupffer cells is protective against ALD (Wan et al. 2014). Furthermore, it was demonstrated that tissue-restorative Kupffer cells promote apoptosis of pro-inflammatory Kupffer cells thereby limiting ALD progression (Wan et al. 2014). Both resident Kupffer cells and infiltrating macrophages are involved in ALD pathogenesis in animal studies. The infiltrating macrophages consist of the $Ly6C^{low}$ and $Ly6C^{+}$ subsets. Upon phagocytosis, the pro-inflammatory $Ly6C^{hi}$ macrophages appear to switch phenotype to the tissue-restorative $Ly6C^{low}$ phenotype (Wang et al. 2014). In the context of ALD, macrophages actively participate in tissue inflammation and damage as well as the resolution of fibrosis rendering therapeutic targeting of macrophage populations difficult. However, further understanding of the microenvironmental cues which govern the switch between tissue-destructive and tissue-restorative macrophages may elucidate additional therapeutic avenues.

4.5.2 Liver Fibrosis

Chronic liver inflammation ultimately leads to liver fibrosis, and macrophages have been shown to exert dual and opposing roles in fibrosis signaling. During the early phase of fibrosis initiation, Ly6C^{hi} monocytes differentiate into pro-inflammatory Ly6C^{hi} macrophages that interact with HSCs to promote fibrosis. Pro-fibrotic macrophages secrete various pro-fibrotic mediators, such as IL-1β, platelet-derived growth factor (PDGF), connective tissue growth factor (CTGF), and TGFβ (Kolb et al. 2001; Seki et al. 2007; Kinnman et al. 2000; Paradis et al. 2002). By the use of CCR2$^{-/-}$ mice and in mice-administered pharmacological inhibitors of CCL2, it was shown that blockade of the CCL2/CCR2 pathway inhibits Ly6C^{hi} monocyte recruitment and attenuates liver fibrosis (Karlmark et al. 2009; Ehling et al. 2014). In addition to promoting fibrosis progression, macrophages have been shown to be instrumental in fibrosis resolution through the production of IL-10, Arg1, and matrix metalloproteinases (MMP) (Pesce et al. 2009; Wilson et al. 2007; Ramachandran et al. 2012; Wynn et al. 2013a). Ramachandran et al. characterized the macrophage subset that facilitates fibrosis resolution as the Ly6C^{low} macrophage. Furthermore, they demonstrated the role of the tissue-restorative Ly6C^{low} macrophages in fibrosis resolution through the use of transgenic CD11B-DTR mice in which administration of diphtheria toxin selectively ablates Ly6C^{low} macrophages. Depletion of the Ly6C^{low} macrophages in a CCl_4-induced hepatic fibrosis model resulted in impaired fibrosis resolution (Ramachandran et al. 2012). Therefore, further understanding the mechanisms involved in the conversion of pro-fibrogenic macrophages to tissue-restorative macrophages could provide therapeutic insight into managing liver fibrosis through promoting macrophage-elicited fibrosis resolution.

4.5.3 Viral Hepatitis

Infection with either hepatitis B virus (HBV) or hepatitis C virus (HCV) can result in chronic liver disease and the progression to HCC (Ganem and Prince 2004; Lauer and Walker 2001). Macrophages have been shown to play an important role in viral hepatitis-induced liver damage. While only limited studies have been conducted showing the impact of HBV infection of Kupffer cells, HCV and its proteins have been shown to activate Kupffer cells. Following infection with HCV, increases in the number of Kupffer cells as well as the expression level of Kupffer cell activation markers such as CD163 and CD33 have been observed (Burgio et al. 1998; Khakoo et al. 1997; McGuinness et al. 2000). In addition, Kupffer cells release pro-inflammatory cytokines IL-1β, IL-6, IL-18, and TNFα (Tu et al. 2010; Hosomura et al. 2011; Shrivastava et al. 2013). The cytokines released by Kupffer cells exhibit opposing effects on modulating antiviral activities with TNFα promoting cellular entry of HCV (Boltjes et al. 2014) and IL-6, IL-18, and IFNγ inhibiting viral replication (Broering et al. 2008; Zhu and Liu 2003; Zhu et al. 2004). In addition, Kupffer cells have been shown to produce pro-fibrogenic factors which promote the progression of liver injury toward fibrosis/cirrhosis (Wallace et al. 2008). Due to the difficulty in distinguishing Kupffer cells from infiltrating macrophages, the available information in regard to the role of infiltrating macrophages in viral hepatitis liver injury is limited. Therefore, to date, the response observed in viral-induced liver injury has been attributed to that of the Kupffer cells. However, future research to elucidate the contributions of the different macrophage population in viral hepatitis will help understand which immune cells actively participate in disease progression and therefore provide potential therapeutic targets.

4.5.4 Hepatocellular Carcinoma (HCC)

Persistent inflammation and fibrosis in the liver ultimately ends with the development of HCC which is the most common type of primary liver cancer and the third leading cause of cancer-related death worldwide (Capece et al. 2013). During primary tumor development in HCC, Kupffer cells are the predominant macrophages cell type, but once the primary tumor is established, infiltrating macrophages are more instrumental in disease progression (Wu et al. 2012; Ju and Tacke 2016). To facilitate the growth and progression of HCC, macrophages produce a number of key signaling molecules associated with angiogenesis, cell growth, and metastasis. These signaling molecules include growth factors (vascular endothelial growth factor (VEGF), epidermal growth factor receptor (EGFR) ligands, PDGFβ, and TGFβ), cytokines (IL-6, TNFα, and IL-10), factors involved in matrix remodeling, and metastasis (MMPs and cyclooxygenase-2 (COX-2)) (Capece et al. 2013). Effectively targeting macrophages could prove beneficial in HCC management.

4.6 Conclusion

Macrophages represent a diverse and heterogeneous population of innate immune cells which depending on the microenvironment actively facilitate tissue damage as well as tissue restoration. An imbalance in the conversion of pro-inflammatory macrophages to tissue-restorative macrophages promotes a chronic inflammatory phenotype which can develop into fibrosis and cirrhosis and ultimately may progress to liver cancer. Hepatocyte damage caused by various insults lead to tissue inflammation, in which pro-inflammatory macrophages play an important role. In the normal process of wound healing, the microenvironment changes and promotes the phenotypic conversion to the tissue-restorative, inflammation-resolving macrophage subset, which plays a critical role in tissue repair. However, during chronic liver injury and inflammation, an imbalance in the conversion of pro-inflammatory macrophages

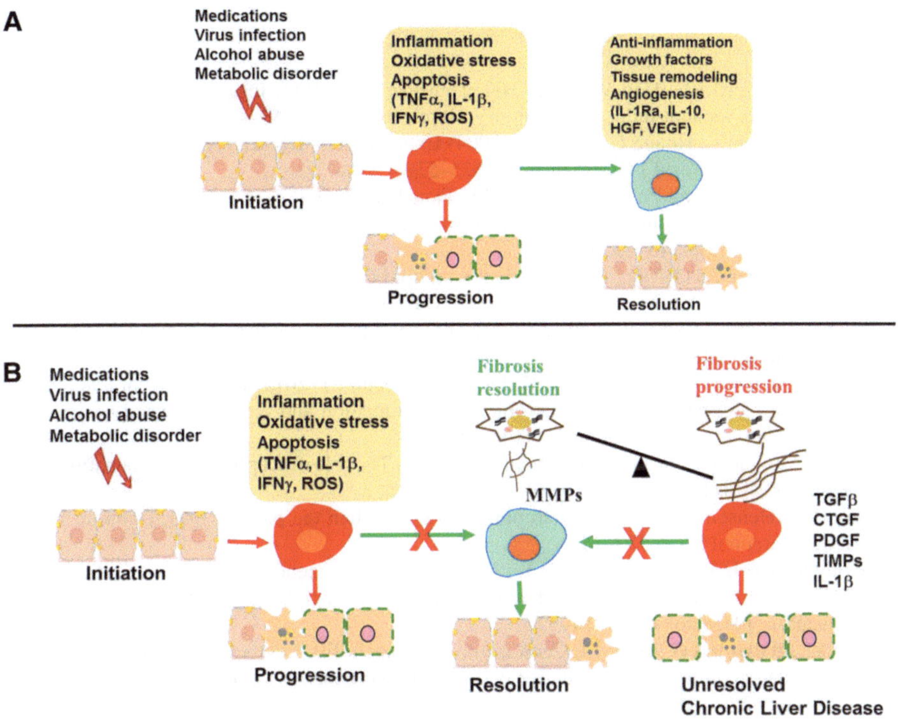

Fig. 4.2 Role of macrophages in inflammation and fibrosis. (**a**) Following insult to hepatocytes, pro-inflammatory macrophages release inflammatory cytokine which facilitates the removal of damage cells and cellular debris. In the normal process of inflammation resolution, pro-inflammatory macrophages undergo a phenotypic conversion to the tissue-restorative, inflammation-resolving macrophage subset. (**b**) When inflammation resolution becomes unbalanced, chronic inflammation can ensue and result in fibrosis development

to tissue-restorative macrophages promotes a chronic inflammatory phenotype which can develop into fibrosis and cirrhosis and ultimately may progress to liver cancer (Fig. 4.2). Potential therapeutic targets to managing excess inflammation in the liver and therefore potentially limiting the development of liver fibrosis consist of blocking recruitment of Ly6C^{hi} monocytes from the vasculature via blockade of the CCL2/CCR2 pathway. Additionally, stimulating the influx of Ly6C^{low} macrophages from the peritoneum could provide another means of modulating the macrophage populations to promote a tissue-restorative environment. Further studies elucidating the molecular pathways governing the conversion of the Ly6C^{hi} pro-inflammatory macrophages to that of the tissue-restorative Ly6C^{low} macrophages will be important in identifying potential therapeutic target to treat liver diseases with exuberant inflammation.

References

Adachi Y, Bradford BU, Gao W, Bojes HK, Thurman RG (1994) Inactivation of Kupffer cells prevents early alcohol-induced liver injury. Hepatology 20(2):453–460

Bangoura G, Yang L, Huang G, Wang W (2004) Expression of HIF-2alpha/EPAS1 in hepatocellular carcinoma. World J Gastroenterol 4:525–530

Bannenberg GL, Chiang N, Ariel A, Arita M, Tjonahen E, Gotlinger KH, Hong S, Serhan CN (2005) Molecular circuits of resolution: formation and actions of Resolvins and Protectins. J Immunol 174(7):4345–4355

Baraona E, Lieber CS (1979) Effects of ethanol on lipid metabolism. J Lipid Res 20(3):289–315

Bergheim I, Weber S, Vos M, Krämer S, Volynets V, Kaserouni S, McClain CJ, Bischoff SC (2008) Antibiotics protect against fructose-induced hepatic lipid accumulation in mice: role of endotoxin. J Hepatol 48(6):983–992

Blériot C, Dupuis T, Jouvion G, Eberl G, Disson O, Lecuit M (2015) Liver-resident macrophage necroptosis orchestrates type 1 microbicidal inflammation and type-2-mediated tissue repair during bacterial infection. Immunity 42(1):145–158

Boltjes A, Movita D, Boonstra A, Woltman AM (2014) The role of Kupffer cells in hepatitis B and hepatitis C virus infections. J Hepatol 61(3):660–671

Bratton DL, Henson PM (2011) Neutrophil clearance: when the party is over, clean-up begins. Trends Immunol 32(8):350–357

Broering R, Wu J, Meng Z, Hilgard P, Lu M, Trippler M, Szczeponek A, Gerken G, Schlaak JF (2008) Toll-like receptor-stimulated non-parenchymal liver cells can Regulate hepatitis C virus replication. J Hepatol 48(6):914–922

Burgio VL, Ballardini G, Artini M, Caratozzolo M, Bianchi FB, Levrero M (1998) Expression of co-stimulatory molecules by Kupffer cells in chronic hepatitis of hepatitis C virus etiology. Hepatology 27(6):1600–1606

Burke B, Tang N, Corke KP, Tazzyman D, Ameri K, Wells M, Lewis CE (2002) Expression of HIF-1alpha by human macrophages: implications for the use of macrophages in hypoxia-regulated cancer Gene therapy. J Pathol 196(2):204–212

Burzyn D, Kuswanto W, Kolodin D, Shadrach JL, Cerletti M, Jang Y, Sefik E, Tan TG, Wagers AJ, Benoist C, Mathis D (2013) A special population of regulatory T cells potentiates muscle repair. Cell 155(6):1282–1295

Capece D, Fischietti M, Verzella D, Gaggiano A, Cicciarelli G, Tessitore A, Zazzeroni F, Alesse E (2013) The inflammatory microenvironment in hepatocellular carcinoma: a pivotal role for tumor-associated macrophages Biomed Res Int 2013:187204

Carlin LM, Auffray C, Geissmann F (2013) Measuring intravascular migration of mouse Ly6Clow monocytes in vivo using Intravital microscopy. Curr Protoc Immunol 33:1–16

Corpechot C, Barbu V, Wendum D, Kinnman N, Rey C, Poupon R, Housset C, Rosmorduc O (2002) Hypoxia-induced VEGF and collagen I expressions are associated with angiogenesis and fibrogenesis in experimental cirrhosis. Hepatology 35(5):1010–1021

Dalli J, Serhan C (2016) Macrophage proresolving mediators-the when and where. Microbiol Spectr 4(3):37–54

Dal-Secco D, Wang J, Zeng Z, Kolaczkowska E, Wong CHY, Petri B, Ransohoff RM, Charo IF, Jenne CN, Kubes P (2015) A dynamic Spectrum of monocytes arising from the in situ reprogramming of CCR2+ monocytes at a site of sterile injury. J Exp Med 212(4):447–456

Dixon LJ, Barnes M, Tang H, Pritchard MT, Nagy LE (2013) Kupffer cells in the liver. Compr Physiol 3(2):785–797

Ehling J, Bartneck M, Wei X, Gremse F, Fech V, Möckel D, Baeck C, Hittatiya K, Eulberg D, Luedde T, Kiessling F, Trautwein C, Lammers T, Tacke F (2014) CCL2-Dependent infiltrating macrophages promote angiogenesis in progressive liver fibrosis. Gut 63(12):1960–1971

Eltzschig HK, Carmeliet P (2011) Hypoxia and inflammation. N Engl J Med 364(20):1976–1977

Enomoto N, Ikejima K, Bradford BU, Rivera CA, Kono H, Goto M, Yamashina S, Schemmer P, Kitamura T, Oide H, Takei Y, Hirose M, Shimizu H, Miyazaki A, Brenner DA, Sato N,

Thurman RG (2000) Role of Kupffer cells and gut-derived endotoxins in alcoholic liver injury. J Gastroenterol Hepatol 15(s1):20–25
Epelman S, Lavine KJ, Beaudin AE, Sojka DK, Carrero JA, Calderon B, Brija T, Brija T, Gautier EL, Ivanov S, Satpathy AT, Schilling JD, Schwendener R, Sergin I, Razani B, Forsberg EC, Yokoyama WM, Unanue ER, Colonna M, Randolph GJ, Mann DL (2014) Embryonic and adult-derived resident cardiac macrophages are maintained through distinct mechanisms at steady state and during inflammation. Immunity 40(1):91–104
Fadok VA, Voelker DR, Campbell PA, Cohen JJ, Bratton DL, Henson PM (1992) Exposure of Phosphatidylserine on the surface of apoptotic lymphocytes triggers specific recognition and removal by macrophages. J Immunol 148(7):2207–2216
Feldstein AE, Canbay A, Angulo P, Taniai M, Burgart LJ, Lindor KD, Gores GJ (2003) Hepatocyte apoptosis and fas expression are prominent features of human nonalcoholic steatohepatitis. Gastroenterology 125(2):437–443
Fernandez-Boyanapalli RF, Frasch SC, McPhillips K, Vandivier RW, Harry BL, Riches DWH, Henson PM, Bratton DL (2009) Impaired apoptotic cell clearance in CGD due to altered macrophage programming is reversed by phosphatidylserine-dependent production of IL-4. Blood 113(9):2047–2055
Ganem D, Prince AM (2004) Hepatitis B virus infection — natural history and clinical consequences. N Engl J Med 350(11):1118–1129
Gao B, Bataller R (2011) Alcoholic liver disease: pathogenesis and new therapeutic targets. Gastroenterology 141(5):1572–1585
Gautier EL, Shay T, Miller J, Greter M, Jakubzick C, Ivanov S, Helft J, Chow A, Elpek KG, Gordonoc S, Mazloon AR, Ma'ayan A, Chua WJ, Hansen TH, Turley SJ, Merad M, Randolph GJ (2012) Gene-expression profiles and transcriptional regulatory pathways that underlie the identity and diversity of mouse tissue macrophages. Nat Immunol 13(11):1118–1128
Geissmann F, Jung S, Littman DR (2003) Blood monocytes consist of two principal subsets with distinct migratory properties. Immunity 19(1):71–82
Ginhoux F, Jung S (2014) Monocytes and macrophages: developmental pathways and tissue homeostasis TL-14. Nat Rev Immunol 14 VN-r(6):392–404
Ginhoux F, Greter M, Leboeuf M, Nandi S, See P, Gokhan S, Mehler MF, Conway SJ, Ng LG, Stanley ER, Samokhvalov IM, Merad M (2010) Fate mapping analysis reveals that adult microglia derive from primitive macrophages. Science 330(6005):841–845
Gobejishvili L, Barve S, Joshi-Barve S, Uriarte S, Song Z, McClain C (2006) Chronic ethanol-mediated decrease in cAMP primes macrophages to enhanced LPS-inducible NF-kappaB activity and TNF expression: relevance to alcoholic liver disease. Am J Physiol Gastrointest Liver Physiol 291(4):G681–G688
Gomez Perdiguero E, Klapproth K, Schulz C, Busch K, Azzoni E, Crozet L, Garner H, Trouillet C, de Bruijn MF, Geissmann F, Rodewald HR (2014) Tissue-resident macrophages originate from yolk-sac-derived Erythro-myeloid progenitors. Nature 518(7540):547–551
Gordy C, Pua H, Sempowski GD, He YW (2011) Regulation of steady-state neutrophil homeostasis by macrophages. Blood 117(2):618–629
Goren I, Allmann N, Yogev N, Schürmann C, Linke A, Holdener M, Waisman A, Pfeilschifter J, Frank S (2009) A transgenic mouse model of inducible macrophage depletion. Am J Pathol 175(1):132–147
Griffiths L, Binley K, Iqball S, Kan O, Maxwell P, Ratcliffe P, Lewis C, Harris A, Kingsman S, Naylor S (2000) The macrophage - a Novel system to deliver Gene therapy to pathological hypoxia. Gene Ther 7(3):255–262
Guilliams M, Kleer ID, Henri S, Post S, Vanhoutte L, Prijck SD, Deswarte K, Malissen B, Hammad H, Lambrecht BN (2013) Alveolar macrophages develop from fetal monocytes that differentiate into long-lived cells in the first week of life via GM-CSF. J Exp Med 210(10):1977–1992
Hanna RN, Carlin LM, Hubbeling HG, Nackiewicz D, Green AM, Punt JA, Geissmann F, Hedrick CC (2011) The transcription factor NR4A1 (Nur77) controls bone marrow differentiation and the survival of Ly6C- monocytes. Nat Immunol 12(8):778–785

Hashimoto D, Chow A, Noizat C, Teo P, Beasley MB, Leboeuf M, Becker CD, See P, Price J, Lucas D, Grete M, Mortha A, Boyer SW, Forsberg EC, Tanaka M, van Rooijen N, Garcia-Sastre A, Stanley ER, Ginhoux F, Frenette PS, Merad M (2013) Tissue-resident macrophages self-maintain locally throughout adult life with minimal contribution from circulating monocytes. Immunity 38(4):792–804

Helk E, Bernin H, Ernst T, Ittrich H, Jacobs T, Heeren J, Tacke F, Tannich E, Lotter H (2013) TNFalpha-mediated liver destruction by Kupffer cells and Ly6Chi monocytes during *Entamoeba histolytica* infection. *PLoS Pathog* 9(1):e1003096

Hilgendorf I, Gerhardt LMS, Tan TC, Winter C, Holderried TAW, Chousterman BG, Iwamoto Y, Liao R, Sirlik A, Sherer-Crosbie M, Hedrick CC, Libby P, Nahrendorf M, Weissleder R, Swirski FK (2014) Ly-6C high monocytes depend on nr4a1 to balance both inflammatory and reparative phases in the infarcted myocardium. Circ Res 114(10):1611–1622

Hirota K (2002) Hypoxia-inducible factor 1, a master transcription factor of cellular hypoxic Gene expression. J Anesth 16(2):150–159

Hoeffel G, Wang Y, Greter M, See P, Teo P, Malleret B, Leboeuf M, Low D, Oller G, Almeida F, Choy SH, Grisotto M, Renia L, Conway SJ, Stanley ER, Chan JK, Ng LG, Samokhvalov IM, Merad M, Ginhoux F (2012) Adult langerhans cells derive predominantly from embryonic fetal liver monocytes with a minor contribution of yolk sac-derived macrophages. J Exp Med 209(6):1167–1181

Holt MP, Cheng L, Ju C (2008) Identification and characterization of infiltrating macrophages in acetaminophen-induced liver injury. J Leukoc Biol 84(6):1410–1421

Hosomura N, Kono H, Tsuchiya M, Ishii K, Ogiku M, Matsuda M, Fujii H (2011) HCV-related proteins activate Kupffer cells isolated from human liver tissues. Dig Dis Sci 56(4):1057–1064

Huynh MLN, Fadok VA, Henson PM (2002) Phosphatidylserine-dependent ingestion of apoptotic cells promotes TGF-beta 1 secretion and the resolution of inflammation. J Clin Investig 109(1):41–50

Ingersoll MA, Spanbroek R, Lottaz C, Gautier EL, Frankenberger M, Hoffmann R, Lang R, Haniffa M, Collin M, Tacke F, Habenicht AJ, Ziegler-Heitbrock L, Randolph GJ (2010) Comparison of gene expression profiles between human and mouse monocyte subsets. Blood 115(3):e10–e19

Ju C, Tacke F (2016) Hepatic macrophages in homeostasis and liver diseases: from pathogenesis to Novel therapeutic strategies. Cell Mol Immunol 13:1–12

Ju C, McCoy JP, Chung CJ, Graf MLM, Pohl LR (2003) Tolerogenic role of Kupffer cells in allergic reactions. Chem Res Toxicol 16(12):1514–1519

Kamei T, Callery MP, Flye MW (1990) Kupffer cell blockade prevents induction of portal venous tolerance in rat cardiac allograft transplantation. J Surg Res 48(5):393–396

Karlmark KR, Weiskirchen R, Zimmermann HW, Gassler N, Ginhoux F, Weber C, Merad M, Luedde T, Trautwein C, Tacke F (2009) Hepatic recruitment of the inflammatory Gr1+ monocyte subset upon liver injury promotes hepatic fibrosis. Hepatology 50(1):261–274

Khakoo SI, Soni PN, Savage K, Brown D, Dhillon AP, Poulter LW, Dusheiko GM (1997) Lymphocyte and macrophage phenotypes in chronic hepatitis C infection. Correlation with disease activity. Am J Pathol 150(3):963–970

Kinnman N, Hultcrantz R, Barbu V, Rey C, Wendum D, Poupon R, Housset C (2000) PDGF-mediated Chemoattraction of hepatic Stellate cells by bile duct segments in Cholestatic liver injury. Lab Invest J Tech Methods Pathol 80(5):697–707

Kolb M, Margetts PJ, Anthony DC, Pitossi F, Gauldie J (2001) Transient expression of IL-1beta Induces acute lung injury and chronic repair leading to pulmonary fibrosis. J Clin Invest 107(12):1529–1536

Koop DR, Klopfenstein B, Limuro Y, Thurman RG (1997) Gadolinium chloride blocks alcohol-dependent liver toxicity in rats treated chronically with Intragastric alcohol despite the induction of CYP2E1. Mol Pharmacol 51(6):944–950

Korns D, Frasch C, Fernandez-Boyanapalli R, Henson PM, Bratton DL (2011) Modulation of macrophage efferocytosis in inflammation. Front Immunol 2:57

Lauer GM, Walker BD (2001) Hepatitis C virus infection. N Engl J Med 345(1):41–52

Leroux A, Ferrere G, Godie V, Cailleux F, Renoud ML, Gaudin F, Naveau S, Prevot S, Makhzami S, Perlemuter G, Cassard-Doulcier AM (2012) Toxic lipids stored by Kupffer cells correlates with their pro-inflammatory phenotype at an early stage of Steatohepatitis. J Hepatol 57(1):141–149
Levy BD, Clish CB, Schmidt B, Gronert K, Serhan CN (2001) Lipid mediator class switching during acute inflammation: signals in resolution. Nat Immunol 2(7):612–619
Lin SL, Castaño AP, Nowlin BT, Lupher ML, Duffield JS (2009) Bone marrow Ly6Chigh monocytes are selectively recruited to injured kidney and differentiate into functionally distinct populations. J Immunol 183(10):6733–6743
Lucas M, Stuart LM, Savill J, Lacy-Hulbert A (2003) Apoptotic cells and innate immune stimuli combine to regulate macrophage cytokine secretion. J Immunol 171(5):2610–2615
Ludovic A, Henry A, Poron F, Baba-Amer Y, Gherardi RK, van Rooijen N, Plonquet A, Chazaud B (2007) Inflammatory monocytes recruited after skeletal muscle injury switch into antiinflammatory macrophages to support myogenesis. J Exp Med 204(5):1057–1069
Marra F, Gastaldelli A, Svegliati Baroni G, Tell G, Tiribelli C (2008) Molecular basis and mechanisms of progression of non-alcoholic Steatohepatitis. Trends Mol Med 14(2):72–81
Maxwell PH, Wiesener MS, Chang GW, Clifford SC, Vaux EC, Cockman ME, Wykoff CC, Pugh CW, Maher ER, Ratcliffe PJ (1999) The tumour suppressor protein VHL targets hypoxia-inducible factors for oxygen-dependent proteolysis. Nature 399(6733):271–275
McGuinness PH, Painter D, Davies S, McCaughan GW (2000) Increases in intrahepatic CD68 positive cells, MAC387 positive cells, and proinflammatory cytokines (particularly interleukin 18) in chronic hepatitis C infection. Gut 46(2):260–269
Mian MOR, Barhoumi T, Briet M, Paradis P, Schiffrin EL (2016) Deficiency of T-regulatory cells exaggerates angiotensin II-induced Microvascular injury by enhancing immune responses. J Hypertens 34(1):97–108
Moon J, Welch TP, Gonzalez FJ, Copple BL (2009) Reduced liver fibrosis in hypoxia-inducible factor-1alpha-deficient mice. Am J Physiol Gastrointest Liver Physiol 296(3):G582–G592
Morias Y, Abels C, Laoui D, Van Overmeire E, Guilliams M, Schouppe E, Tacke F, deVries CJ, De Baetselier P, Beschin A (2015) $Ly6C^{-}$ monocytes Regulate parasite-induced liver inflammation by inducing the differentiation of pathogenic $Ly6C^{+}$ monocytes into macrophages. PLoS Pathog 11(5):e1004873
Murdoc C, Muthana M, Lewis CE (2005) Hypoxia regulates macrophage functions in inflammation. J Immunol 175(10):6257–6263
Murray PJ, Allen JE, Biswas SK, Fisher EA, Gilroy DW, Goerdt S, Gordon S, Hamilton JA, Ivashkiv LB, Lawrence T, Locati M, Mantovani A, Martinez FO, Mege JL, Mosser DM, Natoli G, Saeji P, Schultze JL, Shirey KA, Sica A, Suttles J, Udalova I, van Ginderachter JA, Vogel SN, Wynn TA (2014) Macrophage activation and polarization: nomenclature and experimental guidelines. Immunity 41(1):14–20
Nahrendorf M, Swirski FK, Aikawa E, Stangenberg L, Wurdinger T, Figueiredo J, Libby P, Weissleder R, Pittet MJ (2007) The healing myocardium sequentially mobilizes two monocyte subsets with divergent and complementary functions. J Exp Med 204(12):3037–3047
Newson J, Stables M, Karra E, Arce-Vargas F, Quezada S, Motwani M, Mack M, Yona S, Audzevich T, Gilroy DW (2014) Resolution of acute inflammation bridges the gap between innate and adaptive Immunity. Blood 124(11):1748–1764
Novo E, Povero D, Busletta C, Paternostro C, Di Bonzo LV, Cannito S, Compagnone A, Bandino A, Marra F, Colombatto S, David D, Pinzani M, Parola M (2012) The biphasic Nature of hypoxia-induced directional migration of activated human hepatic Stellate cells. J Pathol 226(4):588–597
Oda T, Hirota K, Nishi K, Takabuchi S, Oda S, Yamada H, Arai T, Fukuda K, Kita T, Adachi T, Semenza GL, Nohara R (2006) Activation of hypoxia-inducible factor 1 during macrophage differentiation. Am J Physiol Cell Physiol 291(1):C104–C113
Ohh M, Park CW, Ivan M, Hoffman M, Kim TY, Huang LE, Pavletich N, Chau V, Kaelin WG (2000) Ubiquitination of hypoxia-inducible factor requires direct binding to the beta-domain of the von Hippel-Lindau protein. Nat Cell Bio 2(7):423–427

Olefsky JM, Glass CK (2010) Macrophages, inflammation, and insulin resistance. Annu Rev Physiol 72:219–246

Paradis V, Dargere D, Bonvoust F, Vidaud M, Segarini P, Bedossa P (2002) Effects and regulation of connective tissue growth factor on hepatic stellate cells. Lab Investig 82(6):767–774

Pecaric-Petkovic T, Didichenko SA, Kaempfer S, Spiegl N, Dahinden CA (2009) Human basophils and Eosinophils are the direct target leukocytes of the Novel IL-1 family member IL-33. Blood 113(7):1526–1534

Pesce JT, Ramalingam TR, Mentink-Kane MM, Wilson MS, Kasmi K, Smith AM, Thompson RW, Cheever AW, Murray PJ, Wynn TA (2009) Arginase-1-expressing macrophages suppress Th2 cytokine-driven inflammation and fibrosis. PLoS Pathog 5(4):e100371

Petrasek J, Bala S, Csak T, Lippai D, Kodys K, Menashy V, Barrieau M, Min SY, Kurt-Jones EA, Szabo G (2012) IL-1 receptor antagonist ameliorates inflammasome-dependent alcoholic steatohepatitis in mice. J Clin Invest 122(10):3476–3489

Ramachandran P, Pellicoro A, Vernon M, Boulter L, Aucott RL, Ali A, Hartland SN, Snowdon VK, Cappon A, Gordon-Walker TT, Williams MJ, Dunbar DR, Manning JR, van Rooijen N, Fallowfield JA, Forbes SJ, Iredale JP (2012) Differential Ly-6C expression identifies the recruited macrophage phenotype, which orchestrates the regression of murine liver fibrosis. PNAS 109(46):E3186–E3195

Ramadori G, Armbrust T (2001) Cytokines in the liver. Eur J Gastroenterol Hepatol 13:777–784

Rivera CA, Adegboyega P, van Rooijen N, Tagalicud A, Allman M, Wallace M (2007) Toll-like receptor-4 signaling and Kupffer cells play pivotal roles in the pathogenesis of non-alcoholic Steatohepatitis. J Hepatol 47(4):571–579

Robinson N, McComb S, Mulligan R, Dudani R, Krishnan L, Sad S (2012) Type I interferon induces necroptosis in macrophages during infection with salmonella enterica serovar typhimurium TL-13. Nat Immunol 13(10):954–962

Rubio A, Guruceaga E, Vazquez-Chantada M, Sandoval J, Martinez-Cruz LA, Segura V, Sevilla JL, Podhorski A, Corrales FJ, Torres L, Rodriguez M, Aillet F, Ariz U, Arrieta JM, Caballeria J, Martin-Duce A, Lu SC, Martinez-Chantar ML, Mato JM (2007) Identification of a Gene-pathway associated with non-alcoholic steatohepatitis. J Hepatol 46(4):708–718

Sakaguchi S, Takahashi S, Sasaki T, Kumagai T, Nagata K (2011) Progression of alcoholic and non-alcoholic Steatohepatitis: common metabolic aspects of innate immune system and oxidative stress. Drug Metab Pharmacokinet 26(1):30–46

Sato K, Yabuki K, Haba T, Maekawa T (1996) Role of Kupffer cells in the induction of tolerance after liver transplantation. J Surg Res 63(2):433–438

Savill J (1997) Apoptosis in resolution of inflammation. J Leukoc Biol 61(4):375–380

Seki E, De Minicis S, Osterreicher CH, Kluwe J, Osawa Y, Brenner D, Schwabe RF (2007) TLR4 enhances TGF-beta signaling and hepatic fibrosis. Nat Med 13(11):1324–1332

Semenza GL (2001) Hypoxia-inducible factor 1: oxygen homeostasis and disease pathophysiology. Trends Mol Med 7(8):345–340

Semenza GL (2014) Oxygen sensing, hypoxia-inducible factors, and disease pathophysiology. Ann Rev Pathol 9:47–71

Serbina NV, Shi C, Pamer EG (2012) Monocyte-mediated immune defense against Murine *Listeria monocytogenes* infection. Adv Immunol 113:119–134

Serhan CN, Savill J (2005) Resolution of inflammation: the beginning programs the end. Nat Immunol 6(12):1191–1197

Shrivastava S, Mukherjee A, Ray R, Ray RB (2013) Hepatitis C virus Induces interleukin-1β (IL-1β)/IL-18 in circulatory and resident liver macrophages. J Virol 87(22):12284–12290

Sica A, Erreni M, Allavena P, Porta C (2015) Macrophage polarization in pathology. Cell Mol Life Sci 72(21):4111–4126

Steevels TAM, Meyaard L (2011) Immune inhibitory receptors: essential regulators of phagocyte function. Eur J Immunol 41(3):575–587

Stutchfield BM, Antoine DJ, Mackinnon AC, Gow DJ, Bain CC, Hawley CA, Hughes M, Francis B, Wojtacha D, Man TY, Dear JW, Dever LR, Mowat AM, Pollard JW, Park BK, Jenkins SJ,

Simpson KJ, Hume DA, Wigmore SJ, Forbes SJ (2015) CSF1 restores innate Immunity following liver injury in mice and serum levels indicate outcomes of patients with acute liver failure. Gastroenterology 149(7):1896–1909

Swirski FK, Nahrendorf M, Etzrodt M, Wildgruber M, Cortez-Retamozo V, Panizzi P, Figueiredo JL, Kohler RH, Chudnovskiy A, Waterman P, Aikawa E, Mempel TR, Libber P, Weissleder R, Pittet MJ (2009) Identification of splenic reservoir monocytes and their deployment to inflammatory sites. Science 325(5940):612–616

Tacke F, Randolph GJ (2006) Migratory fate and differentiation of blood monocyte subsets. Immunobiology 211(6–8):609–618

Tacke F, Ginhoux F, Jakubzick C, van Rooijen N, Merad M, Randolph GJ (2006) Immature monocytes acquire antigens from other cells in the bone marrow and present them to T cells after maturing in the periphery. J Exp Med 203(3):583–597

Tacke F, Alvarez D, Kaplan TJ, Jakubzick C, Spanbroek R, Llodra J, Garin A, Liu J, Mack M, van Rooijen N, Lira SA, Habenicht AJ, Randolph GJ (2007) Monocyte subsets differentially employ CCR2, CCR5, and CX3CR1 to accumulate within atherosclerotic plaques. J Clin Invest 117(1):185–194

Takeda N, O'Dea EL, Doedens A, Kim JW, Weidemann A, Stockmann C, Asagiri M, Simon MC, Hoffmann A, Johnson RS (2010) Differential activation and antagonistic function of HIFalpha isoforms in macrophages are essential for NO homeostasis. Genes Dev 24(5):491–501

Talks KL, Turley H, Gatter KC, Maxwell PH, Pugh CW, Ratcliffe PJ, Harris AL (2000) The expression and distribution of the hypoxia-inducible factors HIF- 1alpha and HIF-2alpha in normal human tissues, cancers, and tumor- associated macrophages. Am J Pathol 157(2):411–421

Tapia-Abellán A, Martínez-Esparza M, Ruiz-Alcaraz AJ, Hernández-Caselles T, Martínez-Pascual C, Miras-López M, Such J, Francés R, García-Peñarrubia P (2012) The peritoneal macrophage inflammatory profile in cirrhosis depends on the alcoholic or hepatitis C viral etiology and is related to ERK phosphorylation. BMC Immunol 13:42

Tu Z, Pierce RH, Kurtis J, Kuroki Y, Crispe IN, Orloff MS (2010) Hepatitis C virus core protein subverts the antiviral activities of human Kupffer cells. Gastroenterology 138(1):305–314

Vandivier RW, Henson PM, Douglas IS (2006) Burying the dead: the impact of failed apoptotic cell removal (efferocytosis) on chronic inflammatory lung disease. Chest 129(6):1673–1682

Varga T, Mounier R, Gogolak P, Poliska S, Chazaud B, Nagy L (2013) Tissue LyC6- macrophages are generated in the absence of circulating LyC6- monocytes and Nur77 in a model of muscle regeneration. J Immunol 191(11):5695–5701

Wallace K, Burt AD, Wright MC (2008) Liver Fibrosis. Biochem J 411(1):1–18

Wan J, Benkdane M, Teixeira-Clerc F, Bonnafous S, Louvet A, Lafdil F, Pecker F, Tran A, Gual P, Mallat A, Lotersztajn S, Pavoine C (2014) M2 Kupffer cells promote M1 Kupffer cell apoptosis: a protective mechanism against alcoholic and nonalcoholic fatty liver disease. Hepatology 59(1):130–142

Wang J, Kubes P (2016) A reservoir of mature cavity macrophages that can rapidly invade visceral organs to affect tissue repair. Cell 165(3):668–678

Wang M, You Q, Lor K, Chen F, Gao B, Ju C (2014) Chronic alcohol ingestion modulates hepatic macrophage populations and functions in mice. J Leuko Biol 96(4):657–665

Weirather J, Hofmann UDW, Beyersdorf N, Ramos GC, Vogel B, Frey A, Ertl G, Kerkau T, Frantz S (2014) Foxp3+ CD4+ T cells improve healing after myocardial infarction by modulating monocyte/macrophage differentiation. Circ Res 115(1):55–67

Wenge RH (2002) Cellular adaptation to hypoxia: O2-sensing protein hydroxylases, hypoxia-inducible transcription factors, and O2-regulated gene expression. FASEB J 16(10):1151–1162

Wieckowska A, Zein NN, Yerian LM, Lopez R, McCullough AJ, Feldstein AE (2006) In vivo assessment of liver cell apoptosis as a novel biomarker of disease severity in nonalcoholic fatty liver disease. Hepatology 44(1):27–33

Wildgruber M, Lee H, Chudnovskiy A, Yoon TJ, Etzrodt M, Pittet MJ, Nahrendorf M, Croce K, Libby P, Weissleder R, Swirski FK (2009) Monocyte subset dynamics in human atherosclerosis can be profiled with magnetic nano-sensors. PLoS One 4(5):e5663

Wilson MW, Elnekave E, Mentink-Kane MM, Hodges MG, Pesce JT, Ramalingam TR, Thompson RW, Kamanaka M, Flavell A, Keane-Myers A, Cheever AW, Wynn TA (2007) IL-13Ralpha2 and IL-10 Coordinately suppress airway inflammation, airway-hyperreactivity, and fibrosis in mice. J Clin Invest 117(10):2941–2951

Wu J, Li J, Salcedo R, Mivechi NF, Trinchieri C, Horuzsko A (2012) The proinflammatory myeloid cell receptor TREM-1 controls Kupffer cell activation and development of hepatocellular carcinoma. Cancer Res 72(16):3977–3986

Wynn TA, Yugandhar VG, Clark M (2013a) Cellular and molecular mechanisms of fibrosis. J Pathol 46(2):26–32

Wynn TA, Chawla A, Pollard JW (2013b) Macrophage biology in development, homeostasis and disease. Nature 496(7446):445–455

Yona S, Kim KW, Wolf Y, Mildner A, Varol D, Breker M, Strauss-Ayali D, Viukov S, Guilliams M, Misharim A, Hume DA, Parlman H, Malissen B, Zelzer E, Jung S (2013) Fate mapping reveals origins and dynamics of monocytes and tissue macrophages under homeostasis. Immunity 38(1):79–91

You M, Fischer M, Deeg MA, Crabb DW (2002) Ethanol Induces fatty acid synthesis pathways by activation of sterol regulatory element-binding protein (SREBP). J Biol Chem 277(32):29342–29347

You Q, Cheng L, Kedl RM, Ju C (2008) Mechanism of T cell tolerance induction by Murine hepatic Kupffer cells. Hepatology 48(3):978–990

You Q, Holt M, Yin H, Li G, Hu CJ, Ju C (2013) Role of hepatic resident and infiltrating macrophages in liver repair after acute injury. Biochem Pharmacol 86(6):836–843

Yu S, Rao S, Reddy JK (2003) Peroxisome proliferator-activated receptors, fatty acid oxidation, steatohepatitis and hepatocarcinogenesis. Curr Mol Med 3(6):561–572

Zhang Z, Bagby GJ, Stoltz D, Oliver P, Schwarzenberger PO, Kolls JK (2001) Prolonged ethanol treatment enhances lipopolysaccharide/phorbol myristate acetate-induced tumor necrosis factor-alpha production in human monocytic cells. Alcohol Clin Exp Res 25(3):444–449

Zhu H, Liu C (2003) Interleukin-1 inhibits hepatitis C virus subgenomic RNA replication by activation of extracellular regulated kinase pathway. J Virol 77(9):5493–5498

Zhu H, Shang X, Terada N, Liu C (2004) STAT3 induces anti-hepatitis C viral activity in liver cells. Biochem Biophys Res Commun 324(2):518–528

Zigmond E, Samia-Grinberg S, Mr P-C, Brazowski E, Shibolet O, Halpern Z, Varol C (2014) Infiltrating monocyte-derived macrophages and resident Kupffer cells display different ontogeny and functions in acute liver injury. J Immunol 193(1):344–345

Chapter 5
Regulation of Apoptosis by Bcl-2 Family Proteins in Liver Injury

Hayato Hikita and Tetsuo Takehara

Abbreviations

ALT	Alanine aminotransferase
BH	BCL-2 homology
ER	Endoplasmic reticulum
HCC	Hepatocellular carcinoma
LPS	Lipopolysaccharide
MOMP	Mitochondrial outer membrane permeability
NASH	Non-alcoholic steatohepatitis
TNF	Tumour necrosis factor

5.1 Introduction

Cell death is traditionally divided into three types according to morphology: type 1, type 2 and type 3 (Green and Llambi 2015). Apoptosis, classified as type 1 cell death, is caspase activity-dependent and requires ATP. Apoptotic cells shrink, and their nucleus becomes fragmented. Type 2 and type 3 cell death are known as autophagic cell death and necrotic cell death, respectively (Green and Llambi 2015). Apoptosis is a mechanism of programmed cell death in which cells actively progress to death. In 1985, Bcl-2 was reported as an oncogene (Tsujimoto et al. 1985)

H. Hikita • T. Takehara, MD, PhD (✉)
Department of Gastroenterology and Hepatology, Osaka University Graduate School of Medicine, 2-2 Yamada-oka, Suita, Osaka 565-0871, Japan
e-mail: takehara@gh.med.osaka-u.ac.jp

W.-X. Ding, X.-M. Yin (eds.), *Molecules, Systems and Signaling in Liver Injury*, Cell Death in Biology and Diseases, DOI 10.1007/978-3-319-58106-4_5

X.-M. Yin and Z. Dong (Series eds.), *Cell Death in Biology and Diseases*

and revealed to be an anti-apoptotic protein (Vaux et al. 1988). After Bcl-2 was reported, similar proteins with the same BCL-2 homology (BH) 1–4 domains were successively found and called Bcl-2 family proteins (Aouacheria et al. 2015). These proteins also participate in apoptosis regulation (Tsujimoto 2003). In this section, we describe the regulation of apoptosis by Bcl-2 family proteins in the liver.

5.2 Mechanisms by Which Bcl-2 Family Proteins Control Apoptosis

Apoptosis is regulated by Bcl-2 family proteins (Tsujimoto 2003), which consist of pro-apoptotic and anti-apoptotic proteins. Currently, five anti-apoptotic Bcl-2 family proteins have been reported, including Bcl-2, Bcl-xL, Mcl-1, Bcl-w and Bfl-1. All anti-apoptotic Bcl-2 family proteins have BH1, BH2, BH3 and BH4 domains. In contrast, pro-apoptotic proteins have been further subdivided into multidomain proteins and BH3-only proteins (Aouacheria et al. 2015; Tsujimoto 2003). Among pro-apoptotic proteins, Bak and Bax have BH1, BH2 and BH3 domains and are considered multidomain proteins. BH3-only proteins have only the BH3 domain. The canonical BH3-only proteins, which are confirmed to have pro-apoptotic effects, consist of eight proteins: Bid, Bim, PUMA, Noxa, Bmf, Bad, Bik and Bad (Aouacheria et al. 2015). Various apoptotic stimuli activate Bak/Bax via the activation of BH3-only proteins. Two pathways can activate BH3-only proteins. One is the extrinsic pathway, which is initiated by the activation of death receptors including the tumour necrosis factor (TNF)-α, Fas and TRAIL receptors. The stimulation of cell death receptors activates caspase-8. While activated caspase-8 directly activates caspase-3 in type 1 cells, such as lymphocytes, in type 2 cells, including hepatocytes, activated caspase-8 cleaves Bid to form truncated Bid, an active form of Bid, leading to Bak/Bax activation (Guicciardi et al. 2013; Scaffidi et al. 1998). The other pathway is the intrinsic pathway, which is initiated by various types of intracellular stress. For example, an increase in endoplasmic reticulum (ER) stress activates Bim and PUMA (Cazanave et al. 2010, 2011), which activate Bak/Bax. Genotoxic stress activates PUMA and Noxa via p53, leading to Bak/Bax activation (Green and Llambi 2015; Khoo et al. 2014). Thus, various apoptotic stimuli result in Bak/Bax activation though the action of BH3-only proteins. Inactive Bax exists not only on the outer membrane of mitochondria but also in the cytosolic fraction, and activated Bax moves to the outer membrane of mitochondria. However, both inactive Bak and activated Bak exist on the outer membrane of mitochondria. Anti-apoptotic Bcl-2 family proteins inhibit Bak/Bax activation.

The identity of the mechanism by which BH3-only proteins activate Bak/Bax is an ongoing controversy. The "direct activation model" states that BH3-only proteins directly activate Bak/Bax (Zheng et al. 2016). In this model, the BH3-only proteins Bid, Bim and PUMA are traditionally thought to be capable of Bak/Bax activation (Edwards et al. 2013; Kuwana et al. 2002, 2005), and the other BH3-only proteins act as sensitizers by neutralizing anti-apoptotic Bcl-2 family proteins. Recently, all BH3-only proteins have been shown to be capable of Bak/Bax activation (Du et al. 2011; Vela et al. 2013). The "indirect activation model" states that Bak/Bax can be

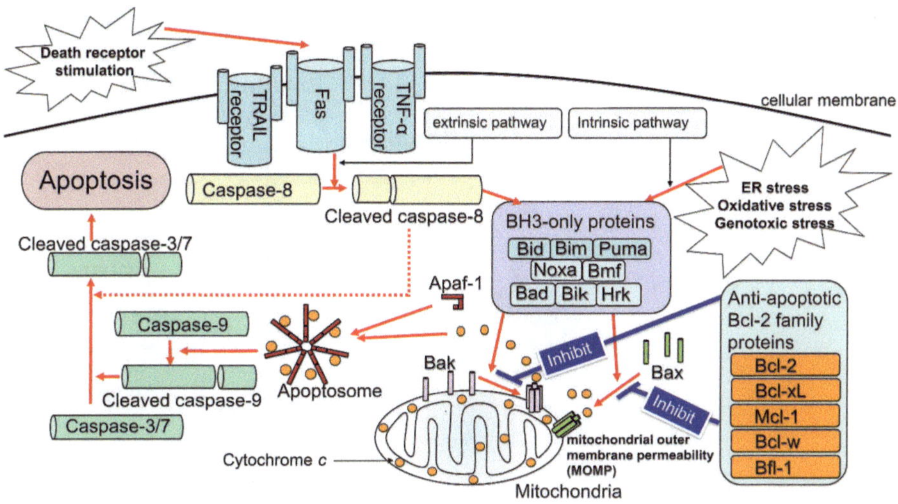

Fig. 5.1 Apoptosis signalling pathway

activated without any stimulation and that their activation is inhibited by anti-apoptotic Bcl-2 family proteins. Activated BH3-only proteins inhibit anti-apoptotic Bcl-2 family proteins, leading to Bak/Bax activation (Willis et al. 2007). Both models are based on scientific evidence, and other developing or combined models have also been proposed (García Sáez and Villunger 2016; Llambi et al. 2011; Lovell et al. 2008). Thus, the precise mechanisms of Bak/Bax activation remain unclear and are the subject of ongoing study (Doerflinger et al. 2015).

When Bak/Bax are activated, they oligomerize to form pores on the outer mitochondrial membrane, resulting in mitochondrial outer membrane permeability (MOMP) (Luna-Vargas and Chipuk 2016). After MOMP occurs, cytochrome c, which is located between the outer and inner mitochondrial membrane, is released into the cytoplasm through the pores. Released cytochrome c forms a complex with Apaf-1, called the apoptosome, in the presence of ATP. The apoptosome activates pro-caspase-9 by cleaving it into caspase-9, which in turn activates caspase-3/7 and CAD to initiate apoptosis (Fig. 5.1). Protein synthesis is not required in the apoptotic signalling cascade that occurs as a result of the sequential activation of Bak/Bax and caspase-3/7. Therefore, apoptosis rapidly occurs after Bak/Bax activation, and it is important to directly or indirectly inhibit anti-apoptotic Bcl-2 family proteins to prevent Bak/Bax activation and subsequent apoptosis.

5.2.1 Bcl-xL and mcl-1 in Hepatocytes

Five Bcl-2 family proteins have been reported to be anti-apoptotic, including Bxl-2, Bcl-xL, Mcl-1, Bcl-w and Bfl-1. In mice, deficiencies in Bcl-2 (Veis et al. 1993), Bcl-w (Print et al. 1998; Ross et al. 1998) or Bfl-1 (Hamasaki et al. 1998) do not produce any phenotype in the liver. In contrast, hepatocyte-specific Bcl-xL

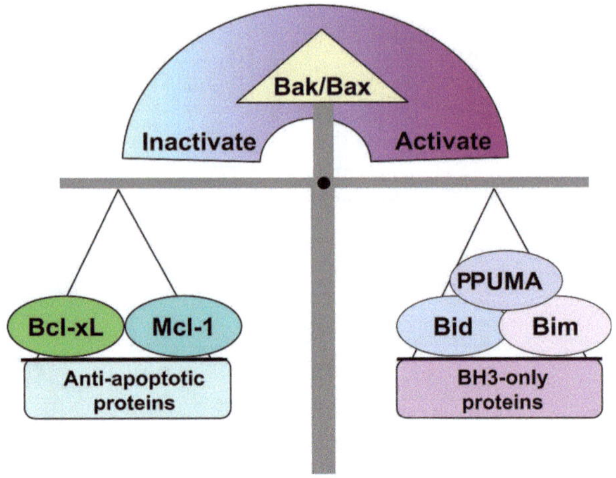

Fig. 5.2 Schematic of hepatocyte apoptosis regulated by Bcl-2 family proteins

knockout mice undergo continuous hepatocyte apoptosis, as evidenced by haematoxylin and eosin (HE) staining, terminal deoxynucleotidyl transferase dUTP nick end labelling (TUNEL) staining and immunohistochemistry of cleaved caspase-3 performed in liver sections (Takehara et al. 2004). Similarly, hepatocyte-specific Mcl-1 knockout mice undergo continuous hepatocyte apoptosis (Hikita et al. 2009b; Vick et al. 2009). Interestingly, serum alanine aminotransferase (ALT) levels in both types of knockout mice increase to approximately 200–800 IU/L, suggesting that not only hepatocyte necrosis but also hepatocyte apoptosis can increase ALT levels. Continuous hepatocyte apoptosis, as well as the increased serum ALT levels, in hepatocyte-specific Bcl-xL knockout mice or hepatocyte-specific Mcl-1 knockout mice was completely inhibited by further knockout of Bak and Bax (Hikita et al. 2009a, b). These results also support the idea that the loss of Bcl-xL or Mcl-1 activates Bak/Bax, leading to hepatocyte apoptosis (Fig. 5.2). Furthermore, hepatocyte-specific Bcl-xL and Mcl-1 double-knockout mice show a loss of hepatocytes in the foetal liver, where most cells are haematopoietic. Mice were born but died within 1 day due to liver failure (Hikita et al. 2009b). While a hetero-deficiency of Bcl-xL or Mcl-1 did not produce any phenotype in the liver, double hetero-deficiency of Bcl-xL and Mcl-1 caused continuous hepatocyte apoptosis, similar to that in Bcl-xL or Mcl-1-deficient mice (Hikita et al. 2009b). Thus, Bcl-xL and Mcl-1 cooperate to protect hepatocytes from apoptosis (Fig. 5.2). In hepatocytes, among the five anti-apoptotic Bcl-2 family proteins, Bcl-xL and Mcl-1 are the key molecules that inhibit apoptosis by producing pro-apoptotic signals under physiological conditions.

5.2.2 BH3-Only Proteins in Hepatocytes

Among the BH3-only proteins, eight have been confirmed to exhibit a pro-apoptotic effect: Bid, Bim, PUMA, Noxa, Bmf, Bad, Bik and Hrk. These proteins act as sensors for various apoptotic stimuli. Bid is activated by caspase-8, which is activated by death receptor signalling mediated by Fas ligand, TRAIL and TNF-α stimulation. Bim is activated by ER stress. However, it is unclear whether sensor proteins are active or inactive under physiological conditions. In mouse experiments, hepatocyte apoptosis observed in Bcl-xL or Mcl-1 knockout mice is suppressed by the further ablation of Bid or Bim (Hikita et al. 2009a, b; Kodama et al. 2013) and is completely suppressed by the ablation of both Bid and Bim (Kodama et al. 2013). These results suggest that Bid and Bim are activated even under physiological conditions and participate in the regulation of apoptosis by the Bcl-2 family proteins. Recently, we found that the ablation of PUMA suppresses hepatocyte apoptosis in Bcl-xL knockout mice (unpublished data), suggesting that PUMA also participates in the regulation of apoptosis under physiological conditions. The liver is located downstream of gut blood flow, which contains lipopolysaccharide (LPS), a component of Gram-negative anaerobic bacteria. The liver contains many resident immune T or NK cells, which stimulate Fas or TRAIL receptors. Therefore, it is logical that Bid is activated under physiological conditions. Hepatocytes synthesize many proteins and suffer from ER stress. Therefore, it is reasonable to conclude that Bim or PUMA are also activated under physiological conditions (Fig. 5.2).

5.2.3 Bcl-2 Family Proteins in Hepatitis

Hepatocyte apoptosis is frequently observed in pathological conditions, and many Bcl-2 family proteins regulate this process. Hepatocyte apoptosis is frequently observed in the livers of patients with chronic hepatitis C or B (Guicciardi and Gores 2005; Hayashi and Mita 1997) and is also detected in the livers of non-alcoholic steatohepatitis (NASH) patients (Feldstein et al. 2003). Caspase inhibitors efficiently decrease serum ALT levels in patients with chronic hepatitis C, chronic hepatitis B and NASH (Anstee et al. 2010; Pockros et al. 2007; Ratziu et al. 2012; Shiffman et al. 2010), suggesting that the increased ALT level in chronic hepatitis is dependent on hepatocyte apoptosis. Many hepatocytes undergoing apoptosis are detected in the livers of patients with fulminant hepatitis (Leifeld et al. 2006). Collectively, hepatocyte apoptosis is one of the characteristic features of hepatitis. Many Bcl-2 family proteins are involved in the mechanisms by which hepatocyte apoptosis occurs.

In viral and autoimmune hepatitis, lymphoid cells in the liver induce hepatocyte apoptosis. T cells or NK cells activate Fas or TRAIL receptors, and monocytes or macrophages activate TNF-α receptors on hepatocytes (Malhi et al. 2010). These death receptor stimuli activate caspase-8, followed by Bak/Bax activation in hepatocytes (Guicciardi et al. 2013). Bak/Bax subsequently activate downstream caspase-3/7, resulting in apoptosis (Fig. 5.1). Indeed, high levels of Fas ligand are detected in the serum of patients with fulminant hepatitis, and Fas is highly expressed in their hepatocytes. Fas agonistic antibodies or Fas ligands consistently induce fulminant hepatitis in mice (Nagata 1996; Ogasawara et al. 1993). Fas activates caspase-8, which in turn activates Bid, leading to apoptosis through Bak/Bax activation (Guicciardi et al. 2013). A Bid deficiency inhibits Fas-induced fulminant hepatitis in mice (Yin et al. 1999); this finding clearly indicates that hepatocytes are type 2 cells. A double deficiency in Bak and Bax also inhibits Fas-induced fulminant hepatitis in mice, suggesting that Fas-mediated apoptosis in hepatocytes is dependent on mitochondria. However, more precise observations have revealed that Fas slowly induces hepatocyte apoptosis in Bak/Bax double-knockout mice (Hikita et al. 2011). Death receptor-mediated apoptosis, which is observed under normal conditions, is thought to be mediated by Bid-Bak/Bax activation in hepatocytes (Scaffidi et al. 1999). However, this result suggests that caspase-8 can directly activate not only Bid but also caspase-3, even in type 2 cells such as hepatocytes. This direct activation of caspase-3 serves as a backup system for the Bak/Bax-mediated apoptosis pathway in the liver (Fig. 5.1).

Hepatocyte apoptosis is induced by several mechanisms in patients with steatohepatitis (Hirsova et al. 2016). One such mechanism is the increase in LPS from the gut, which stimulates toll-like receptors on hepatocytes and macrophages in the liver (Roh and Seki 2013). This mechanism increases the levels of inflammatory cytokines and chemokines and induces the accumulation of monocytes and lymphocytes in the liver, leading to the stimulation of the death receptors on hepatocytes. TNF-α, an inflammatory cytokine, also stimulates TNF-α receptors on hepatocytes. In a mouse model, a TNF-α deficiency (Mukai et al. 2016) or antibiotics (Hu et al. 2015) can suppress steatohepatitis. Lymphocyte activation by inflammation activates Fas or TRAIL receptors on hepatocytes. Free fatty acids increase hepatocyte apoptosis in the presence of insulin, although this increase is blocked by an FAS deficiency (Sommerfeld et al. 2015). In mice, deletion of the TRAIL receptor suppresses steatohepatitis (Werneburg et al. 2007, 2012). Fatty acids can also directly activate TRAIL receptors on hepatocytes (Akazawa et al. 2013). Death receptor activation via various mechanisms followed by Bid activation leads to Bak/Bax activation. Moreover, TRAIL stimulation also activates JNK-dependent Bim, leading to Bak/Bax activation (Cazanave et al. 2010, 2011). Thus, the death receptor-mediated (extrinsic) pathway is one mechanism by which hepatocyte apoptosis is induced in steatohepatitis. However, palmitate and unsaturated free fatty acids inhibit autophagy, leading to increased ER stress in hepatocytes (Tanaka et al. 2016). Increased ER stress activates Bim and PUMA via the activation of CHOP or JNK (Cazanave et al. 2010, 2011). Palmitate also increases the level of reactive oxygen species (ROS) in hepatocytes (Cazanave et al. 2009). This increase in ROS activates JNK (Luedde et al. 2014), leading to Bim or PUMA activation. Thus, Bim or PUMA activation via the intrinsic pathway leads to Bak/Bax activation. PUMA knockout mice are resistant to steatohepatitis (Cazanave et al. 2009). Taken together, the aforementioned findings suggest that the BH3-only proteins Bid, Bim and PUMA are activated by extrinsic and intrinsic pathways and are involved in apoptosis observed in patients with steatohepatitis (Akazawa et al. 2010).

5.2.4 Bcl-2 Family Proteins in Hepatocellular Carcinoma (HCC)

HCC tumours suffer from various stresses, including genotoxic stress, ER stress or death receptor signalling from host immune cells and are required to overexpress anti-apoptotic proteins to evade apoptosis (Hanahan and Weinberg 2011). Indeed, one-third of HCC tumours exhibit Bcl-xL overexpression (Takehara et al. 2001), and Mcl-1 is overexpressed in half of HCCs (Sieghart et al. 2006). The suppression of Bcl-xL deamidation and an increase in let-7 microRNAs are involved in the mechanisms by which functional Bcl-xL is overexpressed in HCC tumours (Shimizu et al. 2010; Takehara and Takahashi 2003). The overexpression of anti-apoptotic proteins in HCC directly increases the growth speed of tumours (Hikita et al. 2010). Interestingly, hepatocyte-specific Bcl-xL or Mcl-1 knockout mice accumulate oxidative stress in the liver and develop HCC (Hikita et al. 2012, 2015; Weber et al. 2010). These HCC tumours in Bcl-xL or Mcl-1 knockout mice maintain the deficiency of Bcl-xL or Mcl-1. However, HCC tumours in Bcl-xL knockout mice overexpress Mcl-1, and HCC tumours in Mcl-1 knockout mice overexpress Bcl-xL (Hikita et al. 2012). These data also support the idea that upregulation of anti-apoptotic proteins is essential for HCC development. Similar to hepatocytes under physiological conditions, HCC tumours are also dependent on the existence of Bcl-xL and Mcl-1. The suppression of Bcl-xL and Mcl-1 is a promising potential HCC treatment. However, Bcl-xL and Mcl-1 knockdown in hepatocytes also results in apoptosis, leading to massive hepatitis, resulting in death in mice (Hikita et al. 2010). An HCC-specific knockdown of Bcl-xL and Mlc-1 is required. Sorafenib, a multi-kinase inhibitor, suppresses Mcl-1, and this phenomenon is HCC specific (Hikita et al. 2010). Combination therapy with sorafenib and an anti-Bcl-xL drug, ABT-737 or ABT-263, is a candidate treatment for HCC. Currently, many small molecules that target anti-apoptotic proteins are under investigation (Delbridge et al. 2016; Luna-Vargas and Chipuk 2016). In addition, death receptor-targeted drugs or these drugs in combination with small molecules that target anti-apoptotic proteins are other potential treatments for HCC. TRAIL activation therapy has recently been reported (Wahl et al. 2013). Thus, targeted therapy of Bcl-2 family proteins is a potential candidate for HCC treatment.

5.3 Conclusion

Hepatocytes suffer from various stresses, including death receptor stimuli and ER stress. These stresses activate corresponding BH3-only proteins, including Bid, Bim and PUMA, which activate Bak/Bax. Bcl-xL and Mcl-1 cooperate to protect hepatocytes from apoptosis. Many Bcl-2 family proteins participate in the regulation of hepatocyte apoptosis not only under pathological liver but also physiological conditions. The balance between anti-apoptotic Bcl-2 family proteins and pro-apoptotic

Bcl-2 family proteins controls Bak/Bax activation to determine the fate of hepatocytes (Fig. 5.2). The modulation of Bcl-2 family proteins has potential for hepatitis treatment, preventative therapy for avoiding liver carcinogenesis and HCC treatment.

References

Akazawa Y, Cazanave S, Mott JL, Elmi N, Bronk SF, Kohno S et al (2010) Palmitoleate attenuates palmitate-induced Bim and PUMA up-regulation and hepatocyte lipoapoptosis. J Hepatol 52(4):586–593

Akazawa Y, Guicciardi ME, Cazanave SC, Bronk SF, Werneburg NW, Kakisaka K et al (2013) Degradation of cIAPs contributes to hepatocyte lipoapoptosis. Am J Physiol Gastrointest Liver Physiol 305(9):G611–G619

Anstee QM, Concas D, Kudo H, Levene A, Pollard J, Charlton P et al (2010) Impact of pan-caspase inhibition in animal models of established steatosis and non-alcoholic steatohepatitis. J Hepatol 53(3):542–550

Aouacheria A, Combet C, Tompa P, Hardwick JM (2015) Redefining the BH3 death domain as a 'short linear motif'. Trends Biochem Sci 40(12):736–748

Cazanave SC, Mott JL, Elmi NA, Bronk SF, Werneburg NW, Akazawa Y et al (2009) JNK1-dependent PUMA expression contributes to hepatocyte lipoapoptosis. J Biol Chem 284(39):26591–26602

Cazanave SC, Elmi NA, Akazawa Y, Bronk SF, Mott JL, Gores GJ (2010) CHOP and AP-1 cooperatively mediate PUMA expression during lipoapoptosis. Am J Physiol Gastrointest Liver Physiol 299(1):G236–G243

Cazanave SC, Mott JL, Bronk SF, Werneburg NW, Fingas CD, Meng XW et al (2011) Death receptor 5 signaling promotes hepatocyte lipoapoptosis. J Biol Chem 286(45):39336–39348

Delbridge AR, Grabow S, Strasser A, Vaux DL (2016) Thirty years of BCL-2: translating cell death discoveries into novel cancer therapies. Nat Rev Cancer 16(2):99–109

Doerflinger M, Glab JA, Puthalakath H (2015) BH3-only proteins: a 20-year stock-take. FEBS J 282(6):1006–1016

Du H, Wolf J, Schafer B, Moldoveanu T, Chipuk JE, Kuwana T (2011) BH3 domains other than Bim and Bid can directly activate Bax/Bak. J Biol Chem 286(1):491–501

Edwards AL, Gavathiotis E, LaBelle JL, Braun CR, Opoku-Nsiah KA, Bird GH et al (2013) Multimodal interaction with BCL-2 family proteins underlies the proapoptotic activity of PUMA BH3. Chem Biol 20(7):888–902

Feldstein AE, Canbay A, Angulo P, Taniai M, Burgart LJ, Lindor KD et al (2003) Hepatocyte apoptosis and fas expression are prominent features of human nonalcoholic steatohepatitis. Gastroenterology 125(2):437–443

García Sáez AJ, Villunger A (2016) MOMP in the absence of BH3-only proteins. Genes Dev 30(8):878–880

Green DR, Llambi F (2015) Cell death signaling. Cold Spring Harb Perspect Biol 7(12):a006080

Guicciardi ME, Gores GJ (2005) Apoptosis: a mechanism of acute and chronic liver injury. Gut 54(7):1024–1033

Guicciardi ME, Malhi H, Mott JL, Gores GJ (2013) Apoptosis and necrosis in the liver. Compr Physiol 3(2):977–1010

Hamasaki A, Sendo F, Nakayama K, Ishida N, Negishi I, Hatakeyama S (1998) Accelerated neutrophil apoptosis in mice lacking A1-a, a subtype of the bcl-2-related A1 gene. J Exp Med 188(11):1985–1992

Hanahan D, Weinberg RA (2011) Hallmarks of cancer: the next generation. Cell 144(5):646–674

Hayashi N, Mita E (1997) Fas system and apoptosis in viral hepatitis. J Gastroenterol Hepatol 12(9–10):S223–S226

Hikita H, Takehara T, Kodama T, Shimizu S, Hosui A, Miyagi T et al (2009a) BH3-only protein bid participates in the Bcl-2 network in healthy liver cells. Hepatology 50(6):1972–1980

Hikita H, Takehara T, Shimizu S, Kodama T, Li W, Miyagi T et al (2009b) Mcl-1 and Bcl-xL cooperatively maintain integrity of hepatocytes in developing and adult murine liver. Hepatology 50(4):1217–1226

Hikita H, Takehara T, Shimizu S, Kodama T, Shigekawa M, Iwase K et al (2010) The Bcl-xL inhibitor, ABT-737, efficiently induces apoptosis and suppresses growth of hepatoma cells in combination with sorafenib. Hepatology 52(4):1310–1321

Hikita H, Takehara T, Kodama T, Shimizu S, Shigekawa M, Hosui A et al (2011) Delayed-onset caspase-dependent massive hepatocyte apoptosis upon Fas activation in Bak/Bax-deficient mice. Hepatology 54(1):240–251

Hikita H, Kodama T, Shimizu S, Li W, Shigekawa M, Tanaka S et al (2012) Bak deficiency inhibits liver carcinogenesis: a causal link between apoptosis and carcinogenesis. J Hepatol 57(1):92–100

Hikita H, Kodama T, Tanaka S, Saito Y, Nozaki Y, Nakabori T et al (2015) Activation of the mitochondrial apoptotic pathway produces reactive oxygen species and oxidative damage in hepatocytes that contribute to liver tumorigenesis. Cancer Prev Res (Phila) 8(8):693–701

Hirsova P, Ibrahim SH, Gores GJ, Malhi H (2016) Lipotoxic lethal and sublethal stress signaling in hepatocytes: relevance to NASH pathogenesis. J Lipid Res 57(10):1758–1770

Hu X, Wang T, Liang S, Li W, Wu X, Jin F (2015) Antibiotic-induced imbalances in gut microbiota aggravates cholesterol accumulation and liver injuries in rats fed a high-cholesterol diet. Appl Microbiol Biotechnol 99(21):9111–9122

Khoo KH, Hoe KK, Verma CS, Lane DP (2014) Drugging the p53 pathway: understanding the route to clinical efficacy. Nat Rev Drug Discov 13(3):217–236

Kodama T, Hikita H, Kawaguchi T, Saito Y, Tanaka S, Shigekawa M et al (2013) The Bcl-2 homology 3 (BH3)-only proteins Bim and Bid are functionally active and restrained by anti-apoptotic B-cell CLL/lymphoma 2 (Bcl-2) family proteins in healthy liver. J Biol Chem 288(42):30009–30018

Kuwana T, Mackey MR, Perkins G, Ellisman MH, Latterich M, Schneiter R et al (2002) Bid, Bax, and lipids cooperate to form supramolecular openings in the outer mitochondrial membrane. Cell 111(3):331–342

Kuwana T, Bouchier-Hayes L, Chipuk JE, Bonzon C, Sullivan BA, Green DR et al (2005) BH3 domains of BH3-only proteins differentially regulate Bax-mediated mitochondrial membrane permeabilization both directly and indirectly. Mol Cell 17(4):525–535

Leifeld L, Nattermann J, Fielenbach M, Schmitz V, Sauerbruch T, Spengler U (2006) Intrahepatic activation of caspases in human fulminant hepatic failure. Liver Int 26(7):872–879

Llambi F, Moldoveanu T, Tait SW, Bouchier-Hayes L, Temirov J, McCormick LL et al (2011) A unified model of mammalian BCL-2 protein family interactions at the mitochondria. Mol Cell 44(4):517–531

Lovell JF, Billen LP, Bindner S, Shamas-Din A, Fradin C, Leber B et al (2008) Membrane binding by tBid initiates an ordered series of events culminating in membrane permeabilization by Bax. Cell 135(6):1074–1084

Luedde T, Kaplowitz N, Schwabe RF (2014) Cell death and cell death responses in liver disease: mechanisms and clinical relevance. Gastroenterology 147(4):765–783.e764

Luna-Vargas MP, Chipuk JE (2016) Physiological and pharmacological control of BAK, BAX, and beyond. Trends Cell Biol 26(12):906–917

Malhi H, Guicciardi M, Gores G (2010) Hepatocyte death: a clear and present danger. Physiol Rev 90(3):1165–1194

Mukai K, Miyagi T, Nishio K, Yokoyama Y, Yoshioka T, Saito Y et al (2016) S100A8 production in CXCR2-expressing CD11b+gr-1high cells aggravates hepatitis in mice fed a high-fat and high-cholesterol diet. J Immunol 196(1):395–406

Nagata S (1996) Apoptosis mediated by the Fas system. Prog Mol Subcell Biol 16:87–103
Ogasawara J, Watanabe-Fukunaga R, Adachi M, Matsuzawa A, Kasugai T, Kitamura Y et al (1993) Lethal effect of the anti-Fas antibody in mice. Nature 364(6440):806–809
Pockros P, Schiff E, Shiffman M, McHutchison J, Gish R, Afdhal N et al (2007) Oral IDN-6556, an antiapoptotic caspase inhibitor, may lower aminotransferase activity in patients with chronic hepatitis C. Hepatology 46(2):324–329
Print CG, Loveland KL, Gibson L, Meehan T, Stylianou A, Wreford N et al (1998) Apoptosis regulator bcl-w is essential for spermatogenesis but appears otherwise redundant. Proc Natl Acad Sci U S A 95(21):12424–12431
Ratziu V, Sheikh MY, Sanyal AJ, Lim JK, Conjeevaram H, Chalasani N et al (2012) A phase 2, randomized, double-blind, placebo-controlled study of GS-9450 in subjects with nonalcoholic steatohepatitis. Hepatology 55(2):419–428
Roh YS, Seki E (2013) Toll-like receptors in alcoholic liver disease, non-alcoholic steatohepatitis and carcinogenesis. J Gastroenterol Hepatol 28(Suppl 1):38–42
Ross AJ, Waymire KG, Moss JE, Parlow AF, Skinner MK, Russell LD et al (1998) Testicular degeneration in Bclw-deficient mice. Nat Genet 18(3):251–256
Scaffidi C, Fulda S, Srinivasan A, Friesen C, Li F, Tomaselli K et al (1998) Two CD95 (APO-1/Fas) signaling pathways. EMBO J 17(6):1675–1687
Scaffidi C, Schmitz I, Zha J, Korsmeyer SJ, Krammer PH, Peter ME (1999) Differential modulation of apoptosis sensitivity in CD95 type I and type II cells. J Biol Chem 274(32):22532–22538
Shiffman ML, Pockros P, McHutchison JG, Schiff ER, Morris M, Burgess G (2010) Clinical trial: the efficacy and safety of oral PF-03491390, a pancaspase inhibitor – a randomized placebo-controlled study in patients with chronic hepatitis C. Aliment Pharmacol Ther 31(9):969–978
Shimizu S, Takehara T, Hikita H, Kodama T, Miyagi T, Hosui A et al (2010) The let-7 family of microRNAs inhibits Bcl-xL expression and potentiates sorafenib-induced apoptosis in human hepatocellular carcinoma. J Hepatol 52(5):698–704
Sieghart W, Losert D, Strommer S, Cejka D, Schmid K, Rasoul-Rockenschaub S et al (2006) Mcl-1 overexpression in hepatocellular carcinoma: a potential target for antisense therapy. J Hepatol 44(1):151–157
Sommerfeld A, Reinehr R, Häussinger D (2015) Free fatty acids shift insulin-induced hepatocyte proliferation towards CD95-dependent apoptosis. J Biol Chem 290(7):4398–4409
Takehara T, Takahashi H (2003) Suppression of Bcl-xL deamidation in human hepatocellular carcinomas. Cancer Res 63(12):3054–3057
Takehara T, Liu X, Fujimoto J, Friedman S, Takahashi H (2001) Expression and role of Bcl-xL in human hepatocellular carcinomas. Hepatology 34(1):55–61
Takehara T, Tatsumi T, Suzuki T, Rucker E r, Hennighausen L, Jinushi M et al (2004) Hepatocyte-specific disruption of Bcl-xL leads to continuous hepatocyte apoptosis and liver fibrotic responses. Gastroenterology 127(4):1189–1197
Tanaka S, Hikita H, Tatsumi T, Sakamori R, Nozaki Y, Sakane S et al (2016) Rubicon inhibits autophagy and accelerates hepatocyte apoptosis and lipid accumulation in nonalcoholic fatty liver disease. Hepatology 64(6):1994–2014
Tsujimoto Y (2003) Cell death regulation by the Bcl-2 protein family in the mitochondria. J Cell Physiol 195(2):158–167
Tsujimoto Y, Cossman J, Jaffe E, Croce CM (1985) Involvement of the bcl-2 gene in human follicular lymphoma. Science 228(4706):1440–1443
Vaux DL, Cory S, Adams JM (1988) Bcl-2 gene promotes haemopoietic cell survival and cooperates with c-myc to immortalize pre-B cells. Nature 335(6189):440–442
Veis DJ, Sorenson CM, Shutter JR, Korsmeyer SJ (1993) Bcl-2-deficient mice demonstrate fulminant lymphoid apoptosis, polycystic kidneys, and hypopigmented hair. Cell 75(2):229–240
Vela L, Gonzalo O, Naval J, Marzo I (2013) Direct interaction of Bax and Bak proteins with Bcl-2 homology domain 3 (BH3)-only proteins in living cells revealed by fluorescence complementation. J Biol Chem 288(7):4935–4946

Vick B, Weber A, Urbanik T, Maass T, Teufel A, Krammer PH et al (2009) Knockout of myeloid cell leukemia-1 induces liver damage and increases apoptosis susceptibility of murine hepatocytes. Hepatology 49(2):627–636

Wahl K, Siegemund M, Lehner F, Vondran F, Nüssler A, Länger F et al (2013) Increased apoptosis induction in hepatocellular carcinoma by a novel tumor-targeted TRAIL fusion protein combined with bortezomib. Hepatology 57(2):625–636

Weber A, Boger R, Vick B, Urbanik T, Haybaeck J, Zoller S et al (2010) Hepatocyte-specific deletion of the antiapoptotic protein myeloid cell leukemia-1 triggers proliferation and hepatocarcinogenesis in mice. Hepatology 51(4):1226–1236

Werneburg NW, Guicciardi ME, Bronk SF, Kaufmann SH, Gores GJ (2007) Tumor necrosis factor-related apoptosis-inducing ligand activates a lysosomal pathway of apoptosis that is regulated by Bcl-2 proteins. J Biol Chem 282(39):28960–28970

Werneburg NW, Bronk SF, Guicciardi ME, Thomas L, Dikeakos JD, Thomas G et al (2012) Tumor necrosis factor-related apoptosis-inducing ligand (TRAIL) protein-induced lysosomal translocation of proapoptotic effectors is mediated by phosphofurin acidic cluster sorting protein-2 (PACS-2). J Biol Chem 287(29):24427–24437

Willis SN, Fletcher JI, Kaufmann T, van Delft MF, Chen L, Czabotar PE et al (2007) Apoptosis initiated when BH3 ligands engage multiple Bcl-2 homologs, not Bax or Bak. Science 315(5813):856–859

Yin X, Wang K, Gross A, Zhao Y, Zinkel S, Klocke B et al (1999) Bid-deficient mice are resistant to Fas-induced hepatocellular apoptosis. Nature 400(6747):886–891

Zheng JH, Viacava Follis A, Kriwacki RW, Moldoveanu T (2016) Discoveries and controversies in BCL-2 protein-mediated apoptosis. FEBS J 283(14):2690–2700

Chapter 6
Beta-Catenin and the Survival of Hepatocytes

Kari Nichole Nejak-Bowen and Satdarshan Pal Singh Monga

Abbreviations

ActD	Actinomycin D
APC	Adenomatous polyposis coli
CBP	CREB-binding protein
CK1	Casein kinase 1
DC	Dendritic cell
DISC	Death-inducing signaling complex
Dkk-1	Dickkopf-related protein 1
Dvl	Dishevelled
FADD	Fas-associated death domain
FGF	Fibroblast growth factor
FRP	Frizzled-related protein
Fz	Frizzled
GalN	D-Galactosamine
GPCR	G-protein-coupled receptor
GS	Glutamine synthetase
GSK3β	Glycogen synthase kinase 3β
HCC	Hepatocellular carcinoma
HGF	Hepatocyte growth factor
HIF-1	Hypoxia-inducible factor-1

K.N. Nejak-Bowen, MBA, PhD (✉) • S.P.S. Monga
Division of Experimental Pathology, Department of Pathology, University of Pittsburgh, School of Medicine, 200 Lothrop Street, S422-BST, Pittsburgh, PA 15261, USA
e-mail: knnst5@pitt.edu; smonga@pitt.edu

W.-X. Ding, X.-M. Yin (eds.), *Molecules, Systems and Signaling in Liver Injury*, Cell Death in Biology and Diseases, DOI 10.1007/978-3-319-58106-4_6

X.-M. Yin and Z. Dong (Series eds.), *Cell Death in Biology and Diseases*

HSC	Hepatic stellate cell
I/R	Ischemia/reperfusion
IAP	X-linked inhibitors of apoptosis
iNOS	Inducible nitric oxide synthase
JNK	c-Jun NH_2-terminal kinase
KO	Knockout
LECT2	Leukocyte cell-derived chemotaxin 2
LEF	Lymphoid enhancer-binding factor
LNA	Locked nucleic antisense
LNP	Lipid nanoparticle
LPS	Lipopolysaccharide
LRP	Low-density lipoprotein receptor-related protein
LSEC	Liver sinusoidal endothelial cell
NFAT	Nuclear factor associated with T cells
NF-κB	Nuclear factor kappa B
PKA	Protein kinase A
PTEN	Phosphatase and tensin homolog deleted on chromosome 10
RNS	Reactive nitrogen species
ROS	Reactive oxygen species
RTK	Receptor tyrosine kinase
Ser	Serine
SMP30	Senescence marker protein-30
T3	Triiodothyronine
TCF	T cell-specific transcription factor
TGF-β	Transforming growth factor-β
Thr	Threonine
TNF	Tumor necrosis factor
TRADD	TNFR-associated death domain
TUNEL	Terminal deoxynucleotidyl transferase (TdT) dUTP nick-end labeling
WIF	Wnt inhibitory factor
Wls	Wntless

6.1 The Wnt/β-Catenin Signaling Pathway

The Wnt/β-catenin signaling pathway is a highly evolutionarily conserved regulatory system that is essential in controlling developmental decisions as well as growth and maintenance of adult tissues. The past three decades' worth of extensive research has increased our understanding of this pathway enormously, demonstrating its indispensable importance in processes such as cell fate, stem cell control, differentiation, growth, survival, proliferation, regeneration, and self-renewal (Cadigan and Nusse 1997; Wodarz and Nusse 1998; Peifer and Polakis 2000; Willert et al. 2003; Logan and Nusse 2004; Nusse 2005; Niehrs 2012). Given its diverse roles, then, it is not surprising that mutations in this signaling pathway have also been associated with various diseases. In particular, dysregulation of Wnt/β-catenin signaling has been observed in the initiation and progression of several different types of cancers, including those in the colon and liver (Peifer and Polakis 2000; Polakis 2000; Paul and Dey 2008).

Although aberrant β-catenin activation was noted in a subset of human hepatocellular carcinomas (HCCs) as early as the late 1990s (de La Coste et al. 1998; Miyoshi et al. 1998; Satoh et al. 2000), in the last decade or so this pathway has become the focus of intense interest and research in the field of hepatic pathophysiology. These studies have contributed much to our understanding of the essential role of Wnt/β-catenin signaling in both liver health and disease (Apte et al. 2008; Monga 2015). This pathway has been shown to be critical in the highly dynamic environment of the developing liver, where it regulates the processes of patterning and specification, as well as hepatoblast proliferation and maturation (Nejak-Bowen and Monga 2008; Monga 2014). Wnt/β-catenin signaling is also active during the postnatal hepatic growth spurt and can be reactivated in an adult liver under conditions of experimentally induced controlled growth, such as in liver regeneration after partial hepatectomy (Nejak-Bowen and Monga 2011). Pathologically, this pathway is also involved in hepatic disorders such as acute liver injury, diet-induced steatohepatitis, fibrosis, cholestasis, and progenitor cell activation (Myung et al. 2007; Apte et al. 2008, 2009; Cheng et al. 2008; Behari et al. 2010; Thompson et al. 2010; Yeh et al. 2010; Nejak-Bowen et al. 2013; Ge et al. 2014). The present chapter will be an in-depth discussion on the recently identified role of Wnt/β-catenin signaling in cell survival and apoptosis during liver injury.

6.1.1 The Canonical Wnt Pathway

Wnt proteins are secreted glycoproteins approximately 40 kDa in size (Tanaka et al. 2002). Importantly, certain posttranslational modifications are required in order for Wnts to be secreted and signal efficiently (Bartscherer and Boutros 2008). For example, Porcupine, located in the endoplasmic reticulum, is essential for palmitoylation and glycosylation of Wnt proteins (Kadowaki et al. 1996; Zhai et al. 2004; Barrott et al. 2011), while Wntless (Wls), a multipass transmembrane protein, conveys the posttranslationally modified Wnts from the Golgi to the membrane for secretion (Banziger et al. 2006; Bartscherer et al. 2006). Once secreted, Wnts can act in either a paracrine or an autocrine fashion, but predominantly Wnts are short-distance signals, meaning that the signal occurs adjacent to the secreting cell, rather than over a long range (Clevers and Nusse 2012).

In the absence of Wnt proteins, Wnt signaling is inactive. This state is characterized by degradation of β-catenin, the central effector in the canonical pathway. A destruction complex, which consists of Axin, adenomatous polyposis coli (APC), casein kinase 1 (CK1), and glycogen synthase kinase 3β (GSK3β), binds free β-catenin in the cytoplasm. When Wnt signaling is absent, β-catenin is phosphorylated by CK1 and GSK3β at the N-terminal region of the protein, specifically at serine 45 (Ser45), threonine 41 (Thr41), Ser37, and Ser37 (Behrens et al. 1998; Amit et al. 2002; Liu et al. 2002; Yanagawa et al. 2002; Clevers 2006). This sequential phosphorylation targets β-catenin for ubiquitination by β-TrC, which ultimately leads to degradation by the proteasome (Aberle et al. 1997; MacDonald et al. 2009). These events maintain the pathway in "OFF" mode by preventing β-catenin nuclear translocation and keeping β-catenin protein levels in the cell low. Inactivation of canonical Wnt signaling can also occur if Wnts are sequestered or prevented from binding to their receptors. One such modulator is Frizzled-related proteins (FRPs), which are smaller proteins (30kD) with Fz-like cysteine-rich domain that bind and sequester Wnts (Rattner et al. 1997). Similarly, Wnt inhibitory factors (WIFs) bind Wnts to inactivate the pathway (Hsieh et al. 1999).

Binding of Wnt to the seven-pass transmembrane Frizzled (Fz) receptor and co-receptor low-density lipoprotein receptor-related protein (LRP) 5/LRP6 on the surface of cells triggers activation of the canonical Wnt pathway (Bhanot et al. 1996; Pinson et al. 2000; Tamai et al. 2000; Wehrli et al. 2000) (Fig. 6.1). Dishevelled (Dvl) interacts with the Frizzled receptor (Wong et al. 2003; Cadigan and Liu 2006), and the Fz/Dvl complex in turn relocates Axin to LRP5/LRP6 (Cliffe et al. 2003). This removes GSK3β and CK1 from the cytoplasmic complex, resulting in hypophosphorylation of β-catenin; these proteins then also phosphorylate LRP5/LRP6 to activate canonical β-catenin signaling (Zeng et al. 2005, 2008). The release of β-catenin from the destruction complex results in the accumulation of cytoplasmic and active β-catenin. β-Catenin then translocates to the nucleus via an as-yet unknown mechanism that may involve Rac1 activation (Esufali and Bapat 2004; Wu et al. 2008; Schlessinger et al. 2009) or direct interaction with components of the nuclear pore complex (Yokoya et al. 1999; Henderson and Fagotto 2002). In the nucleus, β-catenin displaces the transcriptional inhibitor Groucho and binds to lymphoid enhancer-binding factor/T cell-specific transcription factor (LEF/TCF), allowing for initiation of target gene transcription (Molenaar et al. 1996; Cavallo et al. 1998; Logan and Nusse 2004). CREB-binding protein (CBP), which is a known coactivator for several transcription factors, as well as another related acetyltransferase, p300, acts as transcriptional coactivators in the β-catenin-TCF transcription machinery (Hecht et al. 2000; Takemaru and Moon 2000). More recently, P300 and CBP have shown to have differential and sometimes opposite effects on target gene promoters (Ma et al. 2005).

One interesting aspect of β-catenin-TCF target gene expression is that it seems to be tissue specific as well as temporally regulated, which is reflected in the number and diversity of the genes regulated. A partial listing of some notable Wnt/β-catenin-regulated genes is shown in Table 6.1. Importantly, Wnt/β-catenin signaling regulates the expression of many liver-specific genes, which are involved in processes as diverse as metabolism, drug detoxification, zonation, differentiation, and proliferation (Monga 2015).

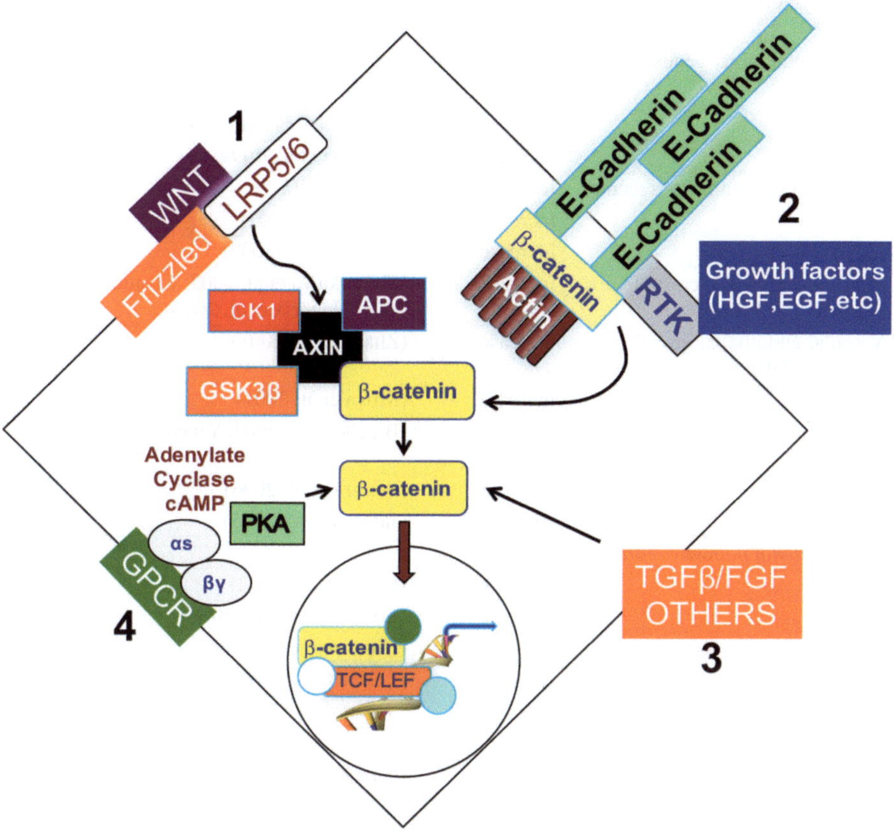

Fig. 6.1 β-catenin signaling in the hepatocyte. (*1*) Canonical Wnt/β-catenin signaling. Upon binding of Wnt to Frizzled receptor and co-receptor LRP5 or LRP6, the signal is transduced to the β-catenin degradation complex comprised of GSK3β, CK1, Axin, and APC to inactivate it. β-Catenin dissociates from the multimeric complex to eventually accumulate in the cytoplasm and eventually translocate to the nucleus where it acts as a cofactor for the TCF/LEF family of transcription factors. (*2*) β-Catenin at the cell surface. β-Catenin mediates cell-cell adhesion by forming a bridge between the actin cytoskeleton and E-cadherin present at the cell surface. β-Catenin also associates with receptor tyrosine kinases (*RTKs*) at the cell surface. In the presence of growth factors, β-catenin is phosphorylated and translocates to the nucleus to activate genes important in proliferation and morphogenesis. (*3*) TGF-β and fibroblast growth factors (*FGFs*) can also activate β-catenin-mediated transcription through multiple mechanisms that are not yet fully elucidated. (*4*) Signaling through G-protein-coupled receptors (*GPCRs*) can stimulate β-catenin activation through production of cyclic AMP (*cAMP*) and PKA

6.1.2 Noncanonical Wnt signaling

Wnt proteins can also activate several pathways that are independent of β-catenin. These noncanonical Wnt signaling pathways utilize other methods of signaling downstream of Wnt/Fz, which may intersect or overlap one another (Behari et al. 2010;

Table 6.1 A partial list of Wnt/β-catenin target genes

General Wnt target genes	Effect of Wnt signaling	Reference
Gastrin	Increase	(Koh et al. 2000)
Survivin	Increase	(Zhang et al. 2001)
MMP7	Increase	(Brabletz et al. 1999)
Epidermal growth factor receptor	Increase	(Tan et al. 2005)
Met	Increase	(Boon et al. 2002)
Jagged	Increase	(Rodilla et al. 2009)
Fibroblast growth factors (FGF)	Increase	(Shimokawa et al. 2003; Chamorro et al. 2005; Hendrix et al. 2006)
Vascular endothelial growth factor (VEGF)	Increase	(Zhang et al. 2001)
Claudin-1	Increase	(Miwa et al. 2001)
Sox9	Increase/ decrease	(Blache et al. 2004; Yano et al. 2005)
Nanog	Decrease	(Pereira et al. 2006)
E-cadherin	Decrease	(Jamora et al. 2003)
Wnt/β-catenin pathway repressors		
Axin2	Increase	(Leung et al. 2002)
βTrCP	Increase	(Spiegelman et al. 2000)
Dkk1	Increase	(Chamorro et al. 2005)
Frp2	Increase	(Lescher et al. 1998)
Naked	Increase	(Wharton et al. 2001)
Wnt/β-catenin positive regulators		
Frizzled	Increase/ decrease	(Cadigan et al. 1998; Willert et al. 2002)
LRP6	Decrease	(Khan et al. 2007)
TCF	Increase	(Roose et al. 1999)
LEF1	Increase	(Hovanes et al. 2001)
Rspo2	Increase	(Kazanskaya et al. 2004)
Cell-cycle regulators/oncogenes		
Cyclin D1	Increase	(He et al. 1998; Shtutman et al. 1999; Tetsu and McCormick 1999)
c-myc	Increase	(He et al. 1998)
Liver-related genes		
Glutamine synthetase	Increase	(Cadoret et al. 2002)
Glutamate transporter-1	Increase	(Cadoret et al. 2002)
Glutathione S-transferase	Increase	(Giera et al. 2010)
Ornithine aminotransferase	Increase	(Cadoret et al. 2002)
Cytochrome P450 enzymes	Increase	(Loeppen et al. 2005; Gougelet et al. 2014)
Regucalcin	Increase	(Nejak-Bowen et al. 2009)
Constitutive androstane receptor	Increase	(Gougelet et al. 2014)
Glucose-6-phosphatase	Increase	(Liu et al. 2011)
Phosphoenolpyruvate carboxykinase (PEPCK)	Increase	(Liu et al. 2011)
Lect2	Increase	(Ovejero et al. 2004)
Nitric oxide synthase 2	Increase	(Du et al. 2006)

Niehrs 2012). Among these alternate signaling mechanisms is the planar cell polarity pathway, which regulates changes in cell polarity and motility through activation of JNK (Boutros et al. 1998; Bartscherer and Boutros 2008; Tada and Kai 2009), and the Wnt/Ca^{2+} pathway, which is mediated by the activation of heterotrimeric G proteins and nuclear factor associated with T cells (NFAT) (Slusarski et al. 1997; De 2011). Additional pathways include the Wnt5-ROR pathway, the Wnt-PKA pathway, the Wnt-mTOR pathway, and others (Semenov et al. 2007; Sugimura and Li 2010; Ho et al. 2012). All of the noncanonical Wnt pathways are discussed further in several recent reviews (Seifert and Mlodzik 2007; De 2011); the remainder of this chapter, however, will focus primarily on the canonical Wnt signaling pathway.

6.1.3 Interaction of β-Catenin with Other Pathways

Apart from its role as a transcriptional cofactor in the canonical Wnt signaling pathway, β-catenin also plays an important role in the maintenance of epithelial cell adherens junctions. β-Catenin forms a bridge between the actin cytoskeleton and E-cadherin present at the cell surface (Aberle et al. 1994; Orsulic et al. 1999; Lilien and Balsamo 2005). The β-catenin/E-cadherin interaction mediates cell-cell adhesion and is regulated by the phosphorylation of β-catenin at tyrosine residue 654 (Roura et al. 1999; Piedra et al. 2001) (Fig. 6.2). Following tyrosine phosphorylation of β-catenin, the β-catenin/E-cadherin complex dissociates, increasing the cytosolic pool of β-catenin, which eventually results in increased transcriptional activity (Piedra et al. 2001; Kam and Quaranta 2009). Specifically, in the liver this dissociation also causes subsequent degradation of E-cadherin, resulting in a loss of adherens junctions and impaired apical trafficking in hepatocytes (Huber and Weis 2001; Theard et al. 2007). Loss of adhesion may also contribute to motility, which is an important component of the cellular response in processes such as development, regeneration, and cancer growth. Recently, however, the role of β-catenin at adherens junctions was reported to be redundant, as liver-specific β-catenin conditional knockout (KO) mice lacking β-catenin in hepatocytes continued to show maintenance of adherens junctions due to an upregulation of γ-catenin or plakoglobin in the KO mice (Wickline et al. 2011, 2013).

β-Catenin also interacts with several receptor tyrosine kinases in the liver. Met is the receptor for hepatocyte growth factor (HGF), a known mitogen, motogen, and morphogen for the liver (Zarnegar 1995). Our laboratory has previously shown that Met and β-catenin associate at the surface of hepatocytes (Monga et al. 2006) (Fig. 6.2). Binding of HGF to Met induces its phosphorylation, which then in turn phosphorylates β-catenin at tyrosine residues 654 and 670, resulting in translocation of β-catenin to the nucleus (Monga et al. 2002; Zeng et al. 2006) (Fig. 6.1). Injecting the human HGF gene into mice also causes dissociation of the Met/β-catenin complex and induction of the β-catenin pathway, resulting in hepatomegaly (Apte et al. 2006). Other reports have demonstrated that HGF positively regulates β-catenin-TCF transactivation, albeit via different mechanisms (Papkoff and Aikawa 1998; Hiscox and

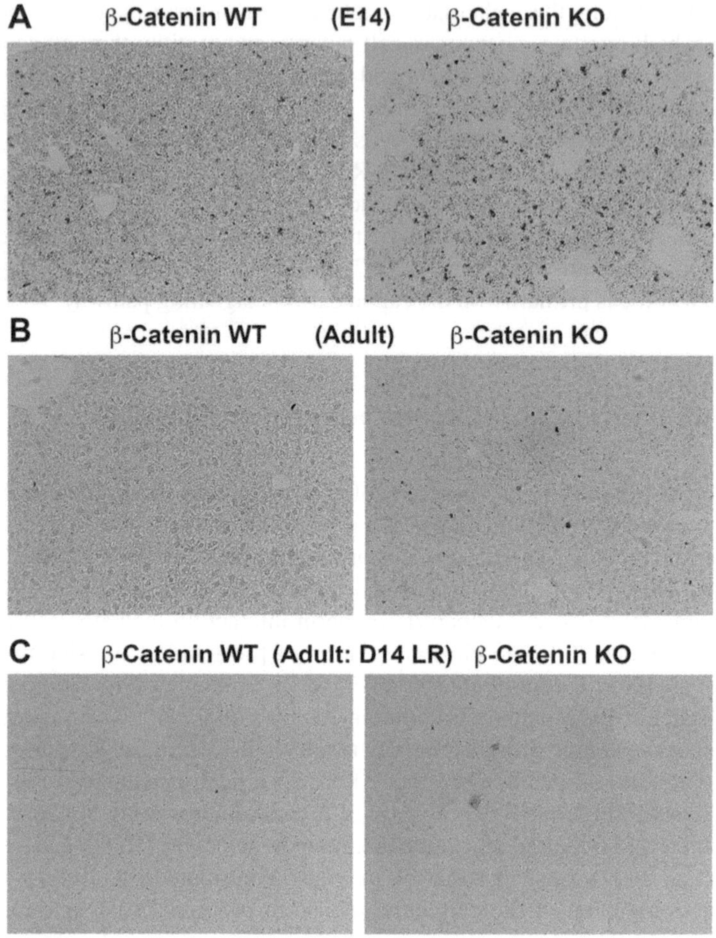

Fig. 6.2 Increased apoptosis in the absence of hepatocyte β-catenin. Terminal deoxynucleotidyl transferase (TdT) dUTP nick-end labeling (TUNEL) immunohistochemistry staining shows increased apoptosis in the livers of mice lacking β-catenin in hepatocytes at E14 (**a**), 4 months after birth at homeostasis (**b**), and 14 days after partial hepatectomy (**c**), compared to wild-type controls

Jiang 1999; Danilkovitch-Miagkova et al. 2001). These findings have important implications for liver cancer. Simultaneous activation of Met and a mutated active form of β-catenin have been found in a subset of human HCC, which has recently been replicated in a genetic model (Tward et al. 2007; Tao et al. 2016a). Hepatoblastomas, which frequently have deletions or missense mutations in exon 3 of the β-catenin gene, also express high levels of Y654 β-catenin, which is induced by HGF/Met signaling (Purcell et al. 2011). Similarly, a direct interaction between β-catenin and the epidermal growth factor pathway has been reported as well (Hoschuetzky et al. 1994; Kanai et al. 1995; Takahashi et al. 1997). ErbB2, a member of this family, can phosphorylate β-catenin at the same tyrosine residue (Y654) as Met, which may increase cell motility (Shibata et al. 1996; Bonvini et al. 2001) (Fig. 6.2).

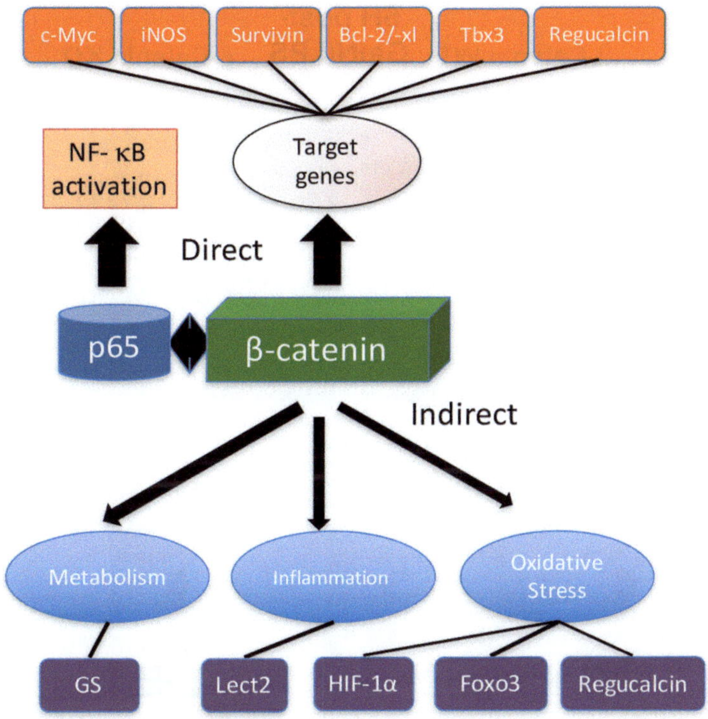

Fig. 6.3 Role of β-catenin in the survival of hepatocytes. β-Catenin can regulate the survival of hepatocytes through multiple mechanisms. It directly regulates the expression of several genes which encode factors with proven roles in regulating cell death. Physical interaction of β-catenin with p65 subunit also directly plays a role in directly regulating NF-κB activity and hence in cell survival. In addition, to β-catenin's role as a cofactor for the TCF family of transcription factors, it can also act as a cofactor for FoxO3 and HIF-1α to regulate the expression of distinct target genes. Through such interactions and regulation of target genes, β-catenin can modulate processes like cell metabolism, inflammation, and oxidative stress and hence can indirectly have an impact on cell survival

Transforming growth factor-β (TGF-β), a known regulator of self-renewal that has also been implicated in the modulation of HCC (Tang et al. 2008), can impact β-catenin signaling as well (Fig. 6.3). TGF-β-mediated loss of E-cadherin results in the release of β-catenin from cell-cell contacts and its subsequent translocation to the cytoplasm. Active β-catenin thus leads to the increased cell motility and invasive phenotype seen in gastrointestinal and liver cancers (Peinado et al. 2003; Katuri et al. 2006). Additionally, the β-catenin-TCF complex physically interacts with Smads 2, 3, and 4, mediators of TGFβ signaling; such interactions prevent β-catenin degradation, facilitate nuclear translocation, or synergistically enhance transcription of a subset of Wnt target genes (Labbe et al. 2000; Lei et al. 2004; Warner et al. 2005; Hirota et al. 2008; Romero et al. 2008; Zhang et al. 2010a).

Lastly, protein kinase A (PKA) has been shown to activate β-catenin downstream of G-protein-coupled receptor (GPCR)/cAMP signaling (Fig. 6.4). PKA has been

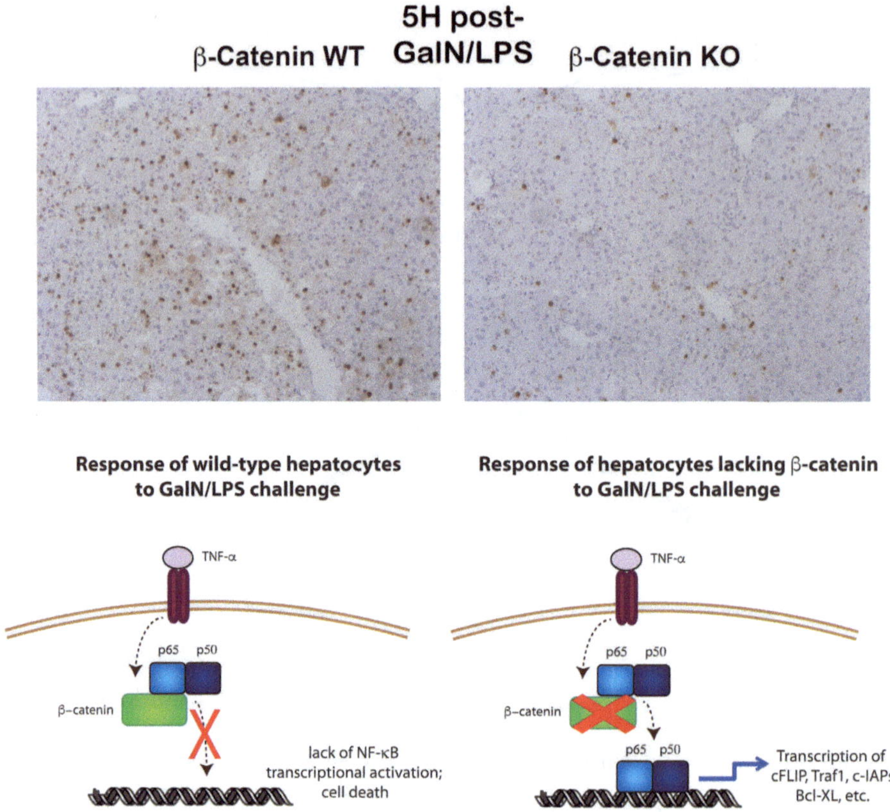

Fig. 6.4 Role of β-catenin in regulating NF-κB/p65 activation in liver injury. *Top*: apoptosis is decreased in livers lacking β-catenin compared to wild-type livers after GalN/LPS injury. *Bottom*: in acute liver injury, administration of LPS results in production of TNF-α; however, the presence of β-catenin prevents p65 from translocating to the nucleus and activating target gene expression. However, in the absence of β-catenin, there is early, robust, and protracted activation of NF-κB when the TNF-α pathway is stimulated, which leads to activation of pro-survival genes and protection from apoptosis

shown to phosphorylate β-catenin at specific residues in the C-terminus (Ser552 and Ser675) to lead to β-catenin nuclear translocation and activation (Taurin et al. 2006).

6.2 Mechanisms of Apoptosis and Cell Survival

Apoptosis, also known as programmed cell death, is a critical component in liver injury and can occur via two major pathways: the death receptor (extrinsic) pathway and the mitochondrial (intrinsic) pathway (Ding and Yin 2004). The death receptor pathway is activated by binding of death receptor ligands to receptors on the cell surface, while the mitochondrial pathway is initiated by intracellular stresses such as DNA damage or changes of intracellular Ca^{2+}. Of these, the death receptor pathway has been extensively studied in hepatocytes, since hepatocytes express an abundance of death receptors (Yoon and Gores 2002). This is presumably due to evolutionary pressure which favored enhanced death receptor expression in the liver to help eliminate pathogens and cells damaged by toxins (Akazawa and Gores 2007). Thus, death receptor-mediated apoptosis in the liver is a critical component of homeostasis, as it targets damaged or abnormal cells for elimination without disrupting liver physiology (Nagata 1997; Akazawa and Gores 2007). Of the six identified death receptors, Fas and TNF-α are considered to have pathologic significance in the liver (Yin and Ding 2003). Importantly, much of our current knowledge on the subject of hepatocyte apoptosis has been derived from experimental induction of liver failure (Galanos et al. 1988).

6.2.1 Fas Pathway

Fas (also known as CD95) is a member of the TNF receptor family which is primarily activated by Fas ligand (FasL), a membrane-spanning protein typically expressed on the surface of activated T cells, although soluble forms of FasL are also present in serum (Berke 1995; Yoon and Gores 2002). Experimentally, the Fas pathway can be induced in mice via intravenous injection of Jo-2 antibody, a Fas agonist which causes death of the animal through massive hepatocyte apoptosis (Ogasawara et al. 1993). Activation of the receptor by FasL or Jo-2 causes trimerization of the receptor, which leads to formation of the death-inducing signaling complex (DISC) (Chinnaiyan et al. 1995; Leist et al. 1996; Ashkenazi and Dixit 1998; Peter and Krammer 2003). This complex consists of the Fas-associated death domain (FADD), which contains a death effector domain (Boldin et al. 1995) that is required for caspase 8 recruitment. Caspase 8, also called the initiator caspase for its ability to activate executioner caspases such as caspases 3, 6, and 7, is activated at the DISC (Thornberry and Lazebnik 1998; Thorburn 2004). In type II cells such as hepatocytes, additional activation of the mitochondrial pathway by caspase 8 is required for apoptosis (Scaffidi et al. 1998). Caspase 8 cleavage of Bid, a proapoptotic protein that translocates to the mitochondria, triggers the release of cytochrome c, which activates caspase 9 and amplifies the apoptotic signal (Luo et al. 1998; Yin et al. 1999).

Hepatocytes can be protected from the effects of Fas by the antiapoptotic Bcl-2 family of proteins, including Bcl-2, myeloid cell leukemia factor-1, and Bcl-x_L, which inhibit the mitochondrial pathway (Lacronique et al. 1996; Rodriguez et al. 1996; de la Coste et al. 1999; Scaffidi et al. 1999). Alternatively, apoptosis can be inhibited at the level of caspase 8 by c-FLIP, a structural mimetic of caspase 8 that lacks proteolytic activity and binds directly to either caspase 8 or FADD (Irmler et al. 1997; Tschopp et al. 1998; Scaffidi et al. 1999). Other inhibitors of Fas-mediated apoptosis include X-linked inhibitors of apoptosis (IAPs) and serine protease inhibitors such as antitrypsin (Deveraux et al. 1997; Daemen et al. 2000).

Recently, it has been shown that Met sequestration of the Fas receptor can also prevent Fas-mediated apoptosis in hepatocytes (Zou et al. 2007). Mice lacking Met were hypersensitive to Fas-induced apoptosis (Huh et al. 2004); similarly, high doses of HGF sensitize apoptosis-resistant bid-deficient hepatocytes to Fas-induced cell death through increased dissociation of Fas from Met (Zhao et al. 2007). Further, YLGA, a peptide derived from the Met α-chain, inhibits Fas-mediated apoptosis in fatty liver disease (Zou et al. 2007). The Met/Fas interaction has been demonstrated to regulate apoptosis in endothelial cells as well (Smyth and Brady 2005).

Upregulation of Fas-mediated apoptosis has been implicated in the progression of many human liver diseases. For example, apoptosis is increased in patients with alcoholic hepatitis, and the number of apoptotic cells correlates with disease severity (Natori et al. 2001). Both the Fas receptor and FasL are strongly expressed in patients with this alcoholic hepatitis, which suggests that FasL-positive hepatocytes can induce apoptosis in adjacent Fas-expressing hepatocytes (Natori et al. 2001; Zhou et al. 2001; Yoon and Gores 2002). Another disease in which apoptosis occurs primarily through Fas upregulation is nonalcoholic steatohepatitis (Feldstein et al. 2003). Toxic bile salt accumulation in cholestatic liver disease results in induction of apoptosis in a Fas-dependent manner independent of FasL (Faubion et al. 1999). Fas signaling is also an essential component in hepatocyte injury and apoptosis in Wilson's disease (Strand et al. 1998). In viral hepatitis, apoptosis, which is regulated either by the pathogen itself or the host cytotoxic T cell-mediated immune response, occurs primarily through the Fas pathway (Kountouras et al. 2003). Fas receptor expression has been correlated with areas of inflammation and necrosis in hepatitis B (Mochizuki et al. 1996) and is also present in livers of patients with hepatitis C (Hiramatsu et al. 1994).

6.2.2 TNF-α Pathway

Tumor necrosis factor (TNF)-α is a pleiotropic cytokine that is critical to many diverse cellular processes in the liver, including proliferation, inflammation, and cell death (Schwabe and Brenner 2006). Execution of these functions occurs when TNF-α binds to its major receptor TNF-R1, causing a conformational change in the receptor that initiates DISC formation similar to that formed during Fas-mediated apoptosis. In the case of TNF-α-mediated signaling, this complex is composed of TNF/TNF-R1 trimers and the TNFR-associated death domain (TRADD) (Hsu et al. 1995, 1996; Chen et al. 2002). TRADD then recruits FADD, TNF-associated factor 2 (Traf-2), and receptor-interacting protein-1 (RIP-1) (Micheau and Tschopp 2003). If antiapoptotic signals are blocked, FADD activates caspase 8, which cleaves other downstream effector caspases like caspase 3 as well as activates proapoptotic members of the Bcl-2 family such as Bax and Bak (Wei et al. 2001). Thus, TNF-α-mediated apoptosis utilizes the same downstream apoptotic machinery as the Fas pathway, albeit independently of Bid (Stennicke et al. 1998; Harper et al. 2003; Chen et al. 2007).

A common means of inducing TNF-α signaling other than administration of TNF-α itself is through the use of lipopolysaccharide or LPS, an endotoxin derived from gram-negative bacteria (Lehmann et al. 1987; Shiratori et al. 1990; Yin and Ding 2003). The interaction of LPS with macrophages stimulates the release of TNF-α, which induces downstream signaling as described before. Unlike Fas-mediated apoptosis, induction of apoptosis via the TNF-α pathway requires hepatocyte sensitization, which is usually accomplished by pretreatment with either D-galactosamine (GalN) or actinomycin D (ActD) before administration of either TNF-α ligand or LPS. GalN is a hepatotoxic agent that depletes UTP, thus inhibiting de novo RNA synthesis and sensitizing hepatocytes to the effects of endotoxin (Lehmann et al. 1987). ActD is a different class of transcriptional inhibitor that binds DNA duplexes and prevents synthesis of protective proteins (Leist et al. 1994). In addition, administration of recombinant TNF-α directly after sensitization causes hepatotoxicity identical to that seen in LPS-treated mice, which also suggests that TNF-α is the major mediator of endotoxin lethality (Lehmann et al. 1987). Cell death caused by activation of the TNF-α pathway is believed to occur primarily by apoptosis, followed by necrosis.

In addition to initiating the cell death machinery, binding of TNF-α to its receptor stimulates formation of signaling complexes that prevent cell death, culminating in the activation of nuclear factor kappa B (NF-κB) (Guicciardi and Gores 2009). NF-κB is a group of dimeric transcription factors that initiate a wide variety of cellular programs in response to inflammation and injury. Inactive NF-κB is sequestered in the cytoplasm by members of the IκB family, which become phosphorylated and degraded by the IKK kinase complex in response to TNF-α signaling (Wajant et al. 2003; Israel 2010). Loss of the IκB complex liberates NF-κB, which then translocates to the nucleus, where it becomes phosphorylated and fully activated (Chen and Greene 2004). NF-κB has also been recognized as a key apoptosis inhibitor through its activation of genes including IAPs, c-FLIP, TRAFs, and Bcl family members, among others (Wang et al. 1998; Karin and Lin 2002). Expression and activation of these proteins suppress caspase activation and thus allow the cell to avoid undergoing apoptosis. Activation of NF-κB by LPS or TNF-α can directly antagonize the proapoptotic effects of the latter (Karin and Lin 2002), as demonstrated in experimental models where p65 is required for protection from TNF-α-mediated apoptosis (Beg and Baltimore 1996; Van Antwerp et al. 1996; Wallach et al. 1999). However, treatment with LPS or TNF-α alone does not cause any hepatocyte injury and thus requires agents like GalN, whose pretreatment prevents NF-κB-mediated pro-survival target gene expression for up to 3–4 hours. In summary, signaling through the TNF-α pathway can result in either life or death for the cell depending on physiological context and pathway cross talk, which favors activation of one pathway over another (Kyriakis 2001).

Like Fas, the TNF-α pathway is also dysregulated in a wide range of liver diseases. In addition to Fas, TNF-R1 is upregulated in the livers of patients with alcoholic liver disease (Ramalho et al. 2006; Akazawa and Gores 2007). Serum levels of TNF-α are also increased in patients with alcoholic steatohepatitis and correlate with disease progression and mortality, suggesting that this pathway plays a role in the pathogenesis of this disease (Bird et al. 1990). Oxidative stress caused by glutathione depletion can exacerbate the severity of alcoholic steatohepatitis by sensitizing hepatocytes to TNF-α (Colell et al. 1998). TNF-R1 and TNF-α are upregulated in both hepatitis B and hepatitis C, probably as a result of inflammation due to viral infection, and are associated with disease progression and severity (Marinos et al. 1995; Fang et al. 1996; Kallinowski et al. 1998; Tokushige et al. 2000). TNF-α-mediated apoptosis also plays a role in ischemia/reperfusion (I/R) injury and fulminant hepatic failure, likely through overaction of the immune system and generation of reactive oxygen intermediaries (Malhi et al. 2010).

6.2.3 Other Cell Signaling Pathways Involved in Cell Survival and Apoptosis

It is also worth mentioning that dysregulation of cell growth and cell cycle checkpoint pathways also contribute to hepatocyte survival and/or apoptosis, especially in HCC. The proto-oncogene c-Myc is a transcription factor that has an important role in regulating apoptotic signaling at the mitochondria, as well as through the death receptor pathways mentioned above (Hoffman and Liebermann 2008). c-Myc also regulates p53, a transcription factor that represses antiapoptotic proteins such as survivin and Bcl-2, resulting in growth arrest (Wu et al. 2001; Hoffman et al. 2002). This pathway can be suppressed by growth factors and is also commonly mutated, deleted, or otherwise activated in HCC, resulting in resistance to apoptosis (Muller et al. 1997; Hainaut and Hollstein 2000; Henriksson et al. 2001; Kodama et al. 2011). Prolonged activation of c-Jun NH_2-terminal kinase (JNK) can also contribute to hepatocyte cell death (Chen et al. 1996; Chang et al. 2006; Das et al. 2009), although its target, the transcription factor c-Jun, protects hepatocytes from apoptosis (Eferl et al. 2003; Fuest et al. 2012). Activation of the TGF-β pathway is generally proapoptotic, although TGF-β can also activate survival signals such as Akt and EGFR ligands (Ohno et al. 1995; Sanchez et al. 1996; Samson et al. 2002; Valdes et al. 2004; Yang et al. 2006; Caja et al. 2007). Finally, dysregulation of growth factor receptor pathways can result in constitutive activation of the signaling cascade, conferring resistance to apoptosis. For example, hepatitis viruses protect hepatocytes from apoptotic cell death by promoting the activation of Ras/PI3K/Akt/mTOR survival pathway (Steelman et al. 2011), and overexpression of components of these growth factor pathways can promote cell survival (Huynh et al. 2003; Liu et al. 2006).

6.3 β-Catenin in Cell Survival and Apoptosis

6.3.1 Cell Death in the Absence of β-Catenin

β-Catenin signaling appears to play a role in modulating cell survival especially in the liver. Deletion of β-catenin or inhibition of Wnt signaling has been shown to induce apoptosis in several in vivo and in vitro models. We will specifically discuss examples during prenatal and postnatal hepatic development.

β-Catenin gene and protein expression peaks at embryonic days E10–E14 in mouse embryonic liver, and during this time β-catenin is localized to the nucleus, cytoplasm, and membrane in different epithelial cells and coincided with ongoing cell proliferation (Micsenyi et al. 2004). Mouse embryonic livers from E9.5–E10 stages cultured in the presence of a β-catenin antisense oligonucleotides showed decreased proliferation and a simultaneous increase in apoptosis, two processes vital to hepatic morphogenesis that follows hepatic specification and induction (Monga et al. 2003). These findings were corroborated by a subsequent study that demonstrated overexpression of β-catenin in developing chicken livers leads to a threefold increase in liver size, which is due at least in part to an expanded hepatoblast population (Suksaweang et al. 2004). In the same study, blocking β-catenin expression through overexpression of pathway inhibitors resulted in decreased liver size and altered liver shape. Characterization of embryos where β-catenin was deleted from hepatoblasts using β-catenin-floxed mice and Foxa3-cre verified the above findings as β-catenin-deficient livers showed a notable decrease in liver size and associated mortality at around E16–E18 (Tan et al. 2008). Among other defects, a dramatic increase in death of resident hepatoblasts was apparent in knockouts (KOs) as shown by immunohistochemistry for TUNEL (Fig. 6.2a). Interestingly, an earlier study employing GSK3β gene knockouts also demonstrated a phenotype of increased liver cell death and liver degeneration that resulted in embryonic lethality (Hoeflich et al. 2000). This seems paradoxical since GSK3β loss would lead to β-catenin activation, which appears to be mitogenic and pro-survival. However, the liver degeneration evident in this model was consistent with excessive TNF-α toxicity and was rescued by neutralizing TNF-α. Thus, GSK3β loss prevented NF-κB activation and led to enhanced cell death in this model and was not due to β-catenin activation. Of further relevance here is that untimely and excessive β-catenin stabilization during hepatic development also has deleterious consequences. APC deletion during liver development also led to a dramatic increase in cell death and a counterintuitive decrease in cell proliferation; however, this was associated with untimely differentiation of hepatoblasts into biliary cells (Decaens et al. 2008). Although not directly tested, this could be a consequence of enhanced β-catenin levels which could modulate cell death through NF-κB modulation as discussed in the forthcoming section.

Our laboratory has also previously shown an increase in hepatocyte apoptosis in adult livers following hepatocyte-specific deletion of β-catenin (Tan et al. 2006). We observed a basal increase in hepatocyte survival as detected by TUNEL staining

(Fig. 6.2b). The exact basis of this observation still remains unclear. Similarly, we have reported an increase in hepatocyte cell death during liver regeneration at both early and late times (Fig. 6.2c) (Tan et al. 2006). Therefore, overall β-catenin appears to contribute to the anti-cell death phenotype in basal hepatocytes.

6.3.2 Mechanisms by Which Wnt/β-Catenin Signaling Regulates Hepatocyte Survival

The regulation of cell survival by Wnt/β-catenin signaling can occur at multiple levels, as described in the next several sections. β-Catenin can exert an inhibitory effect on the transcriptional activity of the NF-κB subunit p65 through physical association, which has implications in both acute liver injury and cancer. Several genes involved in cell survival, such as survivin and regucalcin, are direct transcriptional targets of Wnt/β-catenin signaling. Finally, β-catenin can also play an indirect role in cell survival through regulation of cell metabolism, inflammation, and oxidative stress (Fig. 6.3).

6.3.2.1 Direct Mechanisms

Physical Association

Many studies have addressed the complex cross regulation of the NF-κB and Wnt/β-catenin pathways in other cell and tissue types. For example, loss of E-cadherin in melanoma cells leads to an increase in the amount of cytoplasmic β-catenin, which in turn induces activation of NF-κB through a p38-mediated mechanism (Kuphal et al. 2004). Another study showed that β-catenin exists in a complex with IKKα, a key regulator of NF-κB activity, and that this kinase can phosphorylate β-catenin (Lamberti et al. 2001). Furthermore, inhibition or downregulation of IKKα inhibits growth of prostate cancer and multiple myeloma cell lines through decreased expression and activity of β-catenin (Albanese et al. 2003; Hideshima et al. 2009). Activation of β-catenin is downstream of the IKKα/β/NF-κB pathway in proximal colonic crypts as well (Umar et al. 2009). Other studies have identified the ubiquitin ligase receptor β-TrCP as a convergence point between these two pathways, as it regulates turnover of both β-catenin and NF-κB. Induction of β-TrCP by β-catenin, for example, leads to increased NF-κB transactivation; conversely, inhibition of β-TrCP impairs NF-κB activity while simultaneously causing nuclear accumulation of β-catenin (Spiegelman et al. 2000; Nakayama et al. 2003). From these studies, then, it can be concluded that in hepatocytes, β-catenin may interact with the NF-κB pathway at multiple points in the signaling cascade as well.

The physical association between the β-catenin and the NF-κB survival pathways was first described in colon and breast cancer. In that study, direct binding of β-catenin to NF-κB prevented its translocation and target gene activation in a colon cancer cell line (Deng et al. 2002). Furthermore, a correlation was found between activated β-catenin and inhibition of NF-κB target gene expression in colon and breast cancer tissues, highlighting the clinical relevance of this cross regulation between the β-catenin and NF-κB pathways. Mechanistically, another group has demonstrated that E-cadherin mediates the association of p65 with β-catenin, and thus NF-κB transcriptional activity is inhibited not by the nuclear pool of β-catenin but by the β-catenin associated with the adherens junction complex (Solanas et al. 2008). A direct interaction between the NF-κB p50 subunit and β-catenin was also described in bacterial-colonized HCT116 intestinal epithelial cells. Interestingly, the expression of mutated, constitutively active β-catenin also prevents IκBα degradation, thus inhibiting NF-κB activity by both direct and indirect mechanisms (Sun et al. 2005). Finally, co-immunoprecipitation in colon cancer cells showed a direct interaction between β-catenin and both p50 and p65 proteins. Further, as in the other tissue and cell types mentioned above, manipulation of β-catenin levels inversely influences NF-κB-mediated target gene expression (Du et al. 2009). Collectively, these findings support previous studies in which overexpression of β-catenin is pro-apoptotic and that this property is independent of TCF4 and the transcriptional function of β-catenin (Kim et al. 2000; Deng et al. 2004).

Conversely, p65 can also downregulate the β-catenin pathway by suppressing β-catenin nuclear translocation or activation (Masui et al. 2002; Saegusa et al. 2007). In colon cancer cells, treatment with a nonsteroidal anti-inflammatory drug inhibits Wnt/β-catenin signaling via NF-κB activation and its subsequent physical association with β-catenin (Cho et al. 2005). p65 was also shown to repress β-catenin-dependent cyclin D1 transcription, possibly through a protein-protein interaction (Hwang et al. 2010). Indeed, another group has validated this hypothesis by showing that activation of NF-κB in colon cancer cells blocks β-catenin binding to the cyclin D1 promoter through increased association between β-catenin and p65 (Abe et al. 2014). Thus, p65 inhibits β-catenin activity in much the same way as β-catenin inhibits p65 activity.

In the liver, the importance of this complex was first demonstrated in human HCC tissues. β-Catenin expression inversely correlated with the expression of either Fas and inducible nitric oxide synthase (iNOS), which are both targets of the NF-κB signaling pathway, in liver tumors (Du et al. 2009). Our laboratory has recently demonstrated that liver-specific β-catenin KO mice unexpectedly exhibit prolonged survival and reduced injury to TNF-α induced apoptosis. We elucidated this to be due to the existence of an inhibitory p65/β-catenin complex, which resides in the cytoplasm of hepatocytes. During the process of liver injury, this complex undergoes dynamic regulation, and its absence in the β-catenin KO mice led to earlier and robust activation of NF-κB target gene expression (Nejak-Bowen et al. 2013) (Fig. 6.4). In recent work from our laboratory, we also found the converse – that inhibition of p65 in culture can activate β-catenin – although the implications for cell survival after liver injury are unknown.

Adding to the complexity of these pathway interactions, another group found that GSK3β and APC, which are major upstream regulators of β-catenin, can regulate

NF-κB in colon and breast cancer cells and that this effect was mediated by β-catenin (Deng et al. 2004). In hepatocytes, however, inhibition of GSK3β, which would normally suppress NF-κB activity and promote cell death, instead fails to induce apoptosis after TNF-α challenge. This protection from apoptosis is due to induction of Wnt/β-catenin signaling and production of antiapoptotic factors, which ultimately lead to cell survival (Gotschel et al. 2008). Importantly, the authors found that contrary to previous publications, β-catenin does not directly mediate suppression of NF-κB activity. The differences between this finding and the studies mentioned in the previous paragraphs may be the level at which β-catenin and/or NF-κB is activated or suppressed, as well as the selectivity of the inhibitors or activators for their target protein(s).

The recent investigations into the function of the β-catenin/NF-κB complex and its role in cell survival have implications beyond LPS/TNF-α-induced apoptosis, as the role and regulation of this complex in other acute liver injuries such as acetaminophen toxicity and alcoholic liver injury remain to be explored. Additionally, it will be important to elucidate the role of this complex in cell survival and apoptosis during physiological process such as liver development, growth, and regeneration.

Transcriptional Targets of β-Catenin

Apoptosis can also be suppressed through transcriptional activation of the Wnt/β-catenin pathway. Chen et al. first demonstrated that Wnt/β-catenin signaling inhibits apoptosis in fibroblast cells, but experiments in hepatocytes and HCC cell lines soon followed (Chen et al. 2001). For example, the expression of hepatitis B x antigen in hepatoma cells was found to correlate with upregulation of β-catenin and enhanced cell survival (Lian et al. 2006). Stabilization of β-catenin in AML12 cells also resulted in protection from TNF-α-induced apoptosis (Shang et al. 2004). On the other hand, degradation of β-catenin was observed in hepatoma cells treated with chemotherapeutic agents designed to induce apoptosis (Emanuele et al. 2004). Transfection of hepatoma cells with a tumor-suppressing gene also resulted in β-catenin degradation, concomitant with apoptosis (Yoshibayashi et al. 2007). Additionally, overexpression of miR-122 promoted apoptosis in hepatoma cells, likely through downregulation of Wnt1 and β-catenin (Xu et al. 2012). Although none of these studies identified the downstream targets responsible for β-catenin-mediated cell survival, an oligonucleotide array analysis of β-catenin siRNA-treated HeLa cells identified a differential regulation of genes implicated in cell survival pathways such as PI3K/AKT, NF-κB, Fas, and p53 (Huang et al. 2006). Specific to the liver, other investigators have identified a reduction in the expression of angiogenesis genes, matrix metalloproteinase genes, and anti-apoptosis genes when β-catenin is silenced in HCC cells (Wang et al. 2010). Thus, the diversity of survival pathways affected by Wnt/β-catenin activation highlights its global role in cell signaling and provides strong support for its pro-survival function, independent of its role in regulating NF-κB activity.

One of the prototypical β-catenin target genes is c-Myc, which regulates cell proliferation and differentiation, and is also frequently activated in cancer (Dang 1999, 2012, Klaus and Birchmeier 2008). Interestingly, c-Myc can also induce or sensitize cells to apoptosis, which is a fail-safe program to prevent inappropriate cell proliferation (Hoffman and Liebermann 1998, 2008). In hepatocyte cell lines, c-Myc-induced apoptosis was a result of oxidative stress (Xu et al. 1997). c-Myc is also partly responsible for IFNγ-induced apoptosis in cultured hepatocytes (McCullough et al. 2007). Additionally, overexpression of c-Myc in hepatocytes led to an increase in apoptosis (Nevzorova et al. 2013).

On the other hand, deletion of c-Myc perinatally in vivo also induced apoptosis and oxidative stress in hepatocytes (Baena et al. 2005). Inhibition of c-Myc expression in rat hepatocyte cell lines also sensitizes hepatocytes to TNF-α cytotoxicity (Liu et al. 2000). Interestingly NF-κB transcriptional activity was also decreased in the absence of c-Myc, although these two effects were independent. Another study showed that NF-κB cooperates with c-Myc in promoting hepatocyte survival in a mouse hepatocyte cell line (Bellas and Sonenshein 1999). Thus, c-Myc is an important component in altering the balance between apoptosis and survival, although the molecular regulators influencing this cell fate decision are unclear. Additionally, although not all of these effects may be attributable to β-catenin-regulated transcriptional control of c-Myc, there is evidence that Wnt signaling can directly inhibit c-Myc-induced apoptosis in fibroblasts (You et al. 2002).

Inducible nitric oxide synthase (iNOS, also known as NOS2) is induced by proinflammatory cytokines in the setting of oxidative stress. In the liver, the antioxidant properties of iNOS are hepatoprotective and contribute to cell survival (Kuo et al. 1997). Du et al. were the first to identify that the iNOS gene was directly regulated by Wnt/β-catenin signaling in liver cells (Du et al. 2006). Intriguingly, in a follow-up paper, the same group demonstrated that β-catenin can also regulate iNOS expression indirectly through a physical association with NF-κB. Overexpression or mutation of β-catenin in colon and liver cancer cell lines led to a decrease in cytokine-mediated NF-κB transcriptional activity and a subsequent decrease in iNOS promoter activity; conversely, inhibition or lack of β-catenin induced NF-κB activity and iNOS expression (Du et al. 2009). These findings indicate that β-catenin can regulate the oxidative stress response through an inverse relationship with NF-κB. On the other hand, another group found that activation of β-catenin led to increased transcription of β-TrCP, which in turn degrades IκB. This leads to NF-κβ activation, which results in transcription of iNOS (Bandino et al. 2008). This direct relationship between β-catenin activation and NF-κB transcription of iNOS would appear to be in contrast to Du et al.'s findings; however, in the former, activation of β-catenin was the primary event and occurred due to disruption of adherens junctions and loss of association with E-cadherin, whereas in the latter, cytokine production and activation of NF-κB was the initiating stimulus. Thus, whether iNOS is activated or suppressed by β-catenin may be context-dependent.

Survivin, an antiapoptotic protein that inhibits caspase activation, is a β-catenin target gene (Zhang et al. 2001). Deletion of survivin in hepatocytes during postnatal liver development results in mitotic arrest and cell death (Li et al. 2013). Additionally, survivin is frequently overexpressed in HCC and may play an important role in survival and progression of this disease (Ito et al. 2000; Liu et al. 2008; Peroukides et al. 2010; Yang et al. 2011). Treatment of HCC cell lines with the small-molecule β-catenin inhibitor FH535 reduces survivin expression as well as cell proliferation, suggesting that this therapy may be effective at several levels (Gedaly et al. 2014). Small-molecule antagonists of the TCF4/β-catenin complex also resulted in decreased c-Myc and survivin expression, concomitant with decreased cell survival and increased apoptosis in vitro as well as in HCC xenografts (Wei et al. 2010). Additionally, the proteasomal inhibitor MG132 induced apoptosis in HCC cell lines secondary to increased β-catenin degradation and decreased survivin expression (Cervello et al. 2004). Mutated or constitutively active β-catenin also results in upregulation of the antiapoptotic proteins Bcl-2 and Bcl-x_L, albeit through indirect mechanisms (Oltersdorf et al. 2005; Xie et al. 2005; Li et al. 2007a, b; Mott and Gores 2007). Thus, Wnt/β-catenin signaling can positively impact cell survival through direct upregulation of antiapoptotic genes, although due to the diverse biological effects of this signaling pathway, other mechanisms are likely to contribute as well.

Tbx3 is a member of the T-box gene family that plays a role in patterning events during liver development through transcriptional repression (Davenport et al. 2003). Recently, Renard et al. showed that mutated β-catenin can induce the expression of Tbx3 in human and mouse HCC models as well as HCC cell lines (Renard et al. 2007). After confirming that Tbx3 transcription is directly regulated by β-catenin, they investigated the role of Tbx3 in cellular function. They found that inhibiting Tbx3 blocks β-catenin-mediated cell survival and renders cells sensitive to apoptosis, initiating a sequence of events ultimately leading to downregulation of p53 (Renard et al. 2007).

Regucalcin, also known as senescence marker protein-30 (SMP30) (Fujita 1999), is a Ca^{2+}-binding protein that has been implicated in maintaining cell homeostasis and function (Yamaguchi 2000; Yamaguchi 2005). In the liver, the overexpression of regucalcin in hepatoma cells suppresses cell proliferation and the expression of several known oncogenes, such as c-Myc (Tsurusaki and Yamaguchi 2003). Regucalcin overexpression also has a suppressive effect on apoptosis induced by TNF-α or thapsigargin (Izumi and Yamaguchi 2004). An important role of regucalcin has also been reported in vitamin C biosynthesis (Kondo et al. 2006). Through chromatin immunoprecipitation and in vitro promoter deletion mutants, we recently identified regucalcin to be a novel target of β-catenin signaling in the liver (Nejak-Bowen et al. 2009). Indeed β-catenin KO mice show significantly lower ascorbate levels secondary to decreased expression of both regucalcin and L-gulonolactone oxidase, both critical in vitamin C synthesis in murine hepatocytes (Linster and Van Schaftingen 2007). Furthermore, hepatocytes from β-catenin KO mice, which normally undergo massive apoptosis within 48 hours of culture, can be rescued by vitamin C supplementation. These findings are in agreement with previous studies where regucalcin overexpression has been shown to suppress cell death (Izumi and Yamaguchi 2004). Our results are also concordant with previous studies where mice containing a germline null mutation of SMP30 have been shown to be highly susceptible to both TNF-α and Fas-mediated apoptosis (Ishigami et al. 2002; Matsuyama et al. 2004).

Interestingly, we have also shown that loss of regucalcin promotes apoptosis in HepG2 cells, although the increase in apoptosis was not nearly as pronounced as in the murine hepatocytes (Nejak-Bowen et al. 2009). Moreover, the apoptosis in regucalcin siRNA-transfected hepatoma cells was rescued by supplementation with either ascorbic acid or N-acetylcysteine, a potent antioxidant. This is relevant because human cells lack L-gulonolactone oxidase and are unable to synthesize vitamin C; thus, the increased cell death seen in human hepatoma cells with decreased regucalcin expression cannot be due to restoration of vitamin C biosynthesis. These observations suggest that β-catenin might be important in human hepatocyte survival through regucalcin-dependent regulation of the redox state of the cell, independent of regucalcin's vitamin C biosynthetic function. In fact, previous publications from our lab have documented increased oxidative stress and apoptosis in prenatal β-catenin KO livers (Tan et al. 2008), MCD-induced liver injury (Behari et al. 2010), and adult livers from β-catenin conditional KOs (Tan et al. 2006; Zhang et al. 2010b). Thus, Wnt/β-catenin signaling plays a direct role in protection from apoptosis through regulation of the regucalcin gene, which contributes to cell survival by several mechanisms.

6.3.2.2 Indirect Mechanisms

Oxidative Stress

Reactive oxygen species (ROS), reactive nitrogen species (RNS), and free radicals are critical for physiology. They are a normal by-product of aerobic metabolism and also serve important functions as messenger molecules both inside and outside the cell. Reactive oxygen species generated by growth factors and cytokines are implicated in mediating cellular responses through signal transduction (Lander 1997). However, when the balance between ROS generation and antioxidant protection is disrupted, cells undergo oxidative stress (Sies 1997). Oxidative stress can cause damage through inflammation, ischemia, apoptosis, and necrosis and is a common feature of many liver injuries noted in the previous sections (Diesen and Kuo 2010).

The Wnt/β-catenin signaling pathway can regulate cell survival indirectly through regulation of oxidative stress-induced apoptosis. An interaction between β-catenin and hypoxia-inducible factor-1 (HIF-1) has been described in colorectal cancer which promotes adaptation to hypoxia or insufficient oxygen and nutrients (Kaidi et al. 2007). In the liver, loss of β-catenin in hepatocytes renders mice more susceptible to I/R injury, resulting in increased apoptosis; conversely, overexpression of Wnt-1 confers resistance to hypoxia and I/R injury (Lehwald et al. 2011). In both studies, under conditions of oxygen deprivation, HIF-1α competes with TCF4 for direct binding with β-catenin to induce the expression of genes that regulate adaptation and survival under hypoxic conditions. Therefore, interactions between β-catenin and HIF-1α depend on the redox state of a cell and in turn influence adaptation and survival.

Another mechanism of Wnt/β-catenin-regulated protection against oxidative stress involves the FoxO transcription factors, which control processes such as metabolism, proliferation, apoptosis, and survival. β-Catenin binds directly to FoxO in response to oxidative stress and enhances its transcriptional activity in *C. elegans* and in colon carcinoma cells (Essers et al. 2005; Hoogeboom et al. 2008). However, in the liver, Wnt/β-catenin signaling is important for reduction of oxidative stress-induced apoptosis through suppression, rather than enhancement, of FoxO3. During oxidative liver injury, β-catenin in hepatocytes inhibits FoxO-mediated apoptosis through transcription of serum/glucocorticoid-regulated kinase 1, which inactivates FoxO3, thus preventing transcription of such proapoptotic targets as p27 and Bim (Tao et al. 2013). Thus, regulation of oxidative stress by Wnt/β-catenin signaling appears to be context and tissue specific.

Oxidative stress also contributes to the development of HCC. The absence of β-catenin in hepatocytes promotes enhanced hepatocarcinogenesis in a mouse tumor model; this finding was secondary to sustained oxidative stress, which led to inflammation and apoptosis (Zhang et al. 2010b). Treatment of these β-catenin knockout mice with an antioxidant protects these mice from developing tumors, further supporting a role for β-catenin in regulating oxidative injury during tumorigenesis. Furthermore, identification of the association of CTNNB1 mutations with mutations in the gene encoding NRF2, a transcription factor that is critical in cellular redox homeostasis, using high-resolution copy number analysis of tumors not only suggests cooperation between these two pathways in HCC but also a role for Wnt/β-catenin signaling in the management of oxidative stress (Guichard et al. 2012). Thus, these studies identify a role for β-catenin as a potential tumor suppressor through its antioxidant properties.

Inflammation

Like NF-κB activation, inflammation can be a double-edged sword for the liver. Acute inflammation in response to insult or injury is an important part of immunity; however, chronic inflammation, such as that found in hepatitis and cirrhosis, is a risk factor for the development of HCC (Hagemann et al. 2007). Leukocyte cell-derived chemotaxin 2, or LECT2, was recently identified as a direct target of β-catenin and is mainly synthesized in the liver (Yamagoe et al. 1998; Ovejero et al. 2004). LECT2 was first characterized as a chemotactic factor for neutrophils (Yamagoe et al. 1996). However, it appears that in the liver, LECT2 regulates the homeostasis of natural killer T cells, as mice lacking LECT2 have increased hepatic natural killer T cells. Interestingly, these mice display more severe hepatic injury than controls in response to concanavalin A, characterized by increased apoptotic hepatocytes and elevation in secreted cytokine levels (Saito et al. 2004). These findings implicate LECT2, and thus β-catenin, in the pathogenesis of hepatitis and other inflammatory liver diseases. Another study found that oncogenic β-catenin induces a pro-inflammatory program that is permissive to tumorigenesis. In this model, LECT2 was identified as a key effector involved in this process, as deletion of this gene resulted in a dramatic aggravation of the inflammatory response (Anson et al. 2012). Interestingly, these authors also found that tumors with constitutively active β-catenin also had activated NF-κB as well, although whether that was a result of increased cytokine production or through direct regulation of the pathway by β-catenin is unknown. Accordingly, LECT2 has been utilized as a biomarker for β-catenin-activated hepatocytes in mouse models of HCC and in human HCCs irrespective of β-catenin mutational status (Colnot et al. 2004; Okabe et al. 2014). Serum LECT2 levels have also been used as prognostic indicators in recovery from acute liver failure and in liver regeneration after living donor liver transplantation (Sato et al. 2004a, b). These findings once again emphasize the essential role of β-catenin in regulating hepatocyte survival, even through secondary processes such as inflammation.

Cell Metabolism

As shown in Table 6.1, β-catenin is involved in glutamine metabolism through regulation of several genes including the glutamate transporter GLT-1 as well as glutamine synthetase (GS) (Cadoret et al. 2002). GS catalyzes conversion of glutamate to glutamine, which is required for protein and nucleotide synthesis. Glutamine can also be an alternative energy source for proliferating cells, particularly in tumor cells where the concentrations of reactive oxygen species are high and the citric acid cycle may be impaired (Cederbaum and Rubin 1976; Fukuda et al. 2002; Wan et al. 2015). In fact, a subset of tumors becomes addicted to glutamine as a substrate for maintenance of cell viability (Wise et al. 2008). Because of its dependence on β-catenin signaling, GS has been shown to be a surrogate target for β-catenin-activating mutations in HCC (Cadoret et al. 2002; Zucman-Rossi et al. 2007; Cieply et al. 2009; Lee et al. 2014). To assess if this had any physiological function, Tardito et al. treated various HCC cell lines with GS inhibitors. They found that only the β-catenin-mutated HCC cells were sensitive to glutamine depletion and the resulting inhibition of GS resulted in increased autophagy and apoptosis (Tardito et al. 2011). A follow-up study by this same group showed that glutamine depletion also hinders the growth of HCC xenografts (Chiu et al. 2014). Thus, β-catenin-dependent dysregulation can cause alterations in cell metabolism that have indirect consequences for cell survival.

6.3.3 Wnt/β-Catenin Signaling and Cell Survival in Other Liver Cell Types

In addition to its well-established contribution to hepatocyte cell survival, the Wnt/β-catenin signaling pathway also plays a role in preventing apoptosis in other liver cell types. Although the role and regulation of Wnt/β-catenin signaling are just beginning to be understood in these populations, hepatic stellate cells, endothelial cells, and inflammatory cells have all been shown to utilize this pathway to promote cell survival during liver injury. Together, these findings illustrate the cross talk and interdependency between cell types during liver injury, as well as the redundancy of critical pathways such as Wnt/β-catenin in regulating cell survival regardless of type.

6.3.3.1 Hepatic Stellate Cells

Hepatic stellate cells (HSCs) are fat-storing mesenchymal cells of the liver and are responsible for diverse functions including vitamin A storage, maintenance of sinusoidal vascular tone, and wound healing. During liver injury, HSCs are activated, resulting in deposition of extracellular matrix; in chronic injury, this process ultimately culminates in hepatic fibrosis. Thus, as the major liver cell type that responds to hepatocellular damage, HSCs are a vital component of the liver's response to injury and repair (Friedman 2008). Survival of quiescent, nonactivated HSC is dependent on Wnt signaling (Subramaniam et al. 2012). Wnt signaling is also involved in HSC activation and subsequent fibrosis, as Wnt antagonism restores quiescence in HSC (Cheng et al. 2008; Zhu et al. 2010). Interestingly, high doses of Dickkopf-related protein 1 (Dkk-1), an inhibitor of the β-catenin pathway, also induce apoptosis in cultured activated HSC. Studies by other groups have shown that treatment of hepatic stellate cells with Wnt3a suppresses TRAIL-induced apoptosis and contributes to cell survival (Myung et al. 2007) and that Wnt signaling is downregulated during HSC apoptosis (Shin et al. 2009), again suggesting a positive role for Wnt/β-catenin signaling in HSC survival and activation. On the other hand, several groups have suggested a role for noncanonical Wnt signaling in HSC activation (Jiang et al. 2006; Corbett et al. 2015). Furthermore, treatment with noncanonical Wnt5a can also protect activated HSC from apoptosis, as well as stimulate the expression of profibrogenic mediators from neighboring Kupffer cells. Thus, noncanonical Wnt signaling is implicated in the survival and profibrogenic phenotype of activated HSC through two mechanisms (Corbett et al. 2015). The complex regulation of HSC by canonical or noncanonical Wnt signaling thus needs to be further elucidated, as do the target genes responsible for HSC activation and liver fibrosis.

Liver Sinusoidal Endothelial Cells

Liver sinusoidal endothelial cells (LSECs), in addition to forming a continuous lining of the liver sinusoids, also contribute to repair after liver injury (DeLeve 2013). Although few direct studies exist on the role of Wnt/β-catenin signaling in hepatic endothelial cells, this pathway has been shown to induce proliferation in other endothelial cell types (Wright et al. 1999; Masckauchan et al. 2005). Further studies have demonstrated that the expression of mutated active β-catenin can promote cell survival after growth factor deprivation; activation of the Wnt/β-catenin pathway also promotes angiogenesis by induction of IL-8 (Masckauchan et al. 2005). Noncanonical Wnt signaling can also promote survival of endothelial cells in the absence of growth factors, albeit through a distinct mechanism (Masckauchan et al. 2006). Finally, Gedaly et al. showed that the β-catenin inhibitor FH535 reduced survivin levels in LSEC as well as in HCC cell lines (Gedaly et al. 2014). As with other cell types in which the activation of β-catenin contributes to survival, it will be important to determine the downstream targets and pathways responsible for this protection.

6.3.3.2 Inflammatory and Immune Cells

I/R injury triggers a complex cascade of inflammatory mediators that encompasses Kupffer cell activation, cytokine release, increased expression of cellular adhesion molecules and chemokines, and neutrophil transmigration, culminating in hepatocellular injury and death (Lentsch et al. 2000; Tsung et al. 2005). Knockdown of phosphatase and tensin homolog deleted on chromosome 10 (PTEN), which regulates the innate immune response, results in activation of β-catenin through Akt, which led to resistance against I/R injury through enhancement of antiapoptotic pathways (Kamo et al. 2013). The regulation of these immune networks by β-catenin may occur primarily in macrophage cells, as these cells secrete many cytokines including TNF-α that contribute to injury and inflammation (Colletti et al. 1994; Meng and Lowell 1997; Bruck et al. 2003; Giakoustidis et al. 2003; Wang et al. 2013). Furthermore, activation of β-catenin after PTEN knockdown also inhibited TLF4/NF-κB signaling, and the subsequent decrease in pro-inflammatory response may play a critical role in promoting survival as well (Kamo et al. 2013). Another mechanism by which Wnt/β-catenin signaling may regulate inflammation is found in a follow-up paper, which describes a role for β-catenin in reprogramming dendritic cells (DCs) during I/R injury. In this case, activated β-catenin in DC inhibits PTEN and promotes PI3K/Akt signaling, which in turn downregulates DC maturation and function through inhibition of TLR4. These findings are linked to a decrease in hepatocellular damage, inflammation, and apoptosis after I/R injury (Ke et al. 2013). Thus, enhancement of Wnt/β-catenin signaling in immune cells may ameliorate liver damage through suppression of inflammation.

6.4 Conclusions

Because apoptosis is a common feature in the pathology of many different liver diseases, therapeutic inhibition of this process has the potential to reduce injury and improve patient survival (Yoon and Gores 2002). Several therapies aimed at reducing apoptosis are already being utilized in preclinical and clinical trials in the treatment of liver disease. Infliximab, a monoclonal antibody raised against TNF-α, has been shown to attenuate hepatic injury in a pilot study (Spahr et al. 2002). Administration of caspase inhibitors lowered aminotransferase activity and decreased liver damage in NASH and hepatitis C patients and also protects against

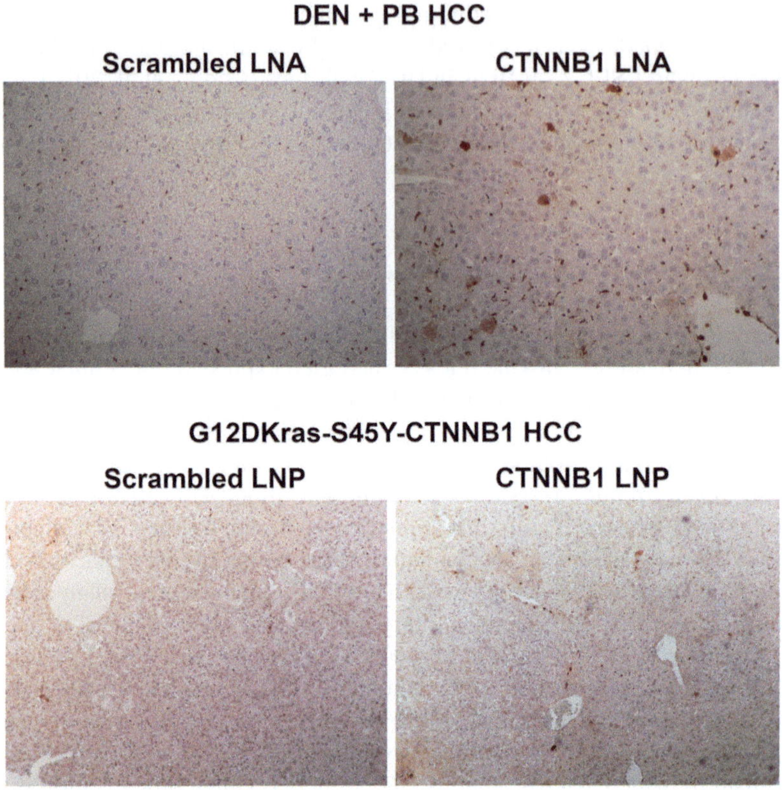

Fig. 6.5 Increased apoptosis in HCC after treatment with β-catenin inhibitors. (**a**) TUNEL staining shows induction of apoptosis in diethylnitrosamine/phenobarbital-induced HCC after treatment with β-catenin-locked nucleic acid antisense (*LNA*). (**b**) Increased apoptosis in Kras/β-catenin-mutated HCC treated with a lipid nanoparticle (*LNP*) loaded with a Dicer substrate siRNA targeting β-catenin

apoptotic injury during liver transplantation (Valentino et al. 2003; Baskin-Bey et al. 2007; Pockros et al. 2007; Masuoka et al. 2009; Ratziu et al. 2012). Chronic cholestatic liver disease is commonly treated with ursodeoxycholic acid, which is believed to attenuate both oxidative stress and apoptosis (Kumar and Tandon 2001; Rodrigues and Steer 2001). Exogenous antioxidant therapy has also been utilized to treat liver injury and apoptosis related to ROS; however, the results of these studies have been mixed (Diesen and Kuo 2010).

Other novel antiapoptotic therapies may include activators of pro-proliferative pathways, which would shift the balance toward cell survival and away from death. The Wnt/β-catenin pathway is one such example of a pro-survival pathway that may be amenable to manipulation. Our group and others have recently shown that triiodothyronine (T3) can activate β-catenin in hepatocytes and induce proliferation; however, further studies will be needed to elucidate the role of T3 in protection from liver injury (Fanti et al. 2014; Alvarado et al. 2016).

An overall useful exploitation of the observation that β-catenin is pro-survival in hepatocytes is to be able to suppress β-catenin signaling in liver tumors to induce tumor cell death. Indeed, β-catenin suppression using modalities like locked nucleic antisense (LNA), or siRNA packaged in liver tumor cell-specific EnCore lipid nanoparticles (LNPs), has shown a dramatic decrease in β-catenin activity and, as a consequence, a profound decrease in tumor burden in various models. For example, using LNA against β-catenin gene (CTNNB1) led to a complete response in a chemical carcinogenesis model where CTNNB1 mutations are drivers of the disease (Delgado et al. 2015). This was in part due to enhanced numbers of apoptotic nuclei as shown by staining for TUNEL (Fig. 6.5a). In another model of HCC, where sleeping beauty transposon and transposase were used to stably express mutant Kras and mutant β-catenin in a subset of hepatocytes, the use of CTNNB1 lipid nanoparticles (LNPs) led to a notable decrease in HCC burden as compared to the controls that received Scrambled LNP (Tao et al. 2016b). Again, the liver sections from the CTNNB1-LNP-treated Kras-β-catenin mice showed enhanced cell death as also detected by increased TUNEL staining (Fig. 6.5b). Thus, anti-β-catenin therapies may be of relevance in the treatment of a subset of HCC cases where unequivocal β-catenin activation is evident due to mutations in CTNNB1, in β-catenin degradation complex components such as AXIN1/AXIN2, or in other mechanisms (Fig. 6.6). The impact of β-catenin suppression on HCC burden is at least in large part predicted to be through the decrease in its direct transcriptional

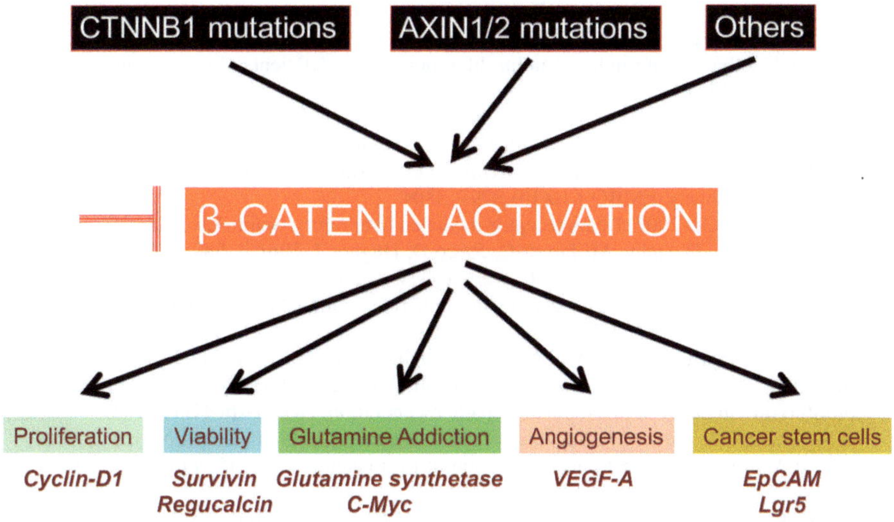

Fig. 6.6 Exploiting β-catenin's pro-survival role in hepatic tumors for therapeutics. β-Catenin activation in tumors like hepatocellular cancers can occur due to mutations in CTNNB1, AXIN1/AXIN2, or other causes. Once activated, β-catenin activates several target genes through its cofactor function and hence can regulate several processes critical in tumor growth and development. Suppression of β-catenin in tumors could lead to tumor cell death due to effects on multiple mechanisms like cell viability, tumor metabolism, and angiogenesis

targets, which have been discussed and hence will impact multiple and critical aspects of tumor biology (Fig. 6.6). In fact, disruption of even one of the various biological functions due to β-catenin downregulation followed by a decrease in its target genes affects another biological function, and hence the effects are multiplied. For example, the effect on tumor metabolism affects tumor cell survival and proliferation, as well as angiogenesis.

Other novel therapies for treatment of acute liver injury may include activators of pro-proliferative pathways such as NF-κB. Paradoxically, in these situations, inhibition of β-catenin, rather than activation, may be necessary. As discussed above, the inhibitory β-catenin/p65 complex prevents NF-κB activation and results in cell death and morbidity after TNF-α treatment, while loss of β-catenin from hepatocytes promotes survival. In recent years, several small-molecule inhibitors of the Wnt/β-catenin pathway have been identified. These antagonists have been touted as the next generation of cancer therapeutics because of their antiproliferative and pro-apoptotic effects in HCC with aberrant Wnt pathway activation (Emami et al. 2004; Lepourcelet et al. 2004; Behari et al. 2007; Handeli and Simon 2008; Delgado et al. 2015; Tao et al. 2016b).

Repurposing these inhibitors of β-catenin to prevent apoptosis in certain cases of liver injury or acute liver failure may thus be a relevant therapeutic strategy.

References

Abe N, Hou DX, Munemasa S, Murata Y, Nakamura Y (2014) Nuclear factor-kappaB sensitizes to benzyl isothiocyanate-induced antiproliferation in p53-deficient colorectal cancer cells. Cell Death Dis 5:e1534

Aberle H, Butz S, Stappert J, Weissig H, Kemler R, Hoschuetzky H (1994) Assembly of the cadherin-catenin complex in vitro with recombinant proteins. J Cell Sci 107(Pt 12):3655–3663

Aberle H, Bauer A, Stappert J, Kispert A, Kemler R (1997) beta-catenin is a target for the ubiquitin-proteasome pathway. EMBO J 16(13):3797–3804

Akazawa Y, Gores GJ (2007) Death receptor-mediated liver injury. Semin Liver Dis 27(4):327–338

Albanese C, Wu K, D'Amico M, Jarrett C, Joyce D, Hughes J, Hulit J, Sakamaki T, Fu M, Ben-Ze'ev A, Bromberg JF, Lamberti C, Verma U, Gaynor RB, Byers SW, Pestell RG (2003) IKKalpha regulates mitogenic signaling through transcriptional induction of cyclin D1 via Tcf. Mol Biol Cell 14(2):585–599

Alvarado TF, Puliga E, Preziosi M, Poddar M, Singh S, Columbano A, Nejak-Bowen K, Monga SP (2016) Thyroid hormone receptor beta agonist induces beta-catenin-dependent hepatocyte proliferation in mice: implications in hepatic regeneration. Gene Expr 17(1):19–34

Amit S, Hatzubai A, Birman Y, Andersen JS, Ben-Shushan E, Mann M, Ben-Neriah Y, Alkalay I (2002) Axin-mediated CKI phosphorylation of beta-catenin at Ser 45: a molecular switch for the Wnt pathway. Genes Dev 16(9):1066–1076

Anson M, Crain-Denoyelle AM, Baud V, Chereau F, Gougelet A, Terris B, Yamagoe S, Colnot S, Viguier M, Perret C, Couty JP (2012) Oncogenic beta-catenin triggers an inflammatory response that determines the aggressiveness of hepatocellular carcinoma in mice. J Clin Invest 122(2):586–599

Apte U, Zeng G, Muller P, Tan X, Micsenyi A, Cieply B, Dai C, Liu Y, Kaestner KH, Monga SP (2006) Activation of Wnt/beta-catenin pathway during hepatocyte growth factor-induced hepatomegaly in mice. Hepatology 44(4):992–1002

Apte U, Thompson MD, Cui S, Liu B, Cieply B, Monga SP (2008) Wnt/beta-catenin signaling mediates oval cell response in rodents. Hepatology 47(1):288–295

Apte U, Singh S, Zeng G, Cieply B, Virji MA, Wu T, Monga SP (2009) Beta-catenin activation promotes liver regeneration after acetaminophen-induced injury. Am J Pathol 175(3):1056–1065

Ashkenazi A, Dixit VM (1998) Death receptors: signaling and modulation. Science 281(5381):1305–1308

Baena E, Gandarillas A, Vallespinos M, Zanet J, Bachs O, Redondo C, Fabregat I, Martinez AC, de Alboran IM (2005) c-Myc regulates cell size and ploidy but is not essential for postnatal proliferation in liver. Proc Natl Acad Sci U S A 102(20):7286–7291

Bandino A, Compagnone A, Bravoco V, Cravanzola C, Lomartire A, Rossetto C, Novo E, Cannito S, Valfre di Bonzo L, Zamara E, Autelli R, Parola M, Colombatto S (2008) Beta-catenin triggers nuclear factor kappaB-dependent up-regulation of hepatocyte inducible nitric oxide synthase. Int J Biochem Cell Biol 40(9):1861–1871

Banziger C, Soldini D, Schutt C, Zipperlen P, Hausmann G, Basler K (2006) Wntless, a conserved membrane protein dedicated to the secretion of Wnt proteins from signaling cells. Cell 125(3):509–522

Barrott JJ, Cash GM, Smith AP, Barrow JR, Murtaugh LC (2011) Deletion of mouse Porcn blocks Wnt ligand secretion and reveals an ectodermal etiology of human focal dermal hypoplasia/Goltz syndrome. Proc Natl Acad Sci U S A 108(31):12752–12757

Bartscherer K, Boutros M (2008) Regulation of Wnt protein secretion and its role in gradient formation. EMBO Rep 9(10):977–982

Bartscherer K, Pelte N, Ingelfinger D, Boutros M (2006) Secretion of Wnt ligands requires Evi, a conserved transmembrane protein. Cell 125(3):523–533

Baskin-Bey ES, Washburn K, Feng S, Oltersdorf T, Shapiro D, Huyghe M, Burgart L, Garrity-Park M, van Vilsteren FG, Oliver LK, Rosen CB, Gores GJ (2007) Clinical trial of the pan-caspase inhibitor, IDN-6556, in human liver preservation injury. Am J Transplant 7(1):218–225

Beg AA, Baltimore D (1996) An essential role for NF-kappaB in preventing TNF-alpha-induced cell death. Science 274(5288):782–784

Behari J, Zeng G, Otruba W, Thompson MD, Muller P, Micsenyi A, Sekhon SS, Leoni L, Monga SP (2007) R-Etodolac decreases beta-catenin levels along with survival and proliferation of hepatoma cells. J Hepatol 46(5):849–857

Behari J, Yeh TH, Krauland L, Otruba W, Cieply B, Hauth B, Apte U, Wu T, Evans R, Monga SP (2010) Liver-specific beta-catenin knockout mice exhibit defective bile acid and cholesterol homeostasis and increased susceptibility to diet-induced steatohepatitis. Am J Pathol 176(2):744–753

Behrens J, Jerchow BA, Wurtele M, Grimm J, Asbrand C, Wirtz R, Kuhl M, Wedlich D, Birchmeier W (1998) Functional interaction of an axin homolog, conductin, with beta-catenin, APC, and GSK3beta. Science 280(5363):596–599

Bellas RE, Sonenshein GE (1999) Nuclear factor kappaB cooperates with c-Myc in promoting murine hepatocyte survival in a manner independent of p53 tumor suppressor function. Cell Growth Differ 10(5):287–294

Berke G (1995) The CTL's kiss of death. Cell 81(1):9–12

Bhanot P, Brink M, Samos CH, Hsieh JC, Wang Y, Macke JP, Andrew D, Nathans J, Nusse R (1996) A new member of the frizzled family from Drosophila functions as a Wingless receptor. Nature 382(6588):225–230

Bird GL, Sheron N, Goka AK, Alexander GJ, Williams RS (1990) Increased plasma tumor necrosis factor in severe alcoholic hepatitis. Ann Intern Med 112(12):917–920

Blache P, van de Wetering M, Duluc I, Domon C, Berta P, Freund JN, Clevers H, Jay P (2004) SOX9 is an intestine crypt transcription factor, is regulated by the Wnt pathway, and represses the CDX2 and MUC2 genes. J Cell Biol 166(1):37–47

Boldin MP, Varfolomeev EE, Pancer Z, Mett IL, Camonis JH, Wallach D (1995) A novel protein that interacts with the death domain of Fas/APO1 contains a sequence motif related to the death domain. J Biol Chem 270(14):7795–7798

Bonvini P, An WG, Rosolen A, Nguyen P, Trepel J, Garcia de Herreros A, Dunach M, Neckers LM (2001) Geldanamycin abrogates ErbB2 association with proteasome-resistant beta-catenin in melanoma cells, increases beta-catenin-E-cadherin association, and decreases beta-catenin-sensitive transcription. Cancer Res 61(4):1671–1677

Boon EM, van der Neut R, van de Wetering M, Clevers H, Pals ST (2002) Wnt signaling regulates expression of the receptor tyrosine kinase met in colorectal cancer. Cancer Res 62(18):5126–5128

Boutros M, Paricio N, Strutt DI, Mlodzik M (1998) Dishevelled activates JNK and discriminates between JNK pathways in planar polarity and wingless signaling. Cell 94(1):109–118

Brabletz T, Jung A, Dag S, Hlubek F, Kirchner T (1999) beta-catenin regulates the expression of the matrix metalloproteinase-7 in human colorectal cancer. Am J Pathol 155(4):1033–1038

Bruck R, Wardi J, Aeed H, Avni Y, Shirin H, Avinoach I, Shahmurov M, Hershkoviz R (2003) Glycine modulates cytokine secretion, inhibits hepatic damage and improves survival in a model of endotoxemia in mice. Liver Int 23(4):276–282

Cadigan KM, Liu YI (2006) Wnt signaling: complexity at the surface. J Cell Sci 119(Pt 3):395–402

Cadigan KM, Nusse R (1997) Wnt signaling: a common theme in animal development. Genes Dev 11(24):3286–3305

Cadigan KM, Fish MP, Rulifson EJ, Nusse R (1998) Wingless repression of Drosophila frizzled 2 expression shapes the Wingless morphogen gradient in the wing. Cell 93(5):767–777

Cadoret A, Ovejero C, Terris B, Souil E, Levy L, Lamers WH, Kitajewski J, Kahn A, Perret C (2002) New targets of beta-catenin signaling in the liver are involved in the glutamine metabolism. Oncogene 21(54):8293–8301

Caja L, Ortiz C, Bertran E, Murillo MM, Miro-Obradors MJ, Palacios E, Fabregat I (2007) Differential intracellular signalling induced by TGF-beta in rat adult hepatocytes and hepatoma cells: implications in liver carcinogenesis. Cell Signal 19(4):683–694

Cavallo RA, Cox RT, Moline MM, Roose J, Polevoy GA, Clevers H, Peifer M, Bejsovec A (1998) Drosophila Tcf and Groucho interact to repress Wingless signalling activity. Nature 395(6702):604–608

Cederbaum AI, Rubin E (1976) Fatty acid oxidation, substrate shuttles, and activity of the citric acid cycle in hepatocellular carcinomas of varying differentiation. Cancer Res 36(9 pt.1):2980–2987

Cervello M, Giannitrapani L, La Rosa M, Notarbartolo M, Labbozzetta M, Poma P, Montalto G, D'Alessandro N (2004) Induction of apoptosis by the proteasome inhibitor MG132 in human HCC cells: Possible correlation with specific caspase-dependent cleavage of beta-catenin and inhibition of beta-catenin-mediated transactivation. Int J Mol Med 13(5):741–748

Chang L, Kamata H, Solinas G, Luo JL, Maeda S, Venuprasad K, Liu YC, Karin M (2006) The E3 ubiquitin ligase itch couples JNK activation to TNFalpha-induced cell death by inducing c-FLIP(L) turnover. Cell 124(3):601–613

Chamorro MN, Schwartz DR, Vonica A, Brivanlou AH, Cho KR, Varmus HE (2005) FGF-20 and DKK1 are transcriptional targets of beta-catenin and FGF-20 is implicated in cancer and development. EMBO J 24(1):73–84

Chen LF, Greene WC (2004) Shaping the nuclear action of NF-kappaB. Nat Rev Mol Cell Biol 5(5):392–401

Chen YR, Wang X, Templeton D, Davis RJ, Tan TH (1996) The role of c-Jun N-terminal kinase (JNK) in apoptosis induced by ultraviolet C and gamma radiation. Duration of JNK activation may determine cell death and proliferation. J Biol Chem 271(50):31929–31936

Chen S, Guttridge DC, You Z, Zhang Z, Fribley A, Mayo MW, Kitajewski J, Wang CY (2001) Wnt-1 signaling inhibits apoptosis by activating beta-catenin/T cell factor-mediated transcription. J Cell Biol 152(1):87–96

Chen YW, Jeng YM, Yeh SH, Chen PJ (2002) P53 gene and Wnt signaling in benign neoplasms: beta-catenin mutations in hepatic adenoma but not in focal nodular hyperplasia. Hepatology 36(4 Pt 1):927–935

Chen X, Ding WX, Ni HM, Gao W, Shi YH, Gambotto AA, Fan J, Beg AA, Yin XM (2007) Bid-independent mitochondrial activation in tumor necrosis factor alpha-induced apoptosis and liver injury. Mol Cell Biol 27(2):541–553

Cheng JH, She H, Han YP, Wang J, Xiong S, Asahina K, Tsukamoto H (2008) Wnt antagonism inhibits hepatic stellate cell activation and liver fibrosis. Am J Physiol Gastrointest Liver Physiol 294(1):G39–G49

Chinnaiyan AM, O'Rourke K, Tewari M, Dixit VM (1995) FADD, a novel death domain-containing protein, interacts with the death domain of Fas and initiates apoptosis. Cell 81(4):505–512

Chiu M, Tardito S, Pillozzi S, Arcangeli A, Armento A, Uggeri J, Missale G, Bianchi MG, Barilli A, Dall'Asta V, Campanini N, Silini EM, Fuchs J, Armeanu-Ebinger S, Bussolati O (2014) Glutamine depletion by crisantaspase hinders the growth of human hepatocellular carcinoma xenografts. Br J Cancer 111(6):1159–1167

Cho M, Gwak J, Park S, Won J, Kim DE, Yea SS, Cha IJ, Kim TK, Shin JG, Oh S (2005) Diclofenac attenuates Wnt/beta-catenin signaling in colon cancer cells by activation of NF-kappaB. FEBS Lett 579(20):4213–4218

Cieply B, Zeng G, Proverbs-Singh T, Geller DA, Monga SP (2009) Unique phenotype of hepatocellular cancers with exon-3 mutations in beta-catenin gene. Hepatology 49(3):821–831

Clevers H (2006) Wnt/beta-catenin signaling in development and disease. Cell 127(3):469–480

Clevers H, Nusse R (2012) Wnt/beta-catenin signaling and disease. Cell 149(6):1192–1205

Cliffe A, Hamada F, Bienz M (2003) A role of Dishevelled in relocating Axin to the plasma membrane during wingless signaling. Curr Biol 13(11):960–966

Colell A, Garcia-Ruiz C, Miranda M, Ardite E, Mari M, Morales A, Corrales F, Kaplowitz N, Fernandez-Checa JC (1998) Selective glutathione depletion of mitochondria by ethanol sensitizes hepatocytes to tumor necrosis factor. Gastroenterology 115(6):1541–1551

Colletti LM, Remick DG, Campbell DA Jr (1994) LPS pretreatment protects from hepatic ischemia/reperfusion. J Surg Res 57(3):337–343

Colnot S, Decaens T, Niwa-Kawakita M, Godard C, Hamard G, Kahn A, Giovannini M, Perret C (2004) Liver-targeted disruption of Apc in mice activates beta-catenin signaling and leads to hepatocellular carcinomas. Proc Natl Acad Sci U S A 101(49):17216–17221

Corbett L, Mann J, Mann DA (2015) Non-canonical Wnt predominates in activated rat hepatic stellate cells, influencing HSC survival and paracrine stimulation of Kupffer cells. PLoS One 10(11):e0142794

de la Coste A, Fabre M, McDonell N, Porteu A, Gilgenkrantz H, Perret C, Kahn A, Mignon A (1999) Differential protective effects of Bcl-xL and Bcl-2 on apoptotic liver injury in transgenic mice. Am J Phys 277(3 Pt 1):G702–G708

Daemen MA, Heemskerk VH, van't Veer C, Denecker G, Wolfs TG, Vandenabeele P, Buurman WA (2000) Functional protection by acute phase proteins alpha(1)-acid glycoprotein and alpha(1)-antitrypsin against ischemia/reperfusion injury by preventing apoptosis and inflammation. Circulation 102(12):1420–1426

Dang CV (1999) c-Myc target genes involved in cell growth, apoptosis, and metabolism. Mol Cell Biol 19(1):1–11

Dang CV (2012) MYC on the path to cancer. Cell 149(1):22–35

Danilkovitch-Miagkova A, Miagkov A, Skeel A, Nakaigawa N, Zbar B, Leonard EJ (2001) Oncogenic mutants of RON and MET receptor tyrosine kinases cause activation of the beta-catenin pathway. Mol Cell Biol 21(17):5857–5868

Das M, Sabio G, Jiang F, Rincon M, Flavell RA, Davis RJ (2009) Induction of hepatitis by JNK-mediated expression of TNF-alpha. Cell 136(2):249–260

Davenport TG, Jerome-Majewska LA, Papaioannou VE (2003) Mammary gland, limb and yolk sac defects in mice lacking Tbx3, the gene mutated in human ulnar mammary syndrome. Development 130(10):2263–2273

De A (2011) Wnt/Ca2+ signaling pathway: a brief overview. Acta Biochim Biophys Sin Shanghai 43(10):745–756

Decaens T, Godard C, de Reynies A, Rickman DS, Tronche F, Couty JP, Perret C, Colnot S (2008) Stabilization of beta-catenin affects mouse embryonic liver growth and hepatoblast fate. Hepatology 47(1):247–258

DeLeve LD (2013) Liver sinusoidal endothelial cells and liver regeneration. J Clin Invest 123(5):1861–1866

Delgado E, Okabe H, Preziosi M, Russell JO, Alvarado TF, Oertel M, Nejak-Bowen KN, Zhang Y, Monga SP (2015) Complete response of Ctnnb1-mutated tumours to beta-catenin suppression by locked nucleic acid antisense in a mouse hepatocarcinogenesis model. J Hepatol 62(2):380–387

Deng J, Miller SA, Wang HY, Xia W, Wen Y, Zhou BP, Li Y, Lin SY, Hung MC (2002) beta-catenin interacts with and inhibits NF-kappa B in human colon and breast cancer. Cancer Cell 2(4):323–334

Deng J, Xia W, Miller SA, Wen Y, Wang HY, Hung MC (2004) Crossregulation of NF-kappaB by the APC/GSK-3beta/beta-catenin pathway. Mol Carcinog 39(3):139–146

Deveraux QL, Takahashi R, Salvesen GS, Reed JC (1997) X-linked IAP is a direct inhibitor of cell-death proteases. Nature 388(6639):300–304

Diesen DL, Kuo PC (2010) Nitric oxide and redox regulation in the liver: Part I. General considerations and redox biology in hepatitis. J Surg Res 162(1):95–109

Ding WX, Yin XM (2004) Dissection of the multiple mechanisms of TNF-alpha-induced apoptosis in liver injury. J Cell Mol Med 8(4):445–454

Du Q, Park KS, Guo Z, He P, Nagashima M, Shao L, Sahai R, Geller DA, Hussain SP (2006) Regulation of human nitric oxide synthase 2 expression by Wnt beta-catenin signaling. Cancer Res 66(14):7024–7031

Du Q, Zhang X, Cardinal J, Cao Z, Guo Z, Shao L, Geller DA (2009) Wnt/beta-catenin signaling regulates cytokine-induced human inducible nitric oxide synthase expression by inhibiting nuclear factor-kappaB activation in cancer cells. Cancer Res 69(9):3764–3771

Eferl R, Ricci R, Kenner L, Zenz R, David JP, Rath M, Wagner EF (2003) Liver tumor development. c-Jun antagonizes the proapoptotic activity of p53. Cell 112(2):181–192

Emami KH, Nguyen C, Ma H, Kim DH, Jeong KW, Eguchi M, Moon RT, Teo JL, Kim HY, Moon SH, Ha JR, Kahn M (2004) A small molecule inhibitor of beta-catenin/CREB-binding protein transcription [corrected]. Proc Natl Acad Sci U S A 101(34):12682–12687

Emanuele S, D'Anneo A, Bellavia G, Vassallo B, Lauricella M, De Blasio A, Vento R, Tesoriere G (2004) Sodium butyrate induces apoptosis in human hepatoma cells by a mitochondria/caspase pathway, associated with degradation of beta-catenin, pRb and Bcl-XL. Eur J Cancer 40(9):1441–1452

Essers MA, de Vries-Smits LM, Barker N, Polderman PE, Burgering BM, Korswagen HC (2005) Functional interaction between beta-catenin and FOXO in oxidative stress signaling. Science 308(5725):1181–1184

Esufali S, Bapat B (2004) Cross-talk between Rac1 GTPase and dysregulated Wnt signaling pathway leads to cellular redistribution of beta-catenin and TCF/LEF-mediated transcriptional activation. Oncogene 23(50):8260–8271

Fang JW, Shen WW, Meager A, Lau JY (1996) Activation of the tumor necrosis factor-alpha system in the liver in chronic hepatitis B virus infection. Am J Gastroenterol 91(4):748–753

Fanti M, Singh S, Ledda-Columbano GM, Columbano A, Monga SP (2014) Tri-iodothyronine induces hepatocyte proliferation by protein kinase A-dependent beta-catenin activation in rodents. Hepatology 59(6):2309–2320

Faubion WA, Guicciardi ME, Miyoshi H, Bronk SF, Roberts PJ, Svingen PA, Kaufmann SH, Gores GJ (1999) Toxic bile salts induce rodent hepatocyte apoptosis via direct activation of Fas. J Clin Invest 103(1):137–145

Feldstein AE, Canbay A, Angulo P, Taniai M, Burgart LJ, Lindor KD, Gores GJ (2003) Hepatocyte apoptosis and fas expression are prominent features of human nonalcoholic steatohepatitis. Gastroenterology 125(2):437–443

Friedman SL (2008) Hepatic stellate cells: protean, multifunctional, and enigmatic cells of the liver. Physiol Rev 88(1):125–172

Fuest M, Willim K, MacNelly S, Fellner N, Resch GP, Blum HE, Hasselblatt P (2012) The transcription factor c-Jun protects against sustained hepatic endoplasmic reticulum stress thereby promoting hepatocyte survival. Hepatology 55(2):408–418

Fujita T (1999) Senescence marker protein-30 (SMP30): structure and biological function. Biochem Biophys Res Commun 254(1):1–4

Fukuda H, Ebara M, Okuyama M, Sugiura N, Yoshikawa M, Saisho H, Shimizu R, Motoji N, Shigematsu A, Watayo T (2002) Increased metabolizing activities of the tricarboxylic acid cycle and decreased drug metabolism in hepatocellular carcinoma. Carcinogenesis 23(12):2019–2023

Galanos C, Freudenberg MA, Matsuura M, Coumbos A (1988) Hypersensitivity to endotoxin and mechanisms of host-response. Prog Clin Biol Res 272:295–308

Ge WS, Wang YJ, Wu JX, Fan JG, Chen YW, Zhu L (2014) beta-catenin is overexpressed in hepatic fibrosis and blockage of Wnt/beta-catenin signaling inhibits hepatic stellate cell activation. Mol Med Rep 9(6):2145–2151

Gedaly R, Galuppo R, Daily MF, Shah M, Maynard E, Chen C, Zhang X, Esser KA, Cohen DA, Evers BM, Jiang J, Spear BT (2014) Targeting the Wnt/beta-catenin signaling pathway in liver cancer stem cells and hepatocellular carcinoma cell lines with FH535. PLoS One 9(6):e99272

Giakoustidis DE, Iliadis S, Tsantilas D, Papageorgiou G, Kontos N, Kostopoulou E, Botsoglou NA, Gerasimidis T, Dimitriadou A (2003) Blockade of Kupffer cells by gadolinium chloride reduces lipid peroxidation and protects liver from ischemia/reperfusion injury. Hepato-Gastroenterology 50(53):1587–1592

Giera S, Braeuning A, Kohle C, Bursch W, Metzger U, Buchmann A, Schwarz M (2010) Wnt/beta-catenin signaling activates and determines hepatic zonal expression of glutathione S-transferases in mouse liver. Toxicol Sci 115(1):22–33

Gotschel F, Kern C, Lang S, Sparna T, Markmann C, Schwager J, McNelly S, von Weizsacker F, Laufer S, Hecht A, Merfort I (2008) Inhibition of GSK3 differentially modulates NF-kappaB, CREB, AP-1 and beta-catenin signaling in hepatocytes, but fails to promote TNF-alpha-induced apoptosis. Exp Cell Res 314(6):1351–1366

Gougelet A, Torre C, Veber P, Sartor C, Bachelot L, Denechaud PD, Godard C, Moldes M, Burnol AF, Dubuquoy C, Terris B, Guillonneau F, Ye T, Schwarz M, Braeuning A, Perret C, Colnot S (2014) T-cell factor 4 and beta-catenin chromatin occupancies pattern zonal liver metabolism in mice. Hepatology 59(6):2344–2357

Guicciardi ME, Gores GJ (2009) Life and death by death receptors. FASEB J 23(6):1625–1637

Guichard C, Amaddeo G, Imbeaud S, Ladeiro Y, Pelletier L, Maad IB, Calderaro J, Bioulac-Sage P, Letexier M, Degos F, Clement B, Balabaud C, Chevet E, Laurent A, Couchy G, Letouze E, Calvo F, Zucman-Rossi J (2012) Integrated analysis of somatic mutations and focal copy-number changes identifies key genes and pathways in hepatocellular carcinoma. Nat Genet 44(6):694–698

Hagemann T, Balkwill F, Lawrence T (2007) Inflammation and cancer: a double-edged sword. Cancer Cell 12(4):300–301

Hainaut P, Hollstein M (2000) p53 and human cancer: the first ten thousand mutations. Adv Cancer Res 77:81–137

Handeli S, Simon JA (2008) A small-molecule inhibitor of Tcf/beta-catenin signaling down-regulates PPARgamma and PPARdelta activities. Mol Cancer Ther 7(3):521–529

Harper N, Hughes M, MacFarlane M, Cohen GM (2003) Fas-associated death domain protein and caspase-8 are not recruited to the tumor necrosis factor receptor 1 signaling complex during tumor necrosis factor-induced apoptosis. J Biol Chem 278(28):25534–25541

He TC, Sparks AB, Rago C, Hermeking H, Zawel L, da Costa LT, Morin PJ, Vogelstein B, Kinzler KW (1998) Identification of c-MYC as a target of the APC pathway. Science 281(5382):1509–1512

Hecht A, Vleminckx K, Stemmler MP, van Roy F, Kemler R (2000) The p300/CBP acetyltransferases function as transcriptional coactivators of beta-catenin in vertebrates. EMBO J 19(8):1839–1850

Henderson BR, Fagotto F (2002) The ins and outs of APC and beta-catenin nuclear transport. EMBO Rep 3(9):834–839

Hendrix ND, Wu R, Kuick R, Schwartz DR, Fearon ER, Cho KR (2006) Fibroblast growth factor 9 has oncogenic activity and is a downstream target of Wnt signaling in ovarian endometrioid adenocarcinomas. Cancer Res 66(3):1354–1362

Henriksson M, Selivanova G, Lindstrom M, Wiman KG (2001) Inactivation of Myc-induced p53-dependent apoptosis in human tumors. Apoptosis 6(1–2):133–137
Hideshima T, Chauhan D, Kiziltepe T, Ikeda H, Okawa Y, Podar K, Raje N, Protopopov A, Munshi NC, Richardson PG, Carrasco RD, Anderson KC (2009) Biologic sequelae of I{kappa}B kinase (IKK) inhibition in multiple myeloma: therapeutic implications. Blood 113(21):5228–5236
Hiramatsu N, Hayashi N, Katayama K, Mochizuki K, Kawanishi Y, Kasahara A, Fusamoto H, Kamada T (1994) Immunohistochemical detection of Fas antigen in liver tissue of patients with chronic hepatitis C. Hepatology 19(6):1354–1359
Hirota M, Watanabe K, Hamada S, Sun Y, Strizzi L, Mancino M, Nagaoka T, Gonzales M, Seno M, Bianco C, Salomon DS (2008) Smad2 functions as a co-activator of canonical Wnt/beta-catenin signaling pathway independent of Smad4 through histone acetyltransferase activity of p300. Cell Signal 20(9):1632–1641
Hiscox S, Jiang WG (1999) Association of the HGF/SF receptor, c-met, with the cell-surface adhesion molecule, E-cadherin, and catenins in human tumor cells. Biochem Biophys Res Commun 261(2):406–411
Ho HY, Susman MW, Bikoff JB, Ryu YK, Jonas AM, Hu L, Kuruvilla R, Greenberg ME (2012) Wnt5a-Ror-Dishevelled signaling constitutes a core developmental pathway that controls tissue morphogenesis. Proc Natl Acad Sci U S A 109(11):4044–4051
Hoeflich KP, Luo J, Rubie EA, Tsao MS, Jin O, Woodgett JR (2000) Requirement for glycogen synthase kinase-3beta in cell survival and NF-kappaB activation. Nature 406(6791):86–90
Hoffman B, Liebermann DA (1998) The proto-oncogene c-myc and apoptosis. Oncogene 17(25):3351–3357
Hoffman B, Liebermann DA (2008) Apoptotic signaling by c-MYC. Oncogene 27(50):6462–6472
Hoffman WH, Biade S, Zilfou JT, Chen J, Murphy M (2002) Transcriptional repression of the anti-apoptotic survivin gene by wild type p53. J Biol Chem 277(5):3247–3257
Hoogeboom D, Essers MA, Polderman PE, Voets E, Smits LM, Burgering BM (2008) Interaction of FOXO with beta-catenin inhibits beta-catenin/T cell factor activity. J Biol Chem 283(14):9224–9230
Hoschuetzky H, Aberle H, Kemler R (1994) Beta-catenin mediates the interaction of the cadherin-catenin complex with epidermal growth factor receptor. J Cell Biol 127(5):1375–1380
Hovanes K, Li TW, Munguia JE, Truong T, Milovanovic T, Lawrence Marsh J, Holcombe RF, Waterman ML (2001) Beta-catenin-sensitive isoforms of lymphoid enhancer factor-1 are selectively expressed in colon cancer. Nat Genet 28(1):53–57
Hsieh JC, Kodjabachian L, Rebbert ML, Rattner A, Smallwood PM, Samos CH, Nusse R, Dawid IB, Nathans J (1999) A new secreted protein that binds to Wnt proteins and inhibits their activities. Nature 398(6726):431–436
Hsu H, Xiong J, Goeddel DV (1995) The TNF receptor 1-associated protein TRADD signals cell death and NF-kappa B activation. Cell 81(4):495–504
Hsu H, Shu HB, Pan MG, Goeddel DV (1996) TRADD-TRAF2 and TRADD-FADD interactions define two distinct TNF receptor 1 signal transduction pathways. Cell 84(2):299–308
Huang M, Wang Y, Sun D, Zhu H, Yin Y, Zhang W, Yang S, Quan L, Bai J, Wang S, Chen Q, Li S, Xu N (2006) Identification of genes regulated by Wnt/beta-catenin pathway and involved in apoptosis via microarray analysis. BMC Cancer 6:221
Huber AH, Weis WI (2001) The structure of the beta-catenin/E-cadherin complex and the molecular basis of diverse ligand recognition by beta-catenin. Cell 105(3):391–402
Huh CG, Factor VM, Sanchez A, Uchida K, Conner EA, Thorgeirsson SS (2004) Hepatocyte growth factor/c-met signaling pathway is required for efficient liver regeneration and repair. Proc Natl Acad Sci U S A 101(13):4477–4482
Huynh H, Nguyen TT, Chow KH, Tan PH, Soo KC, Tran E (2003) Over-expression of the mitogen-activated protein kinase (MAPK) kinase (MEK)-MAPK in hepatocellular carcinoma: its role in tumor progression and apoptosis. BMC Gastroenterol 3:19
Hwang I, Choi YS, Jeon MY, Jeong S (2010) NF-kappaB p65 represses beta-catenin-activated transcription of cyclin D1. Biochem Biophys Res Commun 403(1):79–84

Irmler M, Thome M, Hahne M, Schneider P, Hofmann K, Steiner V, Bodmer JL, Schroter M, Burns K, Mattmann C, Rimoldi D, French LE, Tschopp J (1997) Inhibition of death receptor signals by cellular FLIP. Nature 388(6638):190–195

Ishigami A, Fujita T, Handa S, Shirasawa T, Koseki H, Kitamura T, Enomoto N, Sato N, Shimosawa T, Maruyama N (2002) Senescence marker protein-30 knockout mouse liver is highly susceptible to tumor necrosis factor-alpha- and Fas-mediated apoptosis. Am J Pathol 161(4):1273–1281

Israel A (2010) The IKK complex, a central regulator of NF-kappaB activation. Cold Spring Harb Perspect Biol 2(3):a000158

Ito T, Shiraki K, Sugimoto K, Yamanaka T, Fujikawa K, Ito M, Takase K, Moriyama M, Kawano H, Hayashida M, Nakano T, Suzuki A (2000) Survivin promotes cell proliferation in human hepatocellular carcinoma. Hepatology 31(5):1080–1085

Izumi T, Yamaguchi M (2004) Overexpression of regucalcin suppresses cell death in cloned rat hepatoma H4-II-E cells induced by tumor necrosis factor-alpha or thapsigargin. J Cell Biochem 92(2):296–306

Jamora C, DasGupta R, Kocieniewski P, Fuchs E (2003) Links between signal transduction, transcription and adhesion in epithelial bud development. Nature 422(6929):317–322

Jiang F, Parsons CJ, Stefanovic B (2006) Gene expression profile of quiescent and activated rat hepatic stellate cells implicates Wnt signaling pathway in activation. J Hepatol 45(3):401–409

Kadowaki T, Wilder E, Klingensmith J, Zachary K, Perrimon N (1996) The segment polarity gene porcupine encodes a putative multitransmembrane protein involved in Wingless processing. Genes Dev 10(24):3116–3128

Kaidi A, Williams AC, Paraskeva C (2007) Interaction between beta-catenin and HIF-1 promotes cellular adaptation to hypoxia. Nat Cell Biol 9(2):210–217

Kallinowski B, Haseroth K, Marinos G, Hanck C, Stremmel W, Theilmann L, Singer MV, Rossol S (1998) Induction of tumour necrosis factor (TNF) receptor type p55 and p75 in patients with chronic hepatitis C virus (HCV) infection. Clin Exp Immunol 111(2):269–277

Kam Y, Quaranta V (2009) Cadherin-bound beta-catenin feeds into the Wnt pathway upon adherens junctions dissociation: evidence for an intersection between beta-catenin pools. PLoS One 4(2):e4580

Kamo N, Ke B, Busuttil RW, Kupiec-Weglinski JW (2013) PTEN-mediated Akt/beta-catenin/Foxo1 signaling regulates innate immune responses in mouse liver ischemia/reperfusion injury. Hepatology 57(1):289–298

Kanai Y, Ochiai A, Shibata T, Oyama T, Ushijima S, Akimoto S, Hirohashi S (1995) c-erbB-2 gene product directly associates with beta-catenin and plakoglobin. Biochem Biophys Res Commun 208(3):1067–1072

Karin M, Lin A (2002) NF-kappaB at the crossroads of life and death. Nat Immunol 3(3):221–227

Katuri V, Tang Y, Li C, Jogunoori W, Deng CX, Rashid A, Sidawy AN, Evans S, Reddy EP, Mishra B, Mishra L (2006) Critical interactions between TGF-beta signaling/ELF, and E-cadherin/beta-catenin mediated tumor suppression. Oncogene 25(13):1871–1886

Kazanskaya O, Glinka A, del Barco Barrantes I, Stannek P, Niehrs C, Wu W (2004) R-Spondin2 is a secreted activator of Wnt/beta-catenin signaling and is required for Xenopus myogenesis. Dev Cell 7(4):525–534

Ke B, Shen XD, Kamo N, Ji H, Yue S, Gao F, Busuttil RW, Kupiec-Weglinski JW (2013) beta-catenin regulates innate and adaptive immunity in mouse liver ischemia-reperfusion injury. Hepatology 57(3):1203–1214

Khan Z, Vijayakumar S, de la Torre TV, Rotolo S, Bafico A (2007) Analysis of endogenous LRP6 function reveals a novel feedback mechanism by which Wnt negatively regulates its receptor. Mol Cell Biol 27(20):7291–7301

Kim K, Pang KM, Evans M, Hay ED (2000) Overexpression of beta-catenin induces apoptosis independent of its transactivation function with LEF-1 or the involvement of major G1 cell cycle regulators. Mol Biol Cell 11(10):3509–3523

Klaus A, Birchmeier W (2008) Wnt signalling and its impact on development and cancer. Nat Rev Cancer 8(5):387–398

Kodama T, Takehara T, Hikita H, Shimizu S, Shigekawa M, Tsunematsu H, Li W, Miyagi T, Hosui A, Tatsumi T, Ishida H, Kanto T, Hiramatsu N, Kubota S, Takigawa M, Tomimaru Y, Tomokuni A, Nagano H, Doki Y, Mori M, Hayashi N (2011) Increases in p53 expression induce CTGF synthesis by mouse and human hepatocytes and result in liver fibrosis in mice. J Clin Invest 121(8):3343–3356

Koh TJ, Bulitta CJ, Fleming JV, Dockray GJ, Varro A, Wang TC (2000) Gastrin is a target of the beta-catenin/TCF-4 growth-signaling pathway in a model of intestinal polyposis. J Clin Invest 106(4):533–539

Kondo Y, Inai Y, Sato Y, Handa S, Kubo S, Shimokado K, Goto S, Nishikimi M, Maruyama N, Ishigami A (2006) Senescence marker protein 30 functions as gluconolactonase in L-ascorbic acid biosynthesis, and its knockout mice are prone to scurvy. Proc Natl Acad Sci U S A 103(15):5723–5728

Kountouras J, Zavos C, Chatzopoulos D (2003) Apoptosis in hepatitis C. J Viral Hepat 10(5):335–342

Kumar D, Tandon RK (2001) Use of ursodeoxycholic acid in liver diseases. J Gastroenterol Hepatol 16(1):3–14

Kuo PC, Abe KY, Schroeder RA (1997) Oxidative stress increases hepatocyte iNOS gene transcription and promoter activity. Biochem Biophys Res Commun 234(2):289–292

Kuphal S, Poser I, Jobin C, Hellerbrand C, Bosserhoff AK (2004) Loss of E-cadherin leads to upregulation of NFkappaB activity in malignant melanoma. Oncogene 23(52):8509–8519

Kyriakis JM (2001) Life-or-death decisions. Nature 414(6861):265–266

de La Coste A, Romagnolo B, Billuart P, Renard CA, Buendia MA, Soubrane O, Fabre M, Chelly J, Beldjord C, Kahn A, Perret C (1998) Somatic mutations of the beta-catenin gene are frequent in mouse and human hepatocellular carcinomas. Proc Natl Acad Sci U S A 95(15):8847–8851

Labbe E, Letamendia A, Attisano L (2000) Association of Smads with lymphoid enhancer binding factor 1/T cell-specific factor mediates cooperative signaling by the transforming growth factor-beta and wnt pathways. Proc Natl Acad Sci U S A 97(15):8358–8363

Lacronique V, Mignon A, Fabre M, Viollet B, Rouquet N, Molina T, Porteu A, Henrion A, Bouscary D, Varlet P, Joulin V, Kahn A (1996) Bcl-2 protects from lethal hepatic apoptosis induced by an anti-Fas antibody in mice. Nat Med 2(1):80–86

Lamberti C, Lin KM, Yamamoto Y, Verma U, Verma IM, Byers S, Gaynor RB (2001) Regulation of beta-catenin function by the IkappaB kinases. J Biol Chem 276(45):42276–42286

Lander HM (1997) An essential role for free radicals and derived species in signal transduction. FASEB J 11(2):118–124

Lee JM, Yang J, Newell P, Singh S, Parwani A, Friedman SL, Nejak-Bowen KN, Monga SP (2014) beta-Catenin signaling in hepatocellular cancer: Implications in inflammation, fibrosis, and proliferation. Cancer Lett 343(1):90–97

Lehmann V, Freudenberg MA, Galanos C (1987) Lethal toxicity of lipopolysaccharide and tumor necrosis factor in normal and D-galactosamine-treated mice. J Exp Med 165(3):657–663

Lehwald N, Tao GZ, Jang KY, Sorkin M, Knoefel WT, Sylvester KG (2011) Wnt-beta-catenin signaling protects against hepatic ischemia and reperfusion injury in mice Gastroenterology 141(2):707–718, 718 e701–705

Lei S, Dubeykovskiy A, Chakladar A, Wojtukiewicz L, Wang TC (2004) The murine gastrin promoter is synergistically activated by transforming growth factor-beta/Smad and Wnt signaling pathways. J Biol Chem 279(41):42492–42502

Leist M, Gantner F, Bohlinger I, Germann PG, Tiegs G, Wendel A (1994) Murine hepatocyte apoptosis induced in vitro and in vivo by TNF-alpha requires transcriptional arrest. J Immunol 153(4):1778–1788

Leist M, Gantner F, Kunstle G, Bohlinger I, Tiegs G, Bluethmann H, Wendel A (1996) The 55-kD tumor necrosis factor receptor and CD95 independently signal murine hepatocyte apoptosis and subsequent liver failure. Mol Med 2(1):109–124

Lentsch AB, Kato A, Yoshidome H, McMasters KM, Edwards MJ (2000) Inflammatory mechanisms and therapeutic strategies for warm hepatic ischemia/reperfusion injury. Hepatology 32(2):169–173

Lepourcelet M, Chen YN, France DS, Wang H, Crews P, Petersen F, Bruseo C, Wood AW, Shivdasani RA (2004) Small-molecule antagonists of the oncogenic Tcf/beta-catenin protein complex. Cancer Cell 5(1):91–102

Lescher B, Haenig B, Kispert A (1998) sFRP-2 is a target of the Wnt-4 signaling pathway in the developing metanephric kidney. Dev Dyn 213(4):440–451

Leung JY, Kolligs FT, Wu R, Zhai Y, Kuick R, Hanash S, Cho KR, Fearon ER (2002) Activation of AXIN2 expression by beta-catenin-T cell factor. A feedback repressor pathway regulating Wnt signaling. J Biol Chem 277(24):21657–21665

Li L, Chapman K, Hu X, Wong A, Pasdar M (2007a) Modulation of the oncogenic potential of beta-catenin by the subcellular distribution of plakoglobin. Mol Carcinog 46(10):824–838

Li Q, Dashwood WM, Zhong X, Nakagama H, Dashwood RH (2007b) Bcl-2 overexpression in PhIP-induced colon tumors: cloning of the rat Bcl-2 promoter and characterization of a pathway involving beta-catenin, c-Myc and E2F1. Oncogene 26(42):6194–6202

Li D, Cen J, Chen X, Conway EM, Ji Y, Hui L (2013) Hepatic loss of survivin impairs postnatal liver development and promotes expansion of hepatic progenitor cells in mice. Hepatology 58(6):2109–2121

Lian Z, Liu J, Li L, Li X, Clayton M, Wu MC, Wang HY, Arbuthnot P, Kew M, Fan D, Feitelson MA (2006) Enhanced cell survival of Hep3B cells by the hepatitis B x antigen effector, URG11, is associated with upregulation of beta-catenin. Hepatology 43(3):415–424

Lilien J, Balsamo J (2005) The regulation of cadherin-mediated adhesion by tyrosine phosphorylation/dephosphorylation of beta-catenin. Curr Opin Cell Biol 17(5):459–465

Linster CL, Van Schaftingen E (2007) Vitamin C. Biosynthesis, recycling and degradation in mammals. FEBS J 274(1):1–22

Liu H, Lo CR, Jones BE, Pradhan Z, Srinivasan A, Valentino KL, Stockert RJ, Czaja MJ (2000) Inhibition of c-Myc expression sensitizes hepatocytes to tumor necrosis factor-induced apoptosis and necrosis. J Biol Chem 275(51):40155–40162

Liu C, Li Y, Semenov M, Han C, Baeg GH, Tan Y, Zhang Z, Lin X, He X (2002) Control of beta-catenin phosphorylation/degradation by a dual-kinase mechanism. Cell 108(6):837–847

Liu L, Cao Y, Chen C, Zhang X, McNabola A, Wilkie D, Wilhelm S, Lynch M, Carter C (2006) Sorafenib blocks the RAF/MEK/ERK pathway, inhibits tumor angiogenesis, and induces tumor cell apoptosis in hepatocellular carcinoma model PLC/PRF/5. Cancer Res 66(24):11851–11858

Liu J, Du W, Fan D (2008) Survivin, the promising target in hepatocellular carcinoma gene therapy. Cancer Biol Ther 7(4):555–556

Liu H, Fergusson MM, JJ W, Rovira II, Liu J, Gavrilova O, Lu T, Bao J, Han D, Sack MN, Finkel T (2011) Wnt signaling regulates hepatic metabolism. Sci Signal 4(158):ra6

Loeppen S, Koehle C, Buchmann A, Schwarz M (2005) A beta-catenin-dependent pathway regulates expression of cytochrome P450 isoforms in mouse liver tumors. Carcinogenesis 26(1):239–248

Logan CY, Nusse R (2004) The Wnt signaling pathway in development and disease. Annu Rev Cell Dev Biol 20:781–810

Luo X, Budihardjo I, Zou H, Slaughter C, Wang X (1998) Bid, a Bcl2 interacting protein, mediates cytochrome c release from mitochondria in response to activation of cell surface death receptors. Cell 94(4):481–490

Ma H, Nguyen C, Lee KS, Kahn M (2005) Differential roles for the coactivators CBP and p300 on TCF/beta-catenin-mediated survivin gene expression. Oncogene 24(22):3619–3631

MacDonald BT, Tamai K, He X (2009) Wnt/beta-catenin signaling: components, mechanisms, and diseases. Dev Cell 17(1):9–26

Malhi H, Guicciardi ME, Gores GJ (2010) Hepatocyte death: a clear and present danger. Physiol Rev 90(3):1165–1194

Marinos G, Naoumov NV, Rossol S, Torre F, Wong PY, Gallati H, Portmann B, Williams R (1995) Tumor necrosis factor receptors in patients with chronic hepatitis B virus infection. Gastroenterology 108(5):1453–1463

Masckauchan TN, Shawber CJ, Funahashi Y, Li CM, Kitajewski J (2005) Wnt/beta-catenin signaling induces proliferation, survival and interleukin-8 in human endothelial cells. Angiogenesis 8(1):43–51

Masckauchan TN, Agalliu D, Vorontchikhina M, Ahn A, Parmalee NL, Li CM, Khoo A, Tycko B, Brown AM, Kitajewski J (2006) Wnt5a signaling induces proliferation and survival of endothelial cells in vitro and expression of MMP-1 and Tie-2. Mol Biol Cell 17(12):5163–5172

Masui O, Ueda Y, Tsumura A, Koyanagi M, Hijikata M, Shimotohno K (2002) RelA suppresses the Wnt/beta-catenin pathway without exerting trans-acting transcriptional ability. Int J Mol Med 9(5):489–493

Masuoka HC, Guicciardi ME, Gores GJ (2009) Caspase inhibitors for the treatment of hepatitis C. Clin Liver Dis 13(3):467–475

Matsuyama S, Kitamura T, Enomoto N, Fujita T, Ishigami A, Handa S, Maruyama N, Zheng D, Ikejima K, Takei Y, Sato N (2004) Senescence marker protein-30 regulates Akt activity and contributes to cell survival in Hep G2 cells. Biochem Biophys Res Commun 321(2):386–390

McCullough CT, Tura BJ, Harrison DJ (2007) c-Myc partially mediates IFNgamma-induced apoptosis in the primary hepatocyte. Int J Exp Pathol 88(3):129–136

Meng F, Lowell CA (1997) Lipopolysaccharide (LPS)-induced macrophage activation and signal transduction in the absence of Src-family kinases Hck, Fgr, and Lyn. J Exp Med 185(9):1661–1670

Micheau O, Tschopp J (2003) Induction of TNF receptor I-mediated apoptosis via two sequential signaling complexes. Cell 114(2):181–190

Micsenyi A, Tan X, Sneddon T, Luo JH, Michalopoulos GK, Monga SP (2004) Beta-catenin is temporally regulated during normal liver development. Gastroenterology 126(4):1134–1146

Miwa N, Furuse M, Tsukita S, Niikawa N, Nakamura Y, Furukawa Y (2001) Involvement of claudin-1 in the beta-catenin/Tcf signaling pathway and its frequent upregulation in human colorectal cancers. Oncol Res 12(11-12):469–476

Miyoshi Y, Iwao K, Nagasawa Y, Aihara T, Sasaki Y, Imaoka S, Murata M, Shimano T, Nakamura Y (1998) Activation of the beta-catenin gene in primary hepatocellular carcinomas by somatic alterations involving exon 3. Cancer Res 58(12):2524–2527

Mochizuki K, Hayashi N, Hiramatsu N, Katayama K, Kawanishi Y, Kasahara A, Fusamoto H, Kamada T (1996) Fas antigen expression in liver tissues of patients with chronic hepatitis B. J Hepatol 24(1):1–7

Molenaar M, van de Wetering M, Oosterwegel M, Peterson-Maduro J, Godsave S, Korinek V, Roose J, Destree O, Clevers H (1996) XTcf-3 transcription factor mediates beta-catenin-induced axis formation in Xenopus embryos. Cell 86(3):391–399

Monga SP (2014) Role and regulation of beta-catenin signaling during physiological liver growth. Gene Expr 16(2):51–62

Monga SP (2015) beta-Catenin Signaling and Roles in Liver Homeostasis, Injury, and Tumorigenesis. Gastroenterology 148(7):1294–1310

Monga SP, Mars WM, Pediaditakis P, Bell A, Mule K, Bowen WC, Wang X, Zarnegar R, Michalopoulos GK (2002) Hepatocyte growth factor induces Wnt-independent nuclear translocation of beta-catenin after Met-beta-catenin dissociation in hepatocytes. Cancer Res 62(7):2064–2071

Monga SP, Monga HK, Tan X, Mule K, Pediaditakis P, Michalopoulos GK (2003) Beta-catenin antisense studies in embryonic liver cultures: role in proliferation, apoptosis, and lineage specification. Gastroenterology 124(1):202–216

Monga SP, Micsenyi A, Germinaro M, Apte U, Bell A (2006) beta-Catenin regulation during matrigel-induced rat hepatocyte differentiation. Cell Tissue Res 323(1):71–79

Mott JL, Gores GJ (2007) Piercing the armor of hepatobiliary cancer: Bcl-2 homology domain 3 (BH3) mimetics and cell death. Hepatology 46(3):906–911

Muller M, Strand S, Hug H, Heinemann EM, Walczak H, Hofmann WJ, Stremmel W, Krammer PH, Galle PR (1997) Drug-induced apoptosis in hepatoma cells is mediated by the CD95 (APO-1/Fas) receptor/ligand system and involves activation of wild-type p53. J Clin Invest 99(3):403–413

Myung SJ, Yoon JH, Gwak GY, Kim W, Lee JH, Kim KM, Shin CS, Jang JJ, Lee SH, Lee SM, Lee HS (2007) Wnt signaling enhances the activation and survival of human hepatic stellate cells. FEBS Lett 581(16):2954–2958

Nagata S (1997) Apoptosis by death factor. Cell 88(3):355–365

Nakayama K, Hatakeyama S, Maruyama S, Kikuchi A, Onoe K, Good RA, Nakayama KI (2003) Impaired degradation of inhibitory subunit of NF-kappa B (I kappa B) and beta-catenin as a result of targeted disruption of the beta-TrCP1 gene. Proc Natl Acad Sci U S A 100(15):8752–8757

Natori S, Rust C, Stadheim LM, Srinivasan A, Burgart LJ, Gores GJ (2001) Hepatocyte apoptosis is a pathologic feature of human alcoholic hepatitis. J Hepatol 34(2):248–253

Nejak-Bowen K, Monga SP (2008) Wnt/beta-catenin signaling in hepatic organogenesis. Organogenesis 4(2):92–99

Nejak-Bowen KN, Monga SP (2011) Beta-catenin signaling, liver regeneration and hepatocellular cancer: sorting the good from the bad. Semin Cancer Biol 21(1):44–58

Nejak-Bowen KN, Zeng G, Tan X, Cieply B, Monga SP (2009) Beta-catenin regulates vitamin C biosynthesis and cell survival in murine liver. J Biol Chem 284(41):28115–28127

Nejak-Bowen K, Kikuchi A, Monga SP (2013) Beta-catenin-NF-kappaB interactions in murine hepatocytes: a complex to die for. Hepatology 57(2):763–774

Nevzorova YA, Hu W, Cubero FJ, Haas U, Freimuth J, Tacke F, Trautwein C, Liedtke C (2013) Overexpression of c-myc in hepatocytes promotes activation of hepatic stellate cells and facilitates the onset of liver fibrosis. Biochim Biophys Acta 1832(10):1765–1775

Niehrs C (2012) The complex world of WNT receptor signalling. Nat Rev Mol Cell Biol 13(12):767–779

Nusse R (2005) Wnt signaling in disease and in development. Cell Res 15(1):28–32

Ogasawara J, Watanabe-Fukunaga R, Adachi M, Matsuzawa A, Kasugai T, Kitamura Y, Itoh N, Suda T, Nagata S (1993) Lethal effect of the anti-Fas antibody in mice. Nature 364(6440):806–809

Ohno K, Ammann P, Fasciati R, Maier P (1995) Transforming growth factor beta 1 preferentially induces apoptotic cell death in rat hepatocytes cultured under pericentral-equivalent conditions. Toxicol Appl Pharmacol 132(2):227–236

Okabe H, Delgado E, Lee JM, Yang J, Kinoshita H, Hayashi H, Tsung A, Behari J, Beppu T, Baba H, Monga SP (2014) Role of leukocyte cell-derived chemotaxin 2 as a biomarker in hepatocellular carcinoma. PLoS One 9(6):e98817

Oltersdorf T, Elmore SW, Shoemaker AR, Armstrong RC, Augeri DJ, Belli BA, Bruncko M, Deckwerth TL, Dinges J, Hajduk PJ, Joseph MK, Kitada S, Korsmeyer SJ, Kunzer AR, Letai A, Li C, Mitten MJ, Nettesheim DG, Ng S, Nimmer PM, O'Connor JM, Oleksijew A, Petros AM, Reed JC, Shen W, Tahir SK, Thompson CB, Tomaselli KJ, Wang B, Wendt MD, Zhang H, Fesik SW, Rosenberg SH (2005) An inhibitor of Bcl-2 family proteins induces regression of solid tumours. Nature 435(7042):677–681

Orsulic S, Huber O, Aberle H, Arnold S, Kemler R (1999) E-cadherin binding prevents beta-catenin nuclear localization and beta-catenin/LEF-1-mediated transactivation. J Cell Sci 112(Pt 8):1237–1245

Ovejero C, Cavard C, Perianin A, Hakvoort T, Vermeulen J, Godard C, Fabre M, Chafey P, Suzuki K, Romagnolo B, Yamagoe S, Perret C (2004) Identification of the leukocyte cell-derived chemotaxin 2 as a direct target gene of beta-catenin in the liver. Hepatology 40(1):167–176

Papkoff J, Aikawa M (1998) WNT-1 and HGF regulate GSK3 beta activity and beta-catenin signaling in mammary epithelial cells. Biochem Biophys Res Commun 247(3):851–858

Paul S, Dey A (2008) Wnt signaling and cancer development: therapeutic implication. Neoplasma 55(3):165–176

Peifer M, Polakis P (2000) Wnt signaling in oncogenesis and embryogenesis--a look outside the nucleus. Science 287(5458):1606–1609

Peinado H, Quintanilla M, Cano A (2003) Transforming growth factor beta-1 induces snail transcription factor in epithelial cell lines: mechanisms for epithelial mesenchymal transitions. J Biol Chem 278(23):21113–21123

Pereira L, Yi F, Merrill BJ (2006) Repression of Nanog gene transcription by Tcf3 limits embryonic stem cell self-renewal. Mol Cell Biol 26(20):7479–7491

Peroukides S, Bravou V, Alexopoulos A, Varakis J, Kalofonos H, Papadaki H (2010) Survivin overexpression in HCC and liver cirrhosis differentially correlates with p-STAT3 and E-cadherin. Histol Histopathol 25(3):299–307

Peter ME, Krammer PH (2003) The CD95(APO-1/Fas) DISC and beyond. Cell Death Differ 10(1):26–35

Piedra J, Martinez D, Castano J, Miravet S, Dunach M, de Herreros AG (2001) Regulation of beta-catenin structure and activity by tyrosine phosphorylation. J Biol Chem 276(23):20436–20443

Pinson KI, Brennan J, Monkley S, Avery BJ, Skarnes WC (2000) An LDL-receptor-related protein mediates Wnt signalling in mice. Nature 407(6803):535–538

Pockros PJ, Schiff ER, Shiffman ML, McHutchison JG, Gish RG, Afdhal NH, Makhviladze M, Huyghe M, Hecht D, Oltersdorf T, Shapiro DA (2007) Oral IDN-6556, an antiapoptotic caspase inhibitor, may lower aminotransferase activity in patients with chronic hepatitis C. Hepatology 46(2):324–329

Polakis P (2000) Wnt signaling and cancer. Genes Dev 14(15):1837–1851

Purcell R, Childs M, Maibach R, Miles C, Turner C, Zimmermann A, Sullivan M (2011) HGF/c-Met related activation of beta-catenin in hepatoblastoma. J Exp Clin Cancer Res 30:96

Ramalho RM, Cortez-Pinto H, Castro RE, Sola S, Costa A, Moura MC, Camilo ME, Rodrigues CM (2006) Apoptosis and Bcl-2 expression in the livers of patients with steatohepatitis. Eur J Gastroenterol Hepatol 18(1):21–29

Rattner A, Hsieh JC, Smallwood PM, Gilbert DJ, Copeland NG, Jenkins NA, Nathans J (1997) A family of secreted proteins contains homology to the cysteine-rich ligand-binding domain of frizzled receptors. Proc Natl Acad Sci U S A 94(7):2859–2863

Ratziu V, Sheikh MY, Sanyal AJ, Lim JK, Conjeevaram H, Chalasani N, Abdelmalek M, Bakken A, Renou C, Palmer M, Levine RA, Bhandari BR, Cornpropst M, Liang W, King B, Mondou E, Rousseau FS, McHutchison J, Chojkier M (2012) A phase 2, randomized, double-blind, placebo-controlled study of GS-9450 in subjects with nonalcoholic steatohepatitis. Hepatology 55(2):419–428

Renard CA, Labalette C, Armengol C, Cougot D, Wei Y, Cairo S, Pineau P, Neuveut C, de Reynies A, Dejean A, Perret C, Buendia MA (2007) Tbx3 is a downstream target of the Wnt/beta-catenin pathway and a critical mediator of beta-catenin survival functions in liver cancer. Cancer Res 67(3):901–910

Rodilla V, Villanueva A, Obrador-Hevia A, Robert-Moreno A, Fernandez-Majada V, Grilli A, Lopez-Bigas N, Bellora N, Alba MM, Torres F, Dunach M, Sanjuan X, Gonzalez S, Gridley T, Capella G, Bigas A, Espinosa L (2009) Jagged1 is the pathological link between Wnt and Notch pathways in colorectal cancer. Proc Natl Acad Sci U S A 106(15):6315–6320

Rodrigues CM, Steer CJ (2001) The therapeutic effects of ursodeoxycholic acid as an anti-apoptotic agent. Expert Opin Investig Drugs 10(7):1243–1253

Rodriguez I, Matsuura K, Khatib K, Reed JC, Nagata S, Vassalli P (1996) A bcl-2 transgene expressed in hepatocytes protects mice from fulminant liver destruction but not from rapid death induced by anti-Fas antibody injection. J Exp Med 183(3):1031–1036

Romero D, Iglesias M, Vary CP, Quintanilla M (2008) Functional blockade of Smad4 leads to a decrease in beta-catenin levels and signaling activity in human pancreatic carcinoma cells. Carcinogenesis 29(5):1070–1076

Roose J, Huls G, van Beest M, Moerer P, van der Horn K, Goldschmeding R, Logtenberg T, Clevers H (1999) Synergy between tumor suppressor APC and the beta-catenin-Tcf4 target Tcf1. Science 285(5435):1923–1926

Roura S, Miravet S, Piedra J, Garcia de Herreros A, Dunach M (1999) Regulation of E-cadherin/Catenin association by tyrosine phosphorylation. J Biol Chem 274(51):36734–36740

Saegusa M, Hashimura M, Kuwata T, Hamano M, Okayasu I (2007) Crosstalk between NF-kappaB/p65 and beta-catenin/TCF4/p300 signalling pathways through alterations in GSK-3beta expression during trans-differentiation of endometrial carcinoma cells. J Pathol 213(1):35–45
Saito T, Okumura A, Watanabe H, Asano M, Ishida-Okawara A, Sakagami J, Sudo K, Hatano-Yokoe Y, Bezbradica JS, Joyce S, Abo T, Iwakura Y, Suzuki K, Yamagoe S (2004) Increase in hepatic NKT cells in leukocyte cell-derived chemotaxin 2-deficient mice contributes to severe concanavalin A-induced hepatitis. J Immunol 173(1):579–585
Samson CM, Schrum LW, Bird MA, Lange PA, Brenner DA, Rippe RA, Behrns KE (2002) Transforming growth factor-beta1 induces hepatocyte apoptosis by a c-Jun independent mechanism. Surgery 132(3):441–449
Sanchez A, Alvarez AM, Benito M, Fabregat I (1996) Apoptosis induced by transforming growth factor-beta in fetal hepatocyte primary cultures: involvement of reactive oxygen intermediates. J Biol Chem 271(13):7416–7422
Sato Y, Watanabe H, Kameyama H, Kobayashi T, Yamamoto S, Takeishi T, Hirano K, Oya H, Nakatsuka H, Watanabe T, Kokai H, Yamagoe S, Suzuki K, Oya K, Kojima K, Hatakeyama K (2004a) Changes in serum LECT 2 levels during the early period of liver regeneration after adult living related donor liver transplantation. Transplant Proc 36(8):2357–2358
Sato Y, Watanabe H, Kameyama H, Kobayashi T, Yamamoto S, Takeishi T, Hirano K, Oya H, Nakatsuka H, Watanabe T, Kokai H, Yamagoe S, Suzuki K, Oya K, Kojima K, Hatakeyama K (2004b) Serum LECT2 level as a prognostic indicator in acute liver failure. Transplant Proc 36(8):2359–2361
Satoh S, Daigo Y, Furukawa Y, Kato T, Miwa N, Nishiwaki T, Kawasoe T, Ishiguro H, Fujita M, Tokino T, Sasaki Y, Imaoka S, Murata M, Shimano T, Yamaoka Y, Nakamura Y (2000) AXIN1 mutations in hepatocellular carcinomas, and growth suppression in cancer cells by virus-mediated transfer of AXIN1. Nat Genet 24(3):245–250
Scaffidi C, Fulda S, Srinivasan A, Friesen C, Li F, Tomaselli KJ, Debatin KM, Krammer PH, Peter ME (1998) Two CD95 (APO-1/Fas) signaling pathways. EMBO J 17(6):1675–1687
Scaffidi C, Schmitz I, Krammer PH, Peter ME (1999) The role of c-FLIP in modulation of CD95-induced apoptosis. J Biol Chem 274(3):1541–1548
Schlessinger K, Hall A, Tolwinski N (2009) Wnt signaling pathways meet Rho GTPases. Genes Dev 23(3):265–277
Schwabe RF, Brenner DA (2006) Mechanisms of Liver Injury. I. TNF-alpha-induced liver injury: role of IKK, JNK, and ROS pathways. Am J Physiol Gastrointest Liver Physiol 290(4):G583–G589
Seifert JR, Mlodzik M (2007) Frizzled/PCP signalling: a conserved mechanism regulating cell polarity and directed motility. Nat Rev Genet 8(2):126–138
Semenov MV, Habas R, Macdonald BT, He X (2007) SnapShot: noncanonical Wnt signaling pathways. Cell 131(7):1378
Shang XZ, Zhu H, Lin K, Tu Z, Chen J, Nelson DR, Liu C (2004) Stabilized beta-catenin promotes hepatocyte proliferation and inhibits TNFalpha-induced apoptosis. Lab Investig 84(3):332–341
Shibata T, Ochiai A, Kanai Y, Akimoto S, Gotoh M, Yasui N, Machinami R, Hirohashi S (1996) Dominant negative inhibition of the association between beta-catenin and c-erbB-2 by N-terminally deleted beta-catenin suppresses the invasion and metastasis of cancer cells. Oncogene 13(5):883–889
Shimokawa T, Furukawa Y, Sakai M, Li M, Miwa N, Lin YM, Nakamura Y (2003) Involvement of the FGF18 gene in colorectal carcinogenesis, as a novel downstream target of the beta-catenin/T-cell factor complex. Cancer Res 63(19):6116–6120
Shin HW, Park SY, Lee KB, Jang JJ (2009) Down-regulation of Wnt signaling during apoptosis of human hepatic stellate cells. Hepato-Gastroenterology 56(89):208–212
Shiratori Y, Tanaka M, Hai K, Kawase T, Shirna S, Sugimoto T (1990) Role of endotoxin-responsive macrophages in hepatic injury. Hepatology 11(2):183–192
Sies H (1997) Oxidative stress: oxidants and antioxidants. Exp Physiol 82(2):291–295

Shtutman M, Zhurinsky J, Simcha I, Albanese C, D'Amico M, Pestell R, Ben-Ze'ev A (1999) The cyclin D1 gene is a target of the beta-catenin/LEF-1 pathway. Proc Natl Acad Sci U S A 96(10):5522–5527

Slusarski DC, Corces VG, Moon RT (1997) Interaction of Wnt and a Frizzled homologue triggers G-protein-linked phosphatidylinositol signalling. Nature 390(6658):410–413

Smyth LA, Brady HJ (2005) cMet and Fas receptor interaction inhibits death-inducing signaling complex formation in endothelial cells. Hypertension 46(1):100–106

Solanas G, Porta-de-la-Riva M, Agusti C, Casagolda D, Sanchez-Aguilera F, Larriba MJ, Pons F, Peiro S, Escriva M, Munoz A, Dunach M, de Herreros AG, Baulida J (2008) E-cadherin controls beta-catenin and NF-kappaB transcriptional activity in mesenchymal gene expression. J Cell Sci 121(Pt 13):2224–2234

Spahr L, Rubbia-Brandt L, Frossard JL, Giostra E, Rougemont AL, Pugin J, Fischer M, Egger H, Hadengue A (2002) Combination of steroids with infliximab or placebo in severe alcoholic hepatitis: a randomized controlled pilot study. J Hepatol 37(4):448–455

Spiegelman VS, Slaga TJ, Pagano M, Minamoto T, Ronai Z, Fuchs SY (2000) Wnt/beta-catenin signaling induces the expression and activity of betaTrCP ubiquitin ligase receptor. Mol Cell 5(5):877–882

Steelman LS, Chappell WH, Abrams SL, Kempf RC, Long J, Laidler P, Mijatovic S, Maksimovic-Ivanic D, Stivala F, Mazzarino MC, Donia M, Fagone P, Malaponte G, Nicoletti F, Libra M, Milella M, Tafuri A, Bonati A, Basecke J, Cocco L, Evangelisti C, Martelli AM, Montalto G, Cervello M, McCubrey JA (2011) Roles of the Raf/MEK/ERK and PI3K/PTEN/Akt/mTOR pathways in controlling growth and sensitivity to therapy-implications for cancer and aging. Aging (Albany NY) 3(3):192–222

Stennicke HR, Jurgensmeier JM, Shin H, Deveraux Q, Wolf BB, Yang X, Zhou Q, Ellerby HM, Ellerby LM, Bredesen D, Green DR, Reed JC, Froelich CJ, Salvesen GS (1998) Pro-caspase-3 is a major physiologic target of caspase-8. J Biol Chem 273(42):27084–27090

Strand S, Hofmann WJ, Grambihler A, Hug H, Volkmann M, Otto G, Wesch H, Mariani SM, Hack V, Stremmel W, Krammer PH, Galle PR (1998) Hepatic failure and liver cell damage in acute Wilson's disease involve CD95 (APO-1/Fas) mediated apoptosis. Nat Med 4(5):588–593

Subramaniam N, Sherman MH, Rao R, Wilson C, Coulter S, Atkins AR, Evans RM, Liddle C, Downes M (2012) Metformin-mediated Bambi expression in hepatic stellate cells induces prosurvival Wnt/beta-catenin signaling. Cancer Prev Res (Phila) 5(4):553–561

Sugimura R, Li L (2010) Noncanonical Wnt signaling in vertebrate development, stem cells, and diseases. Birth Defects Res C Embryo Today 90(4):243–256

Suksaweang S, Lin CM, Jiang TX, Hughes MW, Widelitz RB, Chuong CM (2004) Morphogenesis of chicken liver: identification of localized growth zones and the role of beta-catenin/Wnt in size regulation. Dev Biol 266(1):109–122

Sun J, Hobert ME, Duan Y, Rao AS, He TC, Chang EB, Madara JL (2005) Crosstalk between NF-kappaB and beta-catenin pathways in bacterial-colonized intestinal epithelial cells. Am J Physiol Gastrointest Liver Physiol 289(1):G129–G137

Tada M, Kai M (2009) Noncanonical Wnt/PCP signaling during vertebrate gastrulation. Zebrafish 6(1):29–40

Takahashi K, Suzuki K, Tsukatani Y (1997) Induction of tyrosine phosphorylation and association of beta-catenin with EGF receptor upon tryptic digestion of quiescent cells at confluence. Oncogene 15(1):71–78

Takemaru KI, Moon RT (2000) The transcriptional coactivator CBP interacts with beta-catenin to activate gene expression. J Cell Biol 149(2):249–254

Tamai K, Semenov M, Kato Y, Spokony R, Liu C, Katsuyama Y, Hess F, Saint-Jeannet JP, He X (2000) LDL-receptor-related proteins in Wnt signal transduction. Nature 407(6803):530–535

Tan X, Behari J, Cieply B, Michalopoulos GK, Monga SP (2006) Conditional deletion of beta-catenin reveals its role in liver growth and regeneration. Gastroenterology 131(5):1561–1572

Tan X, Yuan Y, Zeng G, Apte U, Thompson MD, Cieply B, Stolz DB, Michalopoulos GK, Kaestner KH, Monga SP (2008) Beta-catenin deletion in hepatoblasts disrupts hepatic morphogenesis and survival during mouse development. Hepatology 47(5):1667–1679

Tanaka K, Kitagawa Y, Kadowaki T (2002) Drosophila segment polarity gene product porcupine stimulates the posttranslational N-glycosylation of wingless in the endoplasmic reticulum. J Biol Chem 277(15):12816–12823
Tan X, Apte U, Micsenyi A, Kotsagrelos E, Luo JH, Ranganathan S, Monga DK, Bell A, Michalopoulos GK, Monga SP (2005) Epidermal growth factor receptor: a novel target of the Wnt/beta-catenin pathway in liver. Gastroenterology 129(1):285–302
Tang Y, Kitisin K, Jogunoori W, Li C, Deng CX, Mueller SC, Ressom HW, Rashid A, He AR, Mendelson JS, Jessup JM, Shetty K, Zasloff M, Mishra B, Reddy EP, Johnson L, Mishra L (2008) Progenitor/stem cells give rise to liver cancer due to aberrant TGF-beta and IL-6 signaling. Proc Natl Acad Sci U S A 105(7):2445–2450
Tao GZ, Lehwald N, Jang KY, Baek J, Xu B, Omary MB, Sylvester KG (2013) Wnt/beta-catenin signaling protects mouse liver against oxidative stress-induced apoptosis through the inhibition of forkhead transcription factor FoxO3. J Biol Chem 288(24):17214–17224
Tao J, Xu E, Zhao Y, Singh S, Li X, Couchy G, Chen X, Zucman-Rossi J, Chikina M, Monga SP (2016a) Modeling a human hepatocellular carcinoma subset in mice through coexpression of met and point-mutant beta-catenin. Hepatology 64(5):1587–1605
Tao J, Zhang R, Singh S, Poddar M, Xu E, Oertel M, Chen X, Ganesh S, Abrams M, Monga SP (2016b) Targeting beta-catenin in hepatocellular cancers induced by co-expression of mutant beta-catenin and K-Ras in mice. Hepatology 65(5):1581–1599.
Tardito S, Chiu M, Uggeri J, Zerbini A, Da Ros F, Dall'Asta V, Missale G, Bussolati O (2011) L-Asparaginase and inhibitors of glutamine synthetase disclose glutamine addiction of beta-catenin-mutated human hepatocellular carcinoma cells. Curr Cancer Drug Targets 11(8):929–943
Taurin S, Sandbo N, Qin Y, Browning D, Dulin NO (2006) Phosphorylation of beta-catenin by cyclic AMP-dependent protein kinase. J Biol Chem 281(15):9971–9976
Tetsu O, McCormick F (1999) Beta-catenin regulates expression of cyclin D1 in colon carcinoma cells. Nature 398(6726):422–426
Theard D, Steiner M, Kalicharan D, Hoekstra D, van Ijzendoorn SC (2007) Cell polarity development and protein trafficking in hepatocytes lacking E-cadherin/beta-catenin-based adherens junctions. Mol Biol Cell 18(6):2313–2321
Thompson MD, Awuah P, Singh S, Monga SP (2010) Disparate cellular basis of improved liver repair in beta-catenin-overexpressing mice after long-term exposure to 3,5-diethoxycarbonyl-1,4-dihydrocollidine. Am J Pathol 177(4):1812–1822
Thorburn A (2004) Death receptor-induced cell killing. Cell Signal 16(2):139–144
Thornberry NA, Lazebnik Y (1998) Caspases: enemies within. Science 281(5381):1312–1316
Tokushige K, Yamaguchi N, Ikeda I, Hashimoto E, Yamauchi K, Hayashi N (2000) Significance of soluble TNF receptor-I in acute-type fulminant hepatitis. Am J Gastroenterol 95(8):2040–2046
Tschopp J, Irmler M, Thome M (1998) Inhibition of fas death signals by FLIPs. Curr Opin Immunol 10(5):552–558
Tsung A, Sahai R, Tanaka H, Nakao A, Fink MP, Lotze MT, Yang H, Li J, Tracey KJ, Geller DA, Billiar TR (2005) The nuclear factor HMGB1 mediates hepatic injury after murine liver ischemia-reperfusion. J Exp Med 201(7):1135–1143
Tsurusaki Y, Yamaguchi M (2003) Overexpression of regucalcin modulates tumor-related gene expression in cloned rat hepatoma H4-II-E cells. J Cell Biochem 90(3):619–626
Tward AD, Jones KD, Yant S, Cheung ST, Fan ST, Chen X, Kay MA, Wang R, Bishop JM (2007) Distinct pathways of genomic progression to benign and malignant tumors of the liver. Proc Natl Acad Sci U S A 104(37):14771–14776
Umar S, Sarkar S, Wang Y, Singh P (2009) Functional cross-talk between beta-catenin and NFkappaB signaling pathways in colonic crypts of mice in response to progastrin. J Biol Chem 284(33):22274–22284
Valdes F, Murillo MM, Valverde AM, Herrera B, Sanchez A, Benito M, Fernandez M, Fabregat I (2004) Transforming growth factor-beta activates both pro-apoptotic and survival signals in fetal rat hepatocytes. Exp Cell Res 292(1):209–218

Valentino KL, Gutierrez M, Sanchez R, Winship MJ, Shapiro DA (2003) First clinical trial of a novel caspase inhibitor: anti-apoptotic caspase inhibitor, IDN-6556, improves liver enzymes. Int J Clin Pharmacol Ther 41(10):441–449
Van Antwerp DJ, Martin SJ, Kafri T, Green DR, Verma IM (1996) Suppression of TNF-alpha-induced apoptosis by NF-kappaB. Science 274(5288):787–789
Wajant H, Pfizenmaier K, Scheurich P (2003) Tumor necrosis factor signaling. Cell Death Differ 10(1):45–65
Wallach D, Varfolomeev EE, Malinin NL, Goltsev YV, Kovalenko AV, Boldin MP (1999) Tumor necrosis factor receptor and Fas signaling mechanisms. Annu Rev Immunol 17:331–367
Wan S, Wu Y, Zhou X, Chen Y, An J, Yu X, Zhang H, Yang H, Xing J (2015) Polymorphisms in Genes of Tricarboxylic Acid Cycle Key Enzymes Are Associated with Early Recurrence of Hepatocellular Carcinoma. PLoS One 10(4):e0124471
Wang CY, Mayo MW, Korneluk RG, Goeddel DV, Baldwin AS Jr (1998) NF-kappaB antiapoptosis: induction of TRAF1 and TRAF2 and c-IAP1 and c-IAP2 to suppress caspase-8 activation. Science 281(5383):1680–1683
Wang XH, Sun X, Meng XW, Lv ZW, Du YJ, Zhu Y, Chen J, Kong DX, Jin SZ (2010) beta-catenin siRNA regulation of apoptosis- and angiogenesis-related gene expression in hepatocellular carcinoma cells: potential uses for gene therapy. Oncol Rep 24(4):1093–1099
Wang B, Zhang Q, Zhu B, Cui Z, Zhou J (2013) Protective effect of gadolinium chloride on early warm ischemia/reperfusion injury in rat bile duct during liver transplantation. PLoS One 8(1):e52743
Warner DR, Greene RM, Pisano MM (2005) Cross-talk between the TGFbeta and Wnt signaling pathways in murine embryonic maxillary mesenchymal cells. FEBS Lett 579(17):3539–3546
Wehrli M, Dougan ST, Caldwell K, O'Keefe L, Schwartz S, Vaizel-Ohayon D, Schejter E, Tomlinson A, DiNardo S (2000) arrow encodes an LDL-receptor-related protein essential for Wingless signalling. Nature 407(6803):527–530
Wei MC, Zong WX, Cheng EH, Lindsten T, Panoutsakopoulou V, Ross AJ, Roth KA, MacGregor GR, Thompson CB, Korsmeyer SJ (2001) Proapoptotic BAX and BAK: a requisite gateway to mitochondrial dysfunction and death. Science 292(5517):727–730
Wei W, Chua MS, Grepper S, So S (2010) Small molecule antagonists of Tcf4/beta-catenin complex inhibit the growth of HCC cells in vitro and in vivo. Int J Cancer 126(10):2426–2436
Wharton KA Jr, Zimmermann G, Rousset R, Scott MP (2001) Vertebrate proteins related to Drosophila Naked Cuticle bind Dishevelled and antagonize Wnt signaling. Dev Biol 234(1):93–106
Wickline ED, Awuah PK, Behari J, Ross M, Stolz DB, Monga SP (2011) Hepatocyte gamma-catenin compensates for conditionally deleted beta-catenin at adherens junctions. J Hepatol 55(6):1256–1262
Wickline ED, Du Y, Stolz DB, Kahn M, Monga SP (2013) gamma-Catenin at adherens junctions: mechanism and biologic implications in hepatocellular cancer after beta-catenin knockdown. Neoplasia 15(4):421–434
Willert J, Epping M, Pollack JR, Brown PO, Nusse R (2002) A transcriptional response to Wnt protein in human embryonic carcinoma cells. BMC Dev Biol 2:8
Willert K, Brown JD, Danenberg E, Duncan AW, Weissman IL, Reya T, Yates JR 3rd, Nusse R (2003) Wnt proteins are lipid-modified and can act as stem cell growth factors. Nature 423(6938):448–452
Wise DR, DeBerardinis RJ, Mancuso A, Sayed N, Zhang XY, Pfeiffer HK, Nissim I, Daikhin E, Yudkoff M, McMahon SB, Thompson CB (2008) Myc regulates a transcriptional program that stimulates mitochondrial glutaminolysis and leads to glutamine addiction. Proc Natl Acad Sci U S A 105(48):18782–18787
Wodarz A, Nusse R (1998) Mechanisms of Wnt signaling in development. Annu Rev Cell Dev Biol 14:59–88
Wong HC, Bourdelas A, Krauss A, Lee HJ, Shao Y, Wu D, Mlodzik M, Shi DL, Zheng J (2003) Direct binding of the PDZ domain of Dishevelled to a conserved internal sequence in the C-terminal region of Frizzled. Mol Cell 12(5):1251–1260

Wright M, Aikawa M, Szeto W, Papkoff J (1999) Identification of a Wnt-responsive signal transduction pathway in primary endothelial cells. Biochem Biophys Res Commun 263(2):384–388
Wu Y, Mehew JW, Heckman CA, Arcinas M, Boxer LM (2001) Negative regulation of bcl-2 expression by p53 in hematopoietic cells. Oncogene 20(2):240–251
Wu X, Tu X, Joeng KS, Hilton MJ, Williams DA, Long F (2008) Rac1 activation controls nuclear localization of beta-catenin during canonical Wnt signaling. Cell 133(2):340–353
Xie H, Huang Z, Sadim MS, Sun Z (2005) Stabilized beta-catenin extends thymocyte survival by up-regulating Bcl-xL. J Immunol 175(12):7981–7988
Xu Y, Nguyen Q, Lo DC, Czaja MJ (1997) c-myc-Dependent hepatoma cell apoptosis results from oxidative stress and not a deficiency of growth factors. J Cell Physiol 170(2):192–199
Xu J, Zhu X, Wu L, Yang R, Yang Z, Wang Q, Wu F (2012) MicroRNA-122 suppresses cell proliferation and induces cell apoptosis in hepatocellular carcinoma by directly targeting Wnt/beta-catenin pathway. Liver Int 32(5):752–760
Yamagoe S, Yamakawa Y, Matsuo Y, Minowada J, Mizuno S, Suzuki K (1996) Purification and primary amino acid sequence of a novel neutrophil chemotactic factor LECT2. Immunol Lett 52(1):9–13
Yamagoe S, Mizuno S, Suzuki K (1998) Molecular cloning of human and bovine LECT2 having a neutrophil chemotactic activity and its specific expression in the liver. Biochim Biophys Acta 1396(1):105–113
Yamaguchi M (2000) The role of regucalcin in nuclear regulation of regenerating liver. Biochem Biophys Res Commun 276(1):1–6
Yamaguchi M (2005) Role of regucalcin in maintaining cell homeostasis and function (review). Int J Mol Med 15(3):371–389
Yanagawa S, Matsuda Y, Lee JS, Matsubayashi H, Sese S, Kadowaki T, Ishimoto A (2002) Casein kinase I phosphorylates the Armadillo protein and induces its degradation in Drosophila. EMBO J 21(7):1733–1742
Yang YA, Zhang GM, Feigenbaum L, Zhang YE (2006) Smad3 reduces susceptibility to hepatocarcinoma by sensitizing hepatocytes to apoptosis through downregulation of Bcl-2. Cancer Cell 9(6):445–457
Yang Y, Zhu J, Gou H, Cao D, Jiang M, Hou M (2011) Clinical significance of Cox-2, Survivin and Bcl-2 expression in hepatocellular carcinoma (HCC). Med Oncol 28(3):796–803
Yano F, Kugimiya F, Ohba S, Ikeda T, Chikuda H, Ogasawara T, Ogata N, Takato T, Nakamura K, Kawaguchi H, Chung UI (2005) The canonical Wnt signaling pathway promotes chondrocyte differentiation in a Sox9-dependent manner. Biochem Biophys Res Commun 333(4):1300–1308
Yeh TH, Krauland L, Singh V, Zou B, Devaraj P, Stolz DB, Franks J, Monga SP, Sasatomi E, Behari J (2010) Liver-specific beta-catenin knockout mice have bile canalicular abnormalities, bile secretory defect, and intrahepatic cholestasis. Hepatology 52(4):1410–1419
Yin XM, Ding WX (2003) Death receptor activation-induced hepatocyte apoptosis and liver injury. Curr Mol Med 3(6):491–508
Yin XM, Wang K, Gross A, Zhao Y, Zinkel S, Klocke B, Roth KA, Korsmeyer SJ (1999) Bid-deficient mice are resistant to Fas-induced hepatocellular apoptosis. Nature 400(6747):886–891
Yokoya F, Imamoto N, Tachibana T, Yoneda Y (1999) beta-catenin can be transported into the nucleus in a Ran-unassisted manner. Mol Biol Cell 10(4):1119–1131
Yoon JH, Gores GJ (2002) Death receptor-mediated apoptosis and the liver. J Hepatol 37(3):400–410
Yoshibayashi H, Okabe H, Satoh S, Hida K, Kawashima K, Hamasu S, Nomura A, Hasegawa S, Ikai I, Sakai Y (2007) SIAH1 causes growth arrest and apoptosis in hepatoma cells through beta-catenin degradation-dependent and -independent mechanisms. Oncol Rep 17(3):549–556
You Z, Saims D, Chen S, Zhang Z, Guttridge DC, Guan KL, MacDougald OA, Brown AM, Evan G, Kitajewski J, Wang CY (2002) Wnt signaling promotes oncogenic transformation by inhibiting c-Myc-induced apoptosis. J Cell Biol 157(3):429–440
Zarnegar R (1995) Regulation of HGF and HGFR gene expression. EXS 74:33–49
Zeng X, Tamai K, Doble B, Li S, Huang H, Habas R, Okamura H, Woodgett J, He X (2005) A dual-kinase mechanism for Wnt co-receptor phosphorylation and activation. Nature 438(7069):873–877

Zeng G, Apte U, Micsenyi A, Bell A, Monga SP (2006) Tyrosine residues 654 and 670 in beta-catenin are crucial in regulation of Met-beta-catenin interactions. Exp Cell Res 312(18):3620–3630

Zeng X, Huang H, Tamai K, Zhang X, Harada Y, Yokota C, Almeida K, Wang J, Doble B, Woodgett J, Wynshaw-Boris A, Hsieh JC, He X (2008) Initiation of Wnt signaling: control of Wnt coreceptor Lrp6 phosphorylation/activation via frizzled, dishevelled and axin functions. Development 135(2):367–375

Zhai L, Chaturvedi D, Cumberledge S (2004) Drosophila wnt-1 undergoes a hydrophobic modification and is targeted to lipid rafts, a process that requires porcupine. J Biol Chem 279(32):33220–33227

Zhang T, Otevrel T, Gao Z, Gao Z, Ehrlich SM, Fields JZ, Boman BM (2001) Evidence that APC regulates survivin expression: a possible mechanism contributing to the stem cell origin of colon cancer. Cancer Res 61(24):8664–8667

Zhang M, Wang M, Tan X, Li TF, Zhang YE, Chen D (2010a) Smad3 prevents beta-catenin degradation and facilitates beta-catenin nuclear translocation in chondrocytes. J Biol Chem 285(12):8703–8710

Zhang XF, Tan X, Zeng G, Misse A, Singh S, Kim Y, Klaunig JE, Monga SP (2010b) Conditional beta-catenin loss in mice promotes chemical hepatocarcinogenesis: role of oxidative stress and platelet-derived growth factor receptor alpha/phosphoinositide 3-kinase signaling. Hepatology 52(3):954–965

Zhao Y, Difrancesca D, Wang X, Zarnegar R, Michalopoulos GK, Yin XM (2007) Promotion of Fas-mediated apoptosis in Type II cells by high doses of hepatocyte growth factor bypasses the mitochondrial requirement. J Cell Physiol 213(2):556–563

Zhou Z, Sun X, Kang YJ (2001) Ethanol-induced apoptosis in mouse liver: Fas- and cytochrome c-mediated caspase-3 activation pathway. Am J Pathol 159(1):329–338

Zhu NL, Wang J, Tsukamoto H (2010) The Necdin-Wnt pathway causes epigenetic peroxisome proliferator-activated receptor gamma repression in hepatic stellate cells. J Biol Chem 285(40):30463–30471

Zou C, Ma J, Wang X, Guo L, Zhu Z, Stoops J, Eaker AE, Johnson CJ, Strom S, Michalopoulos GK, DeFrances MC, Zarnegar R (2007) Lack of Fas antagonism by Met in human fatty liver disease. Nat Med 13(9):1078–1085

Zucman-Rossi J, Benhamouche S, Godard C, Boyault S, Grimber G, Balabaud C, Cunha AS, Bioulac-Sage P, Perret C (2007) Differential effects of inactivated Axin1 and activated beta-catenin mutations in human hepatocellular carcinomas. Oncogene 26(5):774–780

Chapter 7
Nuclear DAMPs in Hepatic Injury and Inflammation

Rui Kang and Daolin Tang

7.1 Introduction

The liver is composed of hepatocytes (parenchymal cells) and non-parenchymal cells (e.g., Kupffer cells, sinusoidal endothelial cells, and stellate cells). As the largest gland and solid organ in the body, the liver plays a critical role not only in the metabolism of carbohydrates, proteins, and fat but also in the clearance of drugs and other harmful substances from the blood. Abnormal cell death, including regulated and nonregulated forms, is implicated in the pathologic presentation of almost all types of liver disease (Luedde et al. 2014). Nonregulated cell death is uncontrolled, whereas regulated cell death (RCD) is controlled by intracellular signals or proteins (Ashkenazi and Salvesen 2014). According to the current classification by the Nomenclature Committee on Cell Death, there are 11 forms of RCD (viz., anoikis, autophagic cell death, apoptosis, cornification, entosis, ferroptosis, mitotic catastrophe, necroptosis, NETosis, parthanatos, and pyroptosis) with different morphological, biochemical, and genetic features (Galluzzi et al. 2015).

In addition to the direct loss of normal cell numbers, abnormal RCD can drive inflammatory injury or immune suppression by releasing damage-associated molecular pattern molecules (DAMPs) (Green et al. 2009). Different from pathogen-associated molecular pattern molecules (PAMPs) generated from foreign pathogens, DAMPs are endogenous molecules released from dead or dying cells (Bianchi 2007; Tang et al. 2012b). The list of DAMPs is rapidly increasing and can be divided into nuclear (e.g., high-mobility group box 1 [HMGB1], histone, and nuclear DNA), mitochondrial (e.g., mitochondrial transcription factor A and mito-

R. Kang (✉) • D. Tang (✉)
Department of Surgery, University of Pittsburgh, Pittsburgh, PA 15213, USA
e-mail: kangr@upmc.edu; tangd2@upmc.edu

W.-X. Ding, X.-M. Yin (eds.), *Molecules, Systems and Signaling in Liver Injury*,
Cell Death in Biology and Diseases, DOI 10.1007/978-3-319-58106-4_7

X.-M. Yin and Z. Dong (Series eds.), *Cell Death in Biology and Diseases*

chondrial DNA), and cytosolic DAMPs (e.g., heat shock protein and S100) and others. PAMPs trigger infection, whereas DAMPs induce sterile inflammation. Both PAMPs and DAMPs can act as danger signals to regulate the inflammatory response by binding to pattern recognition receptors (PRRs), mainly in the immune system (Takeuchi and Akira 2010). These PRRs include toll-like receptors (TLRs), NOD-like receptors, RIG-I-like receptors, and the receptor for advanced glycation end products (RAGE). Meanwhile, the interplay between PAMPs and DAMPs may accelerate and amply the inflammatory response. Thus, danger signals and their receptors are emerging targets in inflammation-associated human diseases, including liver disease (Chen et al. 2013). In this chapter, we will focus on nuclear DAMPs (HMGB1 and histone) in RCD and their roles in liver injury and inflammation.

7.2 Regulated Cell Death in Liver Injury and Inflammation

7.2.1 Apoptosis

7.2.1.1 The Process of Apoptosis

The term "apoptosis" was first used by three pathologists, Kerr, Wyllie, and Currie, in 1972 to describe morphologically distinct ultrastructural changes of RCD (Kerr et al. 1972). As a naturally occurring type of physiological cell death, apoptosis is required for the elimination of unnecessary and unwanted cells to maintain tissue homeostasis. In contrast, abnormal or excessive apoptosis contributes to tissue injury under various pathological conditions. Two apoptotic processes, including the extrinsic and intrinsic pathways, are well documented at molecular levels (Elmore 2007). The extrinsic pathway is mostly mediated by death receptors (DRs), which belong to the tumor necrosis factor (TNF) receptor gene superfamily. After binding to ligands, DRs including the Fas receptor (FasR/CD95), TNF receptor 1, lymphotoxin receptor, DR3, and DR4/DR5, can activate caspase-8 and subsequently caspase-3 to induce apoptosis. Unlike the DR-mediated extrinsic apoptosis pathway, the intrinsic pathway is associated with dynamic mitochondrial changes, including mitochondrial membrane potential loss and subsequent mitochondrial apoptosis-inducing protein release (Wang and Youle 2009). In particular, release of cytochrome c (Liu et al. 1996), second mitochondria-derived activator of caspases (Du et al. 2000), and HtrA serine peptidase 2 (van Loo et al. 2002) from the mitochondria to the cytoplasm can activate caspase-9 and subsequently caspase-3 to induce apoptosis. Thus, caspases (a family of protease enzymes) play central roles in triggering apoptosis, although a caspase-independent apoptotic pathway exists (McIlwain et al. 2015). In addition to caspase, the pro- (e.g., Bax, Bak, Bid, and PUMA) and anti-(e.g., Bcl-2, Bcl-xL, and Mcl-1) apoptotic Bcl-2 family members control mitochondrial membrane potential in apoptosis (Shamas-Din et al. 2013). Notably, caspase 8-mediated cleavage of Bid promotes intrinsic mitochondrial pathway activation, suggesting a direct molecular mechanism linking the extrinsic and intrinsic pathways in the induction of apoptosis (Gross et al. 1999; Li et al. 1998).

7.2.1.2 Apoptosis in Hepatic Pathology

Apoptosis is observed in almost all types of liver injury caused by extrinsic (e.g., viruses, alcohol, and drugs) and intrinsic (e.g., toxic bile acids, free fatty acids, and ischemia/reperfusion [I/R]) factors (Malhi and Gores 2008; Wang 2014). Inflammation mediators from activation of lymphocytes, macrophages, and neutrophils in liver injury also trigger apoptosis in hepatocytes (Lawson et al. 1998). Natural killer cells, a type of cytotoxic lymphocyte, can release perforin/granzyme to induce apoptosis in hepatocytes and hepatic stellate cells (Radaeva et al. 2006; Takeda et al. 2001). Excessive apoptotic death of hepatocytes and non-parenchymal cells contributes to liver injury (Guicciardi and Gores 2005). Apoptotic cell death is generally recognized as a noninflammatory process due to an intact apoptotic body plasma membrane and the rapid clearance of apoptotic cells by phagocytes (e.g., macrophages and dendritic cells). However, engulfment of apoptotic bodies by Kupffer cells (the macrophages in the liver) and hepatic stellate cells has been demonstrated to enhance death ligand production and stimulate the fibrogenic activity (Canbay et al. 2003; Oberhammer et al. 1992; Takehara et al. 2004). These changes exacerbate hepatic fibrosis. In addition, nuclear DAMPs such as histone, nuclear DNA, and HMGB1 can be released by apoptotic cells, which mediates inflammation, the immune response, and cell proliferation (Bell et al. 2006; Chen et al. 2014).

Hepatocellular carcinoma (HCC) is the most common primary malignant tumor of the liver and usually develops from hepatitis, hepatic fibrosis, and cirrhosis. The regulation of apoptosis plays crucial roles not only in the chemotherapeutic treatment of HCC, but also in tumorigenesis of HCC. One hallmark of cancers, including HCC, is intrinsic or acquired resistance to apoptosis (Fabregat 2009). Pharmacologic induction of apoptosis inhibits HCC tumor growth and advancement. Surprisingly, recent studies of $CD95^{-/-}$ and $PUMA^{-/-}$ mice indicate that apoptosis may promote tumor formation at the early stage by promoting stem cell proliferation or activation of the JNK pathway in the tumor microenvironment (Chen et al. 2010; Labi et al. 2010; Michalak et al. 2010; Ranger et al. 2003). Similarly, hepatocyte-specific BID-deficient mice are resistant to diethylnitrosamine (DEN)-induced hepatocarcinogenesis by suppressing inflammation-related compensatory proliferation (Wree et al. 2015). Hepatocyte-specific Bcl-xL or Mcl-1 knockout mice display increased hepatocyte apoptosis and spontaneous HCC formation (Hikita et al. 2012; Weber et al. 2010). One possible underlying mechanism of apoptosis-mediated tumorigenesis is that oxidative injury from apoptotic hepatocytes may promote gene mutation in liver tumorigenesis (Hikita et al. 2015). Collectively, dysregulated apoptosis plays a dual role in HCC tumorigenesis, depending on the tumor stage (Tang et al. 2011c).

7.2.2 Necroptosis

7.2.2.1 The Process of Necroptosis

Necrosis, also termed "oncosis," has been long considered a passive and unregulated form of cell death characterized by nuclear/plasma membrane leakage and adenosine triphosphate (ATP) depletion (Zong and Thompson 2006). Recent studies have highlighted that some cases of necrosis are also regulated and programmed (Galluzzi et al. 2012; Silke et al. 2015; Vanden et al. 2014). This form of necrosis is termed necroptosis and can be traced back to the finding in 1965 that TNF (also called cytotoxin) can induce cell death (Kolb and Granger 1968). We now know that TNF and other DR ligands (e.g., TNF-related apoptosis-inducing ligand) trigger necroptosis when the expression or activity of caspase-8 is inhibited by genetic (e.g., knockout of caspase-8 or Fas-associated protein with death domain) and pharmacological (e.g., ZVAD-FMK) methods (Kaiser et al. 2011; Welz et al. 2011). Additionally, interferons (IFNs), viral nucleic acid, and murine cytomegalovirus can induce necroptosis by IFN-receptor, TLR3, and Z-DNA-binding protein 1, respectively (Pasparakis and Vandenabeele 2015). These stimuli can finally activate receptor-interacting serine-threonine kinase (RIPK)-3 and subsequently target mixed lineage kinase domain-like (MLKL) by protein phosphorylation in the necrosome (He et al. 2009; Murphy et al. 2013; Newton et al. 2014; Sun et al. 2012). Finally, MLKL translocates to the plasma and cytoplasmic membranes, where it triggers necroptosis by modulating ion channel activities and Ca^{2+} influx (Cai et al. 2014). Of note, TNF-mediated RIPK1 phosphorylation is required for RIPK3 activation in necroptosis. In contrast, TLR adaptor molecule 1 (but not RIPK1) is required for viral nucleic acid-mediated RIPK3 activation in necroptosis. The underlying mechanism of IFN-mediated RIPK3 activation remains elusive.

7.2.2.2 Necroptosis in Hepatic Pathology

A number of DAMPs, such as HMGB1, S100A9, IL-33, and mitochondrial DNA, are released under necroptotic conditions and contribute to the subsequent inflammatory response in vitro and in vivo (Duprez et al. 2011; Kovalenko et al. 2009; Lau et al. 2013). Genetic and pharmacologic inhibition of necroptosis by using knockout mice (e.g., RIPK1−/−, RIPK3−/−, and MLKL−/−), necrostatin-1 (a RIPK1 inhibitor), GSK843/872 (a RIPK3 inhibitor), and necrosulfonamide (a MLKL inhibitor) have been demonstrated to provide protection against tissue injury in multiple inflammatory-associated diseases (Linkermann and Green 2014; Zhou and Yuan 2014). Of note, ethanol and acetaminophen increase RIPK3 expression in mouse and human livers. RIPK3−/− and RIPK1−/− mice are more resistant to ethanol-, acetaminophen-, high-fat choline-, or methionine and choline-deficient diet-induced liver injury with decreased proinflammatory cytokine expression (Afonso et al. 2015; Deutsch et al. 2015; Ramachandran et al. 2013; Roychowdhury et al. 2013). Necrostatin-1 not only reduces ethanol- and acetaminophen-induced liver injury but also limits concanavalin A-induced hepatitis in mouse models. However, RIPK1- and RIPK3-deficient mice display a different phenotype in concanavalin A-induced experimental hepatitis. One study showed that loss of RIPK3 limits, whereas loss of RIPK1 exacerbates concanavalin A-induced hepatitis and animal death (Deutsch et al. 2015). Other studies have shown that RIPK3−/− mice are not protected against concanavalin A-induced experimental hepatitis (Weinlich et al. 2013). The reason for these inconsistencies is still unresolved. Liver Kupffer cells can release DAMPs and cytokines in necroptosis to induce type 1 microbicidal inflammation and type-2-mediated liver repair upon infection (Bleriot et al. 2015). Thus, necroptosis pathway impairment contributes to alcohol-related liver disease, drug-induced liver injury, and hepatitis (Bakhautdin et al. 2014; Csak et al. 2011; Deutsch et al. 2015; Li et al. 2014a; Liedtke et al. 2011; Roychowdhury et al. 2013; Wang et al. 2016). Induction of necroptosis can improve anticancer therapies in apoptosis-resistant cancer cells (Chen et al. 2016). A recent study indicated that necroptosis-mediated CXCL1 release promotes pancreatic cancer tumorigenesis through inhibition of antitumor immunity (Seifert et al. 2016). However, the role of necroptosis in HCC tumorigenesis remains unclear.

7.2.3 Autophagic Cell Death

7.2.3.1 Autophagy and Autophagic Cell Death

Autophagy is a highly-conserved and lysosome-dependent degradation pathway in various species (Mizushima and Levine 2010). Currently, it includes three major forms, namely, macroautophagy, microautophagy, and chaperone-mediated autophagy. Among them, macroautophagy (hereafter referred to as autophagy) is well studied and divided into selective (e.g., mitophagy and xenophagy) and nonselective autophagy. Transmission electron microscopy assay has revealed three membrane-associated structures (phagophore, autophagosome, and autolysosome) in the dynamic process of autophagy (Klionsky and Emr 2000). The phagophore is the chief organelle that is vital for initiation of the autophagic cascade. After engulfing the cytosolic materials, the phagophore grows into a sealed autophagosome with the expression of autophagy-related marker microtubule-associated protein light chain 3-II (Mizushima et al. 2010). Finally, autolysosomes are generated after autophagosomes fuse into lysosomes, which facilitates the degradation of the engulfed material. At molecular levels, autophagy-related (ATG) family members can form different protein complexes involved in the regulation of formation and maturation of membrane-associated autophagic structures (Xie et al. 2015). These complexes are also important for the interplay between autophagy and other forms of RCD (Kang et al. 2011; Marino et al. 2014; Wong et al. 2013). In many cases, induction of autophagy promotes cell survival in response to harmful environments (Kroemer et al. 2010). In contrast, excessive autophagy may lead to cell death and tissue injury in some cases (Liu and Levine 2015). In particular, autosis has been recently identified as a specific form of autophagic cell death triggered by activation of Na^+/K^+-ATPase (Liu et al. 2013). The molecular switches between autophagy-mediated survival and death remain largely unidentified (Das et al. 2012).

7.2.3.2 Autophagic Cell Death in Hepatic Pathology

Autophagic dysfunction has been demonstrated to be involved in a number of liver pathologies. Induction of autophagy can maintain the energetic balance of the liver, alleviate endoplasmic reticulum stress, and remove unused or damaged hepatic proteins and organelles (Czaja et al. 2013; Madrigal-Matute and Cuervo 2016; Martinez-Lopez and Singh 2015; Schneider and Cuervo 2014; Yin et al. 2008). Thus, genetic inhibition of autophagy by using liver conditional knockout mice (e.g., ATG5−/− and ATG7−/−) has demonstrated that autophagy deficiency accelerates tissue injury and the inflammatory response in ethanol and acetaminophen-induced liver injury and high-fat diet-induced lipid accumulation, as well as I/R injury (Ding et al. 2010; Ezaki et al. 2011; Hernandez-Gea et al. 2012; Kim et al. 2013; Liu et al. 2015; Ni et al. 2012, 2016, 2014; Seok et al. 2014; Taguchi et al. 2012; Toshima et al. 2014; Xiao et al. 2016; Yang et al. 2010b). Pharmacological induction of autophagy by rapamycin significantly attenuates acetaminophen- and alcohol-induced liver injury (Lin et al. 2013). In contrast to rapamycin, autophagy inhibitors such as chloroquine can cause liver injury (Lin et al. 2013). Of note, the activity of chloroquine may work in an autophagy-independent manner (Maes et al. 2014). Autophagy plays a dual role in cancer biology, depending on tumor stage and type (White 2015). Autophagy inhibits tumorigenesis in HCC, a finding based on studies of autophagy-deficient mice. For example, mice with heterozygous disruption of ATG5 and ATG7 are shown to develop spontaneous liver tumor partly by activation of the nuclear factor (erythroid-derived 2)-like 2 (NRF2) pathway (Ichimura et al. 2013; Inami et al. 2011; Komatsu et al. 2010; Ni et al. 2014; Riley et al. 2010; Taguchi et al. 2012; Takamura et al. 2011). In contrast, autophagy may exert a pro-carcinogenic effect in HCC by promoting cell death or by inducing expression of factors that stimulate cell invasion and metastasis (Di Fazio et al. 2016; Li et al. 2013). Although autophagy has been well studied as a cell survival mechanism providing protection from liver injury, the function of autophagic cell death in liver disease still needs to be better understood.

7.2.4 Pyroptosis

7.2.4.1 The Process of Pyroptosis

The term "pyroptosis" was originally coined to indicate proinflammatory ("pyro") programmed cell death ("ptosis") during bacterial infection in macrophages (Brennan and Cookson 2000). It now refers to a type of RCD in immune cells mediated by the activation of inflammasome. The canonical inflammasomes include NLR (e.g., NLRP1, NLRP3, and NLRC4) and non-NLR (e.g., AIM2) types, which are mediated by activation of caspase-1 in response to PAMPs (e.g., lipopolysaccharide [LPS] and bacterial flagellin) and DAMPs (e.g., ATP, uric acid, DNA, and RNA) (Bergsbaken et al. 2009; Martinon et al. 2009; Rathinam and Fitzgerald 2016; Schroder and Tschopp 2010). In contrast, noncanonical inflammasome is triggered following the binding of intracellular LPS to caspase-11 (in mice) or caspase-4 or caspase-5 (in humans) (Hagar et al. 2013; Kayagaki et al. 2013; Man and Kanneganti 2016; Shi et al. 2014). Gram-negative bacteria secretion of outer membrane vesicles is responsible for the delivery of LPS to the host cell cytosol to bind caspase-11, which will cause cleavage of gasdermin-D (Vanaja et al. 2016). The N-terminal of gasdermin-D can trigger subsequent cell death as well as canonical NLRP3-inflammasome activation (He et al. 2015; Kayagaki et al. 2015; Shi et al. 2015). Thus, cleavage of gasdermin-D links the interplay between noncanonical and canonical inflammasomes (Wallach et al. 2016). Inflammasome activation is usually associated with secretion of proinflammatory cytokines (e.g., IL-1β, IL-1α, IL-18, IL-33, and HMGB1) and thereby triggers injury.

7.2.4.2 Pyroptosis in Hepatic Pathology

Increased caspase-1-dependent pyroptosis has been observed in both Kupffer cells and hepatocytes during the development of nonalcoholic steatohepatitis, alcoholic steatohepatitis, and I/R (Kamo et al. 2013; Petrasek et al. 2012). IL-1β release from inflammasome activation not only directly induces hepatic steatosis and injury but also enhances LPS-mediated TNF-α production, which further exacerbates liver injury (Qiao et al. 2015). Global knockout caspase-1 ameliorates, whereas global knockin NLRP3 increases inflammasome-dependent liver injury in mice (Wree et al. 2014). In contrast, myeloid cell-derived NLRP3 activation resulted in a less severe liver phenotype in the absence of detectable pyroptotic liver cell death (Wree et al. 2014). Thus, NLRP3 inflammasome plays an important role in triggering inflammatory injury and fibrosis in both hepatocytes and non-parenchymal cells (Ganz et al. 2011; Imamura et al. 2009; Zhu et al. 2011). Moreover, heat stroke-induced liver injury requires HMGB1-mediated NLRP3 inflammasome activation (Geng et al. 2015). Caspase-11 also contributes to hepatic inflammation by activation of Kupffer cells (Hendrikx et al. 2013). Autophagy deficiency triggers inflammasome activation in macrophages, which promotes liver injury (Ilyas et al. 2016). Thus, inhibition of pyroptosis with drugs is an alternative approach to the treatment of inflammation in liver disease (Heymann and Tacke 2016; Hoque et al. 2014; Szabo and Petrasek 2015; Yang et al. 2016).

7.2.5 NETosis

7.2.5.1 The Process of NETosis

NETosis is a unique form of RCD that is characterized by the release of decondensed chromatin (e.g., DNA, HMGB1, and histone) and granular contents to the extracellular space (Steinberg and Grinstein 2007; Yipp et al. 2012). This process also termed neutrophil extracellular trapping, which was first observed in neutrophil activation in infection (Brinkmann et al. 2004). In addition to neutrophils, NETosis can be induced in endothelial and cancer cells under pathologic conditions such as hypoxia (Demers and Wagner 2014; Remijsen et al. 2011). NADPH oxidase-mediated reactive oxygen species (ROS) production is an early event in the induction of NETosis. Peptidylarginine deiminase 4 (PAD4) is an enzyme responsible for protein citrullination. PAD4-mediated histone citrullination is a late event in chromatin decondensation and release in NETosis (Branzk and Papayannopoulos 2013). Thus, genetic or pharmacologic inhibition of PAD4 could inhibit NETosis (Branzk and Papayannopoulos 2013; Hahn et al. 2013). However, PAD4-independent NETosis may exist and needs further identification (Metzler et al. 2014).

7.2.5.2 NETosis in Hepatic Pathology

NETosis-associated nuclear DAMP release and tissue damage is implicated in the pathologies of inflammatory and autoimmune disorders such as systemic lupus erythematosus, rheumatoid arthritis, asthma, vessel vasculitis, and psoriasis (Branzk and Papayannopoulos 2013). More recently, formation of NETosis has been reported in liver I/R injury (Huang et al. 2015). Increased NETosis in hepatocytes induces Kupffer cell activation and subsequent proinflammatory cytokine release, which contributes to sterile inflammation. In contrast, blocking NETosis by neutrophil depletion or using PAD4 inhibitor reduces liver I/R injury in a DAMP-dependent manner (Huang et al. 2015). In addition to I/R injury, NETosis-mediated DAMP release also contributes to adhesion, proliferation, migration, and invasion of HCC cells by activation of TLR9 (Tohme et al. 2016). Further studies are needed to elucidate the potential role of NETosis in other types of liver injury such as viral infection.

7.2.6 Ferroptosis

7.2.6.1 Process of Ferroptosis

Ferroptosis is a new form of RCD and was first reported in cancer cells (Dixon et al. 2012). Morphologic, biochemical, and genetic studies highlight the unique aspects of ferroptosis compared to other forms of RCD (e.g., apoptosis, necroptosis, and autophagic cell death). The classical inducer of ferroptosis is erastin, which displayed anticancer activity in an oncogenic Ras-dependent manner in an original study (Dolma et al. 2003). In contrast, other recent studies indicate that erastin triggers ferroptosis in both a RAS-dependent and RAS-independent manner (Nils Eling et al. 2015; Schott et al. 2015; Yagoda et al. 2007; Yu et al. 2015b). In addition to erastin, several clinical drugs (e.g., sulfasalazine, sorafenib, and artesunate) have been demonstrated to trigger ferroptosis not only in cancer cells but also in certain normal cells (e.g., kidney cells, neurons, fibroblasts, and T cells) (Xie et al. 2016; Yang and Stockwell 2015). Oxidative injury from iron accumulation and lipid peroxidation is responsible for ferroptosis induction. Several proteins are involved in the regulation of ferroptosis by targeting two iron metabolism and oxidative stressors, although the molecular mechanism and death-effector remain largely unknown. For example, glutathione peroxidase 4 (Yang et al. 2014), NRF2 (Sun et al. 2016b), metallothionein-1G (MT-1G) (Sun et al. 2016a), and heat shock protein beta-1 (Sun et al. 2015) function as negative regulators of ferroptosis by limiting lipid ROS production or reducing cellular iron uptake. By contrast, NADPH oxidase and p53 (Jiang et al. 2015) act as positive regulators of ferroptosis by promoting lipid ROS production and inhibiting expression of SLC7A11 (a specific light chain subunit of the cystine/glutamate antiporter), respectively. In addition, activation of MAPK (Yagoda et al. 2007) and PKCα (Do Van et al. 2016) also contributes to ferroptosis in different cells. Together, these findings reveal a complex molecular and signaling mechanism in ferroptosis.

7.2.6.2 Ferroptosis in Hepatic Pathology

Ferroptosis has emerged as a new regulatory mechanism for liver injury. For example, acetaminophen has been demonstrated to induce ferroptosis in primary liver cells, and ferroptosis inhibitors such as ferrostain-1 can inhibit acetaminophen-induced liver injury (Lorincz et al. 2015). Ferroptosis has been shown to promote ischemic liver damage, and ferroptosis inhibitor liproxstatin-1 can protect against liver I/R in mice (Friedmann Angeli et al. 2014). Sorafenib, a multiple kinase inhibitor, is now the only approved systemic therapy for advanced HCC. Recent studies indicate that ferroptosis mediates the anticancer activity of sorafenib in HCC cells. Sorafenib-mediated cell death in human (HepG2, Hep3B, and SNU-182) and mouse (Hepal–6) HCC lines is blocked by ferrostatin-1 (a ferroptosis inhibitor), but not ZVAD-FMK (an apoptosis inhibitor) and necrosulfonamide (a necroptosis inhibitor) (Sun et al. 2016b). Activation of the NRF2 pathway protects against ferroptosis in human and mouse HCC cells (Sun et al. 2016b). Upon exposure to ferroptosis-inducing compounds (e.g., erastin, sorafenib, and buthionine sulfoximine), p62/SQSTM1 (an adaptor protein) expression prevents NRF2 degradation and enhances subsequent NRF2 nuclear accumulation through inactivation of Kelch-like ECH-associated protein 1 (Keap1) (Sun et al. 2016b). Genetic or pharmacologic inhibition of NRF2 expression/activity in HCC cells increases the anticancer activity of erastin and sorafenib in HCC cells in vitro and in tumor xenograft models (Sun et al. 2016b). Moreover, Sorafenib specifically induces MT-1G expression by activation of NRF2 in human HCC cells (Sun et al. 2016a). Suppression of MT-1G expression enhances sorafenib sensitivity by induction of ferroptosis (Sun et al. 2016a). These findings not only identify a novel mechanism of sorafenib resistance but also suggest a new link between the NRF2-MT-1G pathway and ferroptosis in human HCC cells.

7.3 Nuclear DAMPs in Hepatic Injury and Inflammation

7.3.1 HMGB1

7.3.1.1 HMGB1 in Cell Death

HMGB1 is the most abundant nonhistone nuclear protein and belongs to the HMG family. In the nucleus, HMGB1 can bind and bend DNA to regulate multiple chromatin-associated events such as gene transcription and DNA repair. As a classical DAMP with cytokine and chemokine functions, HMGB1 is actively secreted by immune cells (e.g., macrophages and neutrophils) or passively released in various types of cell death (e.g., necrosis, apoptosis, necroptosis, autophagic cell death, pyroptosis, NETosis, and ferroptosis) (Kang et al. 2014a). Oxidative stress plays a central role in the regulation of HMGB1 secretion and release (Yu et al. 2015a). Once released, the activity of HMGB1 is regulated at multiple levels such as receptor, partner, and redox status. In general, RAGE (Kokkola et al. 2005), TLR2 (Park et al. 2004), TLR4 (Park et al. 2004), and TLR9 (Tian et al. 2007) are receptors required for the HMGB1-mediated immune response. In contrast, CD24 (Chen et al. 2009) and TIM-3 (Chiba et al. 2012) are negative binding receptors that limit HMGB1 activity in the immune response. Extracellular HMGB1 is a sticky protein and can bind DAMPs (e.g., histone, DNA, and IL-1) and PAMPs (e.g., LPS) to amply the inflammatory response. The oxidative state of released HMGB1 can fine-tune HMGB1 activity (Tang et al. 2012a, b; Venereau et al. 2012; Yang et al. 2010a). In general, deduced all-thiol-HMGB1 displays chemokine activity, whereas disulfide-HMGB1 exhibits cytokine activity. In contrast, oxidized HMGB1 loss displays both chemokine and cytokine activity (Venereau et al. 2012).

Intracellular HMGB1 is generally an antiapoptotic protein in response to ultraviolet radiation, TNF, and TRAIL in vitro (Brezniceanu et al. 2003). Intracellular HMGB1 inhibits apoptosis in both transcriptional-dependent (e.g., regulation of Bcl-2 family protein expression) and transcriptional-independent (e.g., regulation of autophagy and p53 location) manners. In vivo, loss of HMGB1 in the pancreas increases L-arginine-induced apoptosis and necrosis due to oxidative injury (Kang et al. 2014b). Compared with its antiapoptotic role, HMGB1 is a pro-autophagy protein in many cells. HMGB1 promotes autophagy also in both transcriptional-dependent and transcriptional-independent manners. For example, HMGB1 can transcriptionally regulate heat shock protein β-1 (HSPB1) expression, which affects autophagosome and autolysosome formation in response to mitochondrial injury (Tang et al. 2011a). Cytosolic HMGB1 regulates autophagosome formation by direct binding to Beclin-1, a key regulator of autophagy (Tang et al. 2010b). The binding of HMGB1 with Beclin-1 is positively regulated by unc-51-like kinase 1 and nucleus accumbens-1 (Zhang et al. 2012), whereas it is negatively regulated by p53 (Livesey et al. 2012), SNCA/α-synuclein (Song et al. 2014), and lysosomal thiol reductase (Chiang and Maric 2011). Moreover, inhibition of HMGB1 expression by miR34A (Liu et al. 2014a) or miR22 (Li et al. 2014b) limits autophagy in chemotherapy. In contrast, increased HMGB1 translocation from the nucleus to the cytosol by activation of poly [ADP-ribose] polymerase 1 (PARP1) promotes TRAIL resistance by induction of autophagy (Yang et al. 2015). Compared with intracellular HMGB1, extracellular HMGB1-mediated autophagy requires RAGE-mediated PI3KC3 activation (Tang et al. 2010a). In addition to HMGB1-dependent autophagy regulating cell death and inflammation in vitro and in vivo, HMGB1-independent autophagy may exist in the liver and heart (Huebener et al. 2014; Sun and Tang 2014). Induction of proptosis by PAMPs increases HMGB1 release in macrophages (Lu et al. 2012). Released HMGB1 in turn accelerates inflammasome activation and the release of IL-1β and IL-18, again implicating the involvement of DAMPs and PAMPs in systematic inflammation.

7.3.1.2 Role of HMGB1 in Hepatic Pathology

Extracellular HMGB1 plays a significant pathogenic role in liver injury, and serum HMGB1 is a useful biomarker of liver disease. The involvement of extracellular HMGB1 in liver injury was first described in a liver warm I/R mouse model (Tsung et al. 2005; Watanabe et al. 2005). Activation of ROS and the calcium/calmodulin-dependent protein kinase signaling pathway is required for HMGB1 release by hepatocytes in I/R. TLR4 contributes to HMGB1-mediated proinflammatory cytokine generation and organ damage in warm hepatic I/R injury (Tsung et al. 2005). Blocking HMGB1 release and activity protects against I/R injury. In addition to TLR4, TLR9 (Bamboat et al. 2010) and RAGE (Zeng et al. 2009) also mediate liver I/R injury, suggesting that HMGB1 may interact with multiple receptors to mediate warm liver I/R injury. HMGB1 is also released by hepatocytes in hepatitis C virus or hepatitis B virus (HBV) infection, which mediates the antiviral response in a TLR4-dependent pathway (Jung et al. 2011; Zhou et al. 2011). In addition to viral hepatitis, HMGB1 release has been significantly observed in nonalcoholic fatty liver disease, hepatic fibrosis, and alcohol- and drug-induced liver injury. Once released, HMGB1 promotes the inflammation and proliferation response via TLR4-MyD88 or the MAPK-NF-κB or inflammasome pathways (Geng et al. 2015). Similarity, hypoxia-induced HMGB1 release in HCC cell lines can activate TLR4- and RAGE-signaling pathways to induce inflammasome-dependent secretion of IL-1β and IL-18. These changes in the tumor microenvironment promote HCC invasion and metastasis (Yan et al. 2012). Serum HMGB1 levels were significantly higher in HCC and HBV patients and are associated with poor outcomes in HCC patients (Cheng et al. 2008; Jiang et al. 2012).

The role of intracellular HMGB1 in the liver was recently investigated using hepatocyte-specific HMGB1 conditional knockout mice (Sun and Tang 2015). HMGB1 depletion does not affect liver development and function under physiological conditions. However, different groups observed a dual role of HMGB1 in the liver under pathological conditions. Conditional knockout of HMGB1 in the liver increases nuclear and mitochondrial injury and nuclear DAMP release in response to I/R injury (Huang et al. 2014). In contrast, mice with hepatocyte-specific HMGB1 depletion were more resistant to acetaminophen- and alcohol-induced liver injury by inhibition of inflammation, fatty acid synthesis, and fatty acid β-oxidation(Ge et al. 2014; Huebener et al. 2015). Understanding the tissue- and cell-specific roles of HMGB1 in stress is important for HMGB1-based target therapy (Chen et al. 2013; Tang et al. 2014).

7.3.2 *Histone*

7.3.2.1 Role of Histone in Cell Death

Histones are highly alkaline proteins found in eukaryotic cells (Bhasin et al. 2006; Luger et al. 1997). Histones are the fundamental unit of chromatin, namely, the nucleosome. In eukaryotes, the nucleosome is composed of 147 bp DNA wrapped around eight core histone proteins, including two each of H2A, H2B, H3, and H4. The linker histones, H1 and its variant forms, have been implicated in the formation of higher orders of chromatin structure. Posttranslational modifications (PTMs) of histones at the amino-terminal tails, including acetylation, phosphorylation, methylation, and ubiquitination, are changed in response to various stimuli. These changes led to the generation of the histone code hypothesis, which plays a critical role in epigenetic regulation of gene expression in several cellular processes, including various types of cell death (Strahl and Allis 2000). Histone phosphorylation and acetylation have been suggested to affect chromatin function and structure during apoptosis (Hajji and Joseph 2010). For example, histone H2B phosphorylation at Ser14 has been proposed as an epigenetic marker of apoptotic cells associated with DNA degradation (Ajiro et al. 2010). Moreover, phosphorylation of the histone H2AX at Ser139 (termed γ-H2AX) is a hallmark of the DNA-damage response at the early stage of apoptosis (Sharma et al. 2012). Compared with these apoptotic histone codes (Huang et al. 2013; Sanders et al. 2013), recent studies have also described histone codes in autophagy (e.g., H3K56ac, H4K16ac, H3K4me3, H4K20me3, and H3K9me2) (Fullgrabe et al. 2014). The anticancer activity of the histone deacetylase (HDAC) inhibitor depends on induction of forms of RCD such as apoptosis, necroptosis, and autophagic cell death (Xu et al. 2007). HDAC-mediated epigenetic reprogramming also regulates pyroptosis in macrophages (Ha et al. 2014). How these histone PTMs contribute to the interplay between different kinds of RCD remains to be elucidated in detail.

Histones, a new type of nuclear DAMP, can be released by dead and dying cells in apoptosis, necrosis, and NETosis (Chen et al. 2014). In apoptosis, core histones (H2A, H2B, H3, and H4) and link histone (H1) separate from genomic DNA, translocate to the cytosol, and finally release into the extracellular space (Gabler et al. 2004; Gilthorpe et al. 2013; Wu et al. 2002). Both histone release and DNA fragmentation formation are important nuclear events in apoptosis. This process is primarily mediated by CAD (caspase-activated DNase)/DFF-40(DNA fragmentation factor) (Wu et al. 2002). Compared with histone release in apoptotic cells, histone secretion by activated immune cells (e.g., basophils, mast cells, and neutrophils) is mediated by NETosis in response to pathogen infection (Brinkmann et al. 2004). In addition to histones, other nuclear DAMPs, including HMGB1 and DNA, are important components released in NETosis (Brinkmann et al. 2004). These nuclear DAMPs can capture and degrade invading pathogens and play an important role in antibacterial responses. In contrast, extracellular histones have been implicated as inflammatory mediators in lethal infections (e.g., sepsis (Xu et al. 2009)) and sterile inflammation (e.g., I/R injury (Allam et al. 2012; Bosmann et al. 2013; De Meyer et al. 2012; Huang et al. 2011; Wen et al. 2013; Xu et al. 2011) and pancreatitis (Kang et al. 2013)). Extracellular histones also induce disseminated intravascular coagulation by promoting coagulation and thrombosis (Semeraro et al. 2011). TLRs and RAGE are major receptors required for histone activity in tissue injury and inflammation (Kang et al. 2016; Xu et al. 2011). The nuclear DAMP complex can trigger apoptosis, autophagy, and pyroptosis in immune and nonimmune cells, which in turn regulates proinflammatory cytokine release in cell death (Liu et al. 2014b).

7.3.2.2 Role of Histone in Hepatic Pathology

Several studies have shown that extracellular histone plays a significantly pathologic role in liver injury, including concanavalin A-induced hepatitis, acetaminophen-induced hepatotoxicity, liver I/R, and acute liver failure. Elevated serum histones have been observed in animal models with I/R and chemical-induced liver injury (Allam et al. 2012; Huang et al. 2011; Wen et al. 2013; Xu et al. 2011). Exogenous histones can induce proinflammatory cytokine (TNF-α and IL-6) release by binding to TLR2, TLR4, and TLR9 to activate NF-κB and the mitogen-activated protein kinase [MAPK]) pathway in the liver. Blocking histone-TLR signaling by using histone-neutralizing antibody or knockout of TLR2, TLR4, and TLR9 in mice limits liver injury. In addition to directly activating TLR9, extracellular histones also enhance nucleic acid–mediated inflammation through TLR9 (Xu et al. 2011). Downstream of TLRs, myeloid differentiation factor 88 (MyD88) also contributes to histone-mediated liver I/R injury and sterile inflammation (Huang et al. 2011). The extracellular histone-TLR9 pathway promotes ROS production, which is required for NLRP3 inflammasome activation in Kupffer cells during sterile inflammatory liver injury. In contrast, RAGE, but not TLRs, is required for histone/DNA-mediated AIM2 inflammasome activation in macrophages during sterile inflammatory pancreas injury (Liu et al. 2014b). These findings provide different receptor-dependent mechanisms of inflammasome activation by histones in the sterile inflammatory response.

In patients with multiple tumors, especially in advanced cancer, levels of serum nucleosome, including histone and DNA, are significantly higher compared with healthy populations (Holdenrieder et al. 2001; Trejo-Becerril et al. 2003). Recent evidence has shown that release of nuclear DAMPs, including HMGB1, DNA, and histone, promotes tumor metastasis to the liver after surgical stress in NETosis

(Tohme et al. 2016). In contrast, genetic or pharmacologic inhibition of PAD4 or use of DNAse limits nuclear DAMP-mediated tumor metastasis (Tohme et al. 2016). Thus, nuclear DAMPs from abnormal cell death are important contributors to tumor metastasis.

7.4 Concluding Remarks

The liver performs many essential functions related to digestion, metabolism, and immunity. Liver injuries not only cause various types of cell death but also stimulate sterile inflammation. DAMPs such as HMGB1 and histone are critical molecular mediators linking cell death and the sterile inflammatory response in liver injury. Blocking the interaction between DAMPs and their receptors (e.g., TLRs and RAGE) can protect against liver injury from drugs, alcohol, viruses, trauma, and cancer in animal models. Anti-DAMP-based therapeutic strategies may be useful in patients with liver disease. However, many challenges still remain with respect to understanding the structure, modification, and function of DAMPs in liver injury and inflammation. Emerging data also points to the importance of interplay among different types of RCD in the regulation of DAMP release and the inflammatory response. Future studies are required to fully clarify and distinguish between these mechanisms.

Acknowledgments We apologize to the researchers who were not referenced due to space limitations. We thank Christine Heiner (Department of Surgery, University of Pittsburgh) for her critical reading of the manuscript. This work was supported by the National Institutes of Health of the USA (R01GM115366 and R01CA160417 to D.T.), the National Natural Science Foundation of Guangdong (2016A030308 D.T.), and a Research Scholar Grant from the American Cancer Society (RSG-16-014-01-CDD to D.T.).

References

Afonso MB, Rodrigues PM, Carvalho T, Caridade M, Borralho P, Cortez-Pinto H, Castro RE, Rodrigues CM (2015) Necroptosis is a key pathogenic event in human and experimental murine models of non-alcoholic steatohepatitis. Clin Sci (Lond) 129:721–739

Ajiro K, Scoltock AB, Smith LK, Ashasima M, Cidlowski JA (2010) Reciprocal epigenetic modification of histone H2B occurs in chromatin during apoptosis in vitro and in vivo. Cell Death Differ 17:984–993

Allam R, Scherbaum CR, Darisipudi MN, Mulay SR, Hagele H, Lichtnekert J, Hagemann JH, Rupanagudi KV, Ryu M, Schwarzenberger C, Hohenstein B, Hugo C, Uhl B, Reichel CA, Krombach F, Monestier M, Liapis H, Moreth K, Schaefer L, Anders HJ (2012) Histones from dying renal cells aggravate kidney injury via TLR2 and TLR4. J Am Soc Nephrol 23:1375–1388

Ashkenazi A, Salvesen G (2014) Regulated cell death: signaling and mechanisms. Annu Rev Cell Dev Biol 30:337–356

Bakhautdin B, Das D, Mandal P, Roychowdhury S, Danner J, Bush K, Pollard K, Kaspar JW, Li W, Salomon RG, McMullen MR, Nagy LE (2014) Protective role of HO-1 and carbon monoxide in ethanol-induced hepatocyte cell death and liver injury in mice. J Hepatol 61:1029–1037

Bamboat ZM, Balachandran VP, Ocuin LM, Obaid H, Plitas G, DeMatteo RP (2010) Toll-like receptor 9 inhibition confers protection from liver ischemia-reperfusion injury. Hepatology 51:621–632

Bell CW, Jiang W, Reich CF 3rd, Pisetsky DS (2006) The extracellular release of HMGB1 during apoptotic cell death. Am J Physiol Cell Physiol 291:C1318–C1325

Bergsbaken T, Fink SL, Cookson BT (2009) Pyroptosis: host cell death and inflammation. Nat Rev Microbiol 7:99–109

Bhasin M, Reinherz EL, Reche PA (2006) Recognition and classification of histones using support vector machine. J Comput Biol J Comput Mol Cell Biol 13:102–112

Bianchi ME (2007) DAMPs, PAMPs and alarmins: all we need to know about danger. J Leukoc Biol 81:1–5

Bleriot C, Dupuis T, Jouvion G, Eberl G, Disson O, Lecuit M (2015) Liver-resident macrophage necroptosis orchestrates type 1 microbicidal inflammation and type-2-mediated tissue repair during bacterial infection. Immunity 42:145–158

Bosmann M, Grailer JJ, Ruemmler R, Russkamp NF, Zetoune FS, Sarma JV, Standiford TJ, Ward PA (2013) Extracellular histones are essential effectors of C5aR-and C5L2-mediated tissue damage and inflammation in acute lung injury. FASEB J 27:5010–5021

Branzk N, Papayannopoulos V (2013) Molecular mechanisms regulating NETosis in infection and disease. Semin Immunopathol 35:513–530

Brennan MA, Cookson BT (2000) Salmonella induces macrophage death by caspase-1-dependent necrosis. Mol Microbiol 38:31–40

Brezniceanu ML, Volp K, Bosser S, Solbach C, Lichter P, Joos S, Zornig M (2003) HMGB1 inhibits cell death in yeast and mammalian cells and is abundantly expressed in human breast carcinoma. FASEB J 17:1295–1297

Brinkmann V, Reichard U, Goosmann C, Fauler B, Uhlemann Y, Weiss DS, Weinrauch Y, Zychlinsky A (2004) Neutrophil extracellular traps kill bacteria. Science 303:1532–1535

Cai Z, Jitkaew S, Zhao J, Chiang HC, Choksi S, Liu J, Ward Y, Wu LG, Liu ZG (2014) Plasma membrane translocation of trimerized MLKL protein is required for TNF-induced necroptosis. Nat Cell Biol 16:55–65

Canbay A, Feldstein AE, Higuchi H, Werneburg N, Grambihler A, Bronk SF, Gores GJ (2003) Kupffer cell engulfment of apoptotic bodies stimulates death ligand and cytokine expression. Hepatology 38:1188–1198

Chen GY, Tang J, Zheng P, Liu Y (2009) CD24 and Siglec-10 selectively repress tissue damage-induced immune responses. Science 323:1722–1725

Chen L, Park SM, Tumanov AV, Hau A, Sawada K, Feig C, Turner JR, Fu YX, Romero IL, Lengyel E, Peter ME (2010) CD95 promotes tumour growth. Nature 465:492–496

Chen R, Hou W, Zhang Q, Kang R, Fan XG, Tang D (2013) Emerging role of high-mobility group box 1 (HMGB1) in liver diseases. Mol Med 19:357–366

Chen R, Kang R, Fan XG, Tang D (2014) Release and activity of histone in diseases. Cell Death Dis 5:e1370

Chen D, Yu J, Zhang L (2016) Necroptosis: an alternative cell death program defending against cancer. Biochim Biophys Acta 1865:228–236

Cheng BQ, Jia CQ, Liu CT, Lu XF, Zhong N, Zhang ZL, Fan W, Li YQ (2008) Serum high mobility group box chromosomal protein 1 is associated with clinicopathologic features in patients with hepatocellular carcinoma. Dig Liver Dis Off J Italian Soc Gastroenterol Italian Assoc Study Liver 40:446–452

Chiang HS, Maric M (2011) Lysosomal thiol reductase negatively regulates autophagy by altering glutathione synthesis and oxidation. Free Radic Biol Med 51:688–699

Chiba S, Baghdadi M, Akiba H, Yoshiyama H, Kinoshita I, Dosaka-Akita H, Fujioka Y, Ohba Y, Gorman JV, Colgan JD, Hirashima M, Uede T, Takaoka A, Yagita H, Jinushi M (2012) Tumor-

infiltrating DCs suppress nucleic acid-mediated innate immune responses through interactions between the receptor TIM-3 and the alarmin HMGB1. Nat Immunol 13:832–842
Csak T, Dolganiuc A, Kodys K, Nath B, Petrasek J, Bala S, Lippai D, Szabo G (2011) Mitochondrial antiviral signaling protein defect links impaired antiviral response and liver injury in steatohepatitis in mice. Hepatology 53:1917–1931
Czaja MJ, Ding WX, Donohue TM Jr, Friedman SL, Kim JS, Komatsu M, Lemasters JJ, Lemoine A, Lin JD, Ou JH, Perlmutter DH, Randall G, Ray RB, Tsung A, Yin XM (2013) Functions of autophagy in normal and diseased liver. Autophagy 9:1131–1158
Das G, Shravage BV, Baehrecke EH (2012) Regulation and function of autophagy during cell survival and cell death. Cold Spring Harb Perspect Biol 4. pii: a008813. doi:10.1101/cshperspect.a008813
De Meyer SF, Suidan GL, Fuchs TA, Monestier M, Wagner DD (2012) Extracellular chromatin is an important mediator of ischemic stroke in mice. Arterioscler Thromb Vasc Biol 32:1884–1891
Demers M, Wagner DD (2014) NETosis: a new factor in tumor progression and cancer-associated thrombosis. Semin Thromb Hemost 40:277–283
Deutsch M, Graffeo CS, Rokosh R, Pansari M, Ochi A, Levie EM, Van Heerden E, Tippens DM, Greco S, Barilla R, Tomkotter L, Zambirinis CP, Avanzi N, Gulati R, Pachter HL, Torres-Hernandez A, Eisenthal A, Daley D, Miller G (2015) Divergent effects of RIP1 or RIP3 blockade in murine models of acute liver injury. Cell Death Dis 6:e1759
Di Fazio P, Waldegger P, Jabari S, Lingelbach S, Montalbano R, Ocker M, Slater EP, Bartsch DK, Illig R, Neureiter D, Wissniowski TT (2016) Autophagy-related cell death by pan-histone deacetylase inhibition in liver cancer. Oncotarget 7(20):28998–29010
Ding WX, Li M, Chen X, Ni HM, Lin CW, Gao W, Lu B, Stolz DB, Clemens DL, Yin XM (2010) Autophagy reduces acute ethanol-induced hepatotoxicity and steatosis in mice. Gastroenterology 139:1740–1752
Dixon SJ, Lemberg KM, Lamprecht MR, Skouta R, Zaitsev EM, Gleason CE, Patel DN, Bauer AJ, Cantley AM, Yang WS, Morrison B 3rd, Stockwell BR (2012) Ferroptosis: an iron-dependent form of nonapoptotic cell death. Cell 149:1060–1072
Do Van B, Gouel F, Jonneaux A, Timmerman K, Gele P, Petrault M, Bastide M, Laloux C, Moreau C, Bordet R, Devos D, Devedjian JC (2016) Ferroptosis, a newly characterized form of cell death in Parkinson's disease that is regulated by PKC. Neurobiol Dis 94:169–178
Dolma S, Lessnick SL, Hahn WC, Stockwell BR (2003) Identification of genotype-selective antitumor agents using synthetic lethal chemical screening in engineered human tumor cells. Cancer Cell 3:285–296
Du C, Fang M, Li Y, Li L, Wang X (2000) Smac, a mitochondrial protein that promotes cytochrome c-dependent caspase activation by eliminating IAP inhibition. Cell 102:33–42
Duprez L, Takahashi N, Van Hauwermeiren F, Vandendriessche B, Goossens V, Vanden Berghe T, Declercq W, Libert C, Cauwels A, Vandenabeele P (2011) RIP kinase-dependent necrosis drives lethal systemic inflammatory response syndrome. Immunity 35:908–918
Eling N, Reuter L, Hazin J, Hamacher-Brady A, Brady NR (2015) Identification of artesunate as a specific activator of ferroptosis in pancreatic cancer cells. Oncoscience 2:517–532
Elmore S (2007) Apoptosis: a review of programmed cell death. Toxicol Pathol 35:495–516
Ezaki J, Matsumoto N, Takeda-Ezaki M, Komatsu M, Takahashi K, Hiraoka Y, Taka H, Fujimura T, Takehana K, Yoshida M, Iwata J, Tanida I, Furuya N, Zheng DM, Tada N, Tanaka K, Kominami E, Ueno T (2011) Liver autophagy contributes to the maintenance of blood glucose and amino acid levels. Autophagy 7:727–736
Fabregat I (2009) Dysregulation of apoptosis in hepatocellular carcinoma cells. World J Gastroenterol 15:513–520
Friedmann Angeli JP, Schneider M, Proneth B, Tyurina YY, Tyurin VA, Hammond VJ, Herbach N, Aichler M, Walch A, Eggenhofer E, Basavarajappa D, Radmark O, Kobayashi S, Seibt T, Beck H, Neff F, Esposito I, Wanke R, Forster H, Yefremova O, Heinrichmeyer M, Bornkamm GW, Geissler EK, Thomas SB, Stockwell BR, O'Donnell VB, Kagan VE, Schick JA, Conrad M

(2014) Inactivation of the ferroptosis regulator Gpx4 triggers acute renal failure in mice. Nat Cell Biol 16:1180–1191

Fullgrabe J, Heldring N, Hermanson O, Joseph B (2014) Cracking the survival code: autophagy-related histone modifications. Autophagy 10:556–561

Gabler C, Blank N, Hieronymus T, Schiller M, Berden JH, Kalden JR, Lorenz HM (2004) Extranuclear detection of histones and nucleosomes in activated human lymphoblasts as an early event in apoptosis. Ann Rheum Dis 63:1135–1144

Galluzzi L, Vitale I, Abrams JM, Alnemri ES, Baehrecke EH, Blagosklonny MV, Dawson TM, Dawson VL, El-Deiry WS, Fulda S, Gottlieb E, Green DR, Hengartner MO, Kepp O, Knight RA, Kumar S, Lipton SA, Lu X, Madeo F, Malorni W, Mehlen P, Nunez G, Peter ME, Piacentini M, Rubinsztein DC, Shi Y, Simon HU, Vandenabeele P, White E, Yuan J, Zhivotovsky B, Melino G, Kroemer G (2012) Molecular definitions of cell death subroutines: recommendations of the nomenclature committee on cell death 2012. Cell Death Differ 19:107–120

Galluzzi L, Bravo-San Pedro JM, Vitale I, Aaronson SA, Abrams JM, Adam D, Alnemri ES, Altucci L, Andrews D, Annicchiarico-Petruzzelli M, Baehrecke EH, Bazan NG, Bertrand MJ, Bianchi K, Blagosklonny MV, Blomgren K, Borner C, Bredesen DE, Brenner C, Campanella M, Candi E, Cecconi F, Chan FK, Chandel NS, Cheng EH, Chipuk JE, Cidlowski JA, Ciechanover A, Dawson TM, Dawson VL, De Laurenzi V, De Maria R, Debatin KM, Di Daniele N, Dixit VM, Dynlacht BD, El-Deiry WS, Fimia GM, Flavell RA, Fulda S, Garrido C, Gougeon ML, Green DR, Gronemeyer H, Hajnoczky G, Hardwick JM, Hengartner MO, Ichijo H, Joseph B, Jost PJ, Kaufmann T, Kepp O, Klionsky DJ, Knight RA, Kumar S, Lemasters JJ, Levine B, Linkermann A, Lipton SA, Lockshin RA, Lopez-Otin C, Lugli E, Madeo F, Malorni W, Marine JC, Martin SJ, Martinou JC, Medema JP, Meier P, Melino S, Mizushima N, Moll U, Munoz-Pinedo C, Nunez G, Oberst A, Panaretakis T, Penninger JM, Peter ME, Piacentini M, Pinton P, Prehn JH, Puthalakath H, Rabinovich GA, Ravichandran KS, Rizzuto R, Rodrigues CM, Rubinsztein DC, Rudel T, Shi Y, Simon HU, Stockwell BR, Szabadkai G, Tait SW, Tang HL, Tavernarakis N, Tsujimoto Y, Vanden Berghe T, Vandenabeele P, Villunger A, Wagner EF, Walczak H, White E, Wood WG, Yuan J, Zakeri Z, Zhivotovsky B, Melino G, Kroemer G (2015) Essential versus accessory aspects of cell death: recommendations of the NCCD 2015. Cell Death Differ 22:58–73

Ganz M, Csak T, Nath B, Szabo G (2011) Lipopolysaccharide induces and activates the Nalp3 inflammasome in the liver. World J Gastroenterol 17:4772–4778

Ge X, Antoine DJ, Lu Y, Arriazu E, Leung TM, Klepper AL, Branch AD, Fiel MI, Nieto N (2014) High mobility group box-1 (HMGB1) participates in the pathogenesis of alcoholic liver disease (ALD). J Biol Chem 289:22672–22691

Geng Y, Ma Q, Liu YN, Peng N, Yuan FF, Li XG, Li M, Wu YS, Li BL, Song WB, Zhu W, Xu WW, Fan J, Su L (2015) Heatstroke induces liver injury via IL-1beta and HMGB1-induced pyroptosis. J Hepatol 63:622–633

Gilthorpe JD, Oozeer F, Nash J, Calvo M, Bennett DL, Lumsden A, Pini A (2013) Extracellular histone H1 is neurotoxic and drives a pro-inflammatory response in microglia. F1000Res 2:148

Green DR, Ferguson T, Zitvogel L, Kroemer G (2009) Immunogenic and tolerogenic cell death. Nat Rev Immunol 9:353–363

Gross A, Yin XM, Wang K, Wei MC, Jockel J, Milliman C, Erdjument-Bromage H, Tempst P, Korsmeyer SJ (1999) Caspase cleaved BID targets mitochondria and is required for cytochrome c release, while BCL-XL prevents this release but not tumor necrosis factor-R1/Fas death. J Biol Chem 274:1156–1163

Guicciardi ME, Gores GJ (2005) Apoptosis: a mechanism of acute and chronic liver injury. Gut 54:1024–1033

Ha SD, Han CY, Reid C, Kim SO (2014) HDAC8-mediated epigenetic reprogramming plays a key role in resistance to anthrax lethal toxin-induced pyroptosis in macrophages. J Immunol 193:1333–1343

Hagar JA, Powell DA, Aachoui Y, Ernst RK, Miao EA (2013) Cytoplasmic LPS activates caspase-11: implications in TLR4-independent endotoxic shock. Science 341:1250–1253

Hahn S, Giaglis S, Chowdhury CS, Hosli I, Hasler P (2013) Modulation of neutrophil NETosis: interplay between infectious agents and underlying host physiology. Semin Immunopathol 35:439–453

Hajji N, Joseph B (2010) Epigenetic regulation of cell life and death decisions and deregulation in cancer. Essays Biochem 48:121–146

He S, Wang L, Miao L, Wang T, Du F, Zhao L, Wang X (2009) Receptor interacting protein kinase-3 determines cellular necrotic response to TNF-alpha. Cell 137:1100–1111

He WT, Wan H, Hu L, Chen P, Wang X, Huang Z, Yang ZH, Zhong CQ, Han J (2015) Gasdermin D is an executor of pyroptosis and required for interleukin-1beta secretion. Cell Res 25:1285–1298

Hendrikx T, Bieghs V, Walenbergh SM, van Gorp PJ, Verheyen F, Jeurissen ML, Steinbusch MM, Vaes N, Binder CJ, Koek GH, Stienstra R, Netea MG, Hofker MH, Shiri-Sverdlov R (2013) Macrophage specific caspase-1/11 deficiency protects against cholesterol crystallization and hepatic inflammation in hyperlipidemic mice. PLoS One 8:e78792

Hernandez-Gea V, Ghiassi-Nejad Z, Rozenfeld R, Gordon R, Fiel MI, Yue Z, Czaja MJ, Friedman SL (2012) Autophagy releases lipid that promotes fibrogenesis by activated hepatic stellate cells in mice and in human tissues. Gastroenterology 142:938–946

Heymann F, Tacke F (2016) Immunology in the liver – from homeostasis to disease. Nat Rev Gastroenterol Hepatol 13:88–110

Hikita H, Kodama T, Shimizu S, Li W, Shigekawa M, Tanaka S, Hosui A, Miyagi T, Tatsumi T, Kanto T, Hiramatsu N, Morii E, Hayashi N, Takehara T (2012) Bak deficiency inhibits liver carcinogenesis: a causal link between apoptosis and carcinogenesis. J Hepatol 57:92–100

Hikita H, Kodama T, Tanaka S, Saito Y, Nozaki Y, Nakabori T, Shimizu S, Hayashi Y, Li W, Shigekawa M, Sakamori R, Miyagi T, Hiramatsu N, Tatsumi T, Takehara T (2015) Activation of the mitochondrial apoptotic pathway produces reactive oxygen species and oxidative damage in hepatocytes that contribute to liver tumorigenesis. Cancer Prev Res (Phila) 8:693–701

Holdenrieder S, Stieber P, Bodenmuller H, Busch M, Fertig G, Furst H, Schalhorn A, Schmeller N, Untch M, Seidel D (2001) Nucleosomes in serum of patients with benign and malignant diseases. Int J Cancer J Int du cancer 95:114–120

Hoque R, Farooq A, Ghani A, Gorelick F, Mehal WZ (2014) Lactate reduces liver and pancreatic injury in toll-like receptor- and inflammasome-mediated inflammation via GPR81-mediated suppression of innate immunity. Gastroenterology 146:1763–1774

Huang H, Evankovich J, Yan W, Nace G, Zhang L, Ross M, Liao X, Billiar T, Xu J, Esmon CT, Tsung A (2011) Endogenous histones function as alarmins in sterile inflammatory liver injury through toll-like receptor 9 in mice. Hepatology 54:999–1008

Huang SK, Scruggs AM, Donaghy J, Horowitz JC, Zaslona Z, Przybranowski S, White ES, Peters-Golden M (2013) Histone modifications are responsible for decreased Fas expression and apoptosis resistance in fibrotic lung fibroblasts. Cell Death Dis 4:e621

Huang H, Nace GW, McDonald KA, Tai S, Klune JR, Rosborough BR, Ding Q, Loughran P, Zhu X, Beer-Stolz D, Chang EB, Billiar T, Tsung A (2014) Hepatocyte specific HMGB1 deletion worsens the injury in liver ischemia/reperfusion: a role for intracellular HMGB1 in cellular protection. Hepatology 59:1984–1997

Huang H, Tohme S, Al-Khafaji AB, Tai S, Loughran P, Chen L, Wang S, Kim J, Billiar T, Wang Y, Tsung A (2015) Damage-associated molecular pattern-activated neutrophil extracellular trap exacerbates sterile inflammatory liver injury. Hepatology 62:600–614

Huebener P, Gwak GY, Pradere JP, Quinzii CM, Friedman R, Lin CS, Trent CM, Mederacke I, Zhao E, Dapito DH, Lin Y, Goldberg IJ, Czaja MJ, Schwabe RF (2014) High-mobility group box 1 is dispensable for autophagy, mitochondrial quality control, and organ function in vivo. Cell Metab 19:539–547

Huebener P, Pradere JP, Hernandez C, Gwak GY, Caviglia JM, Mu X, Loike JD, Jenkins RE, Antoine DJ, Schwabe RF (2015) The HMGB1/RAGE axis triggers neutrophil-mediated injury amplification following necrosis. J Clin Invest 125:539–550

Ichimura Y, Waguri S, Sou YS, Kageyama S, Hasegawa J, Ishimura R, Saito T, Yang Y, Kouno T, Fukutomi T, Hoshii T, Hirao A, Takagi K, Mizushima T, Motohashi H, Lee MS, Yoshimori T, Tanaka K, Yamamoto M, Komatsu M (2013) Phosphorylation of p62 activates the Keap1-Nrf2 pathway during selective autophagy. Mol Cell 51:618–631

Ilyas G, Zhao E, Liu K, Lin Y, Tesfa L, Tanaka KE, Czaja MJ (2016) Macrophage autophagy limits acute toxic liver injury in mice through down regulation of interleukin-1beta. J Hepatol 64:118–127

Imamura M, Tsutsui H, Yasuda K, Uchiyama R, Yumikura-Futatsugi S, Mitani K, Hayashi S, Akira S, Taniguchi S, Van Rooijen N, Tschopp J, Yamamoto T, Fujimoto J, Nakanishi K (2009) Contribution of TIR domain-containing adapter inducing IFN-beta-mediated IL-18 release to LPS-induced liver injury in mice. J Hepatol 51:333–341

Inami Y, Waguri S, Sakamoto A, Kouno T, Nakada K, Hino O, Watanabe S, Ando J, Iwadate M, Yamamoto M, Lee MS, Tanaka K, Komatsu M (2011) Persistent activation of Nrf2 through p62 in hepatocellular carcinoma cells. J Cell Biol 193:275–284

Jiang W, Wang Z, Li X, Fan X, Duan Y (2012) High-mobility group box 1 is associated with clinicopathologic features in patients with hepatocellular carcinoma. Pathol Oncol Res POR 18:293–298

Jiang L, Kon N, Li T, Wang SJ, Su T, Hibshoosh H, Baer R, Gu W (2015) Ferroptosis as a p53-mediated activity during tumour suppression. Nature 520:57–62

Jung JH, Park JH, Jee MH, Keum SJ, Cho MS, Yoon SK, Jang SK (2011) Hepatitis C virus infection is blocked by HMGB1 released from virus-infected cells. J Virol 85:9359–9368

Kaiser WJ, Upton JW, Long AB, Livingston-Rosanoff D, Daley-Bauer LP, Hakem R, Caspary T, Mocarski ES (2011) RIP3 mediates the embryonic lethality of caspase-8-deficient mice. Nature 471:368–372

Kamo N, Ke B, Ghaffari AA, Shen XD, Busuttil RW, Cheng G, Kupiec-Weglinski JW (2013) ASC/caspase-1/IL-1beta signaling triggers inflammatory responses by promoting HMGB1 induction in liver ischemia/reperfusion injury. Hepatology 58:351–362

Kang R, Zeh HJ, Lotze MT, Tang D (2011) The Beclin 1 network regulates autophagy and apoptosis. Cell Death Differ 18:571–580

Kang R, Zhang Q, Hou W, Yan Z, Chen R, Bonaroti J, Bansal P, Billiar TR, Tsung A, Wang Q, Bartlett DL, Whitcomb DC, Chang EB, Zhu X, Wang H, Lu B, Tracey KJ, Cao L, Fan XG, Lotze MT, Zeh HJ 3rd, Tang D (2013) Intracellular Hmgb1 inhibits inflammatory nucleosome release and limits acute pancreatitis in mice. Gastroenterology 146:1097–1107

Kang R, Chen R, Zhang Q, Hou W, Wu S, Cao L, Huang J, Yu Y, Fan XG, Yan Z, Sun X, Wang H, Wang Q, Tsung A, Billiar TR, Zeh HJ 3rd, Lotze MT, Tang D (2014a) HMGB1 in health and disease. Mol Asp Med 40:1–116

Kang R, Zhang Q, Hou W, Yan Z, Chen R, Bonaroti J, Bansal P, Billiar TR, Tsung A, Wang Q, Bartlett DL, Whitcomb DC, Chang EB, Zhu X, Wang H, Lu B, Tracey KJ, Cao L, Fan XG, Lotze MT, Zeh HJ 3rd, Tang D (2014b) Intracellular Hmgb1 inhibits inflammatory nucleosome release and limits acute pancreatitis in mice. Gastroenterology 146:1097–1107

Kang R, Chen R, Xie M, Cao L, Lotze MT, Tang D, Zeh HJ 3rd (2016) The receptor for advanced glycation end products activates the AIM2 Inflammasome in acute pancreatitis. J Immunol 196:4331–4337

Kayagaki N, Wong MT, Stowe IB, Ramani SR, Gonzalez LC, Akashi-Takamura S, Miyake K, Zhang J, Lee WP, Muszynski A, Forsberg LS, Carlson RW, Dixit VM (2013) Noncanonical inflammasome activation by intracellular LPS independent of TLR4. Science 341:1246–1249

Kayagaki N, Stowe IB, Lee BL, O'Rourke K, Anderson K, Warming S, Cuellar T, Haley B, Roose-Girma M, Phung QT, Liu PS, Lill JR, Li H, Wu J, Kummerfeld S, Zhang J, Lee WP, Snipas SJ, Salvesen GS, Morris LX, Fitzgerald L, Zhang Y, Bertram EM, Goodnow CC, Dixit VM (2015) Caspase-11 cleaves gasdermin D for non-canonical inflammasome signalling. Nature 526:666–671

Kerr JF, Wyllie AH, Currie AR (1972) Apoptosis: a basic biological phenomenon with wide-ranging implications in tissue kinetics. Br J Cancer 26:239–257

Kim KH, Jeong YT, Oh H, Kim SH, Cho JM, Kim YN, Kim SS, Kim do H, Hur KY, Kim HK, Ko T, Han J, Kim HL, Kim J, Back SH, Komatsu M, Chen H, Chan DC, Konishi M, Itoh N, Choi CS, Lee MS (2013) Autophagy deficiency leads to protection from obesity and insulin resistance by inducing Fgf21 as a mitokine. Nat Med 19:83–92

Klionsky DJ, Emr SD (2000) Autophagy as a regulated pathway of cellular degradation. Science 290:1717–1721

Kokkola R, Andersson A, Mullins G, Ostberg T, Treutiger CJ, Arnold B, Nawroth P, Andersson U, Harris RA, Harris HE (2005) RAGE is the major receptor for the proinflammatory activity of HMGB1 in rodent macrophages. Scand J Immunol 61:1–9

Kolb WP, Granger GA (1968) Lymphocyte in vitro cytotoxicity: characterization of human lymphotoxin. Proc Natl Acad Sci U S A 61:1250–1255

Komatsu M, Kurokawa H, Waguri S, Taguchi K, Kobayashi A, Ichimura Y, Sou YS, Ueno I, Sakamoto A, Tong KI, Kim M, Nishito Y, Iemura S, Natsume T, Ueno T, Kominami E, Motohashi H, Tanaka K, Yamamoto M (2010) The selective autophagy substrate p62 activates the stress responsive transcription factor Nrf2 through inactivation of Keap1. Nat Cell Biol 12:213–223

Kovalenko A, Kim JC, Kang TB, Rajput A, Bogdanov K, Dittrich-Breiholz O, Kracht M, Brenner O, Wallach D (2009) Caspase-8 deficiency in epidermal keratinocytes triggers an inflammatory skin disease. J Exp Med 206:2161–2177

Kroemer G, Marino G, Levine B (2010) Autophagy and the integrated stress response. Mol Cell 40:280–293

Labi V, Erlacher M, Krumschnabel G, Manzl C, Tzankov A, Pinon J, Egle A, Villunger A (2010) Apoptosis of leukocytes triggered by acute DNA damage promotes lymphoma formation. Genes Dev 24:1602–1607

Lau A, Wang S, Jiang J, Haig A, Pavlosky A, Linkermann A, Zhang ZX, Jevnikar AM (2013) RIPK3-mediated necroptosis promotes donor kidney inflammatory injury and reduces allograft survival. Am J Transplant 13:2805–2818

Lawson JA, Fisher MA, Simmons CA, Farhood A, Jaeschke H (1998) Parenchymal cell apoptosis as a signal for sinusoidal sequestration and transendothelial migration of neutrophils in murine models of endotoxin and Fas-antibody-induced liver injury. Hepatology 28:761–767

Li H, Zhu H, Xu CJ, Yuan J (1998) Cleavage of BID by caspase 8 mediates the mitochondrial damage in the Fas pathway of apoptosis. Cell 94:491–501

Li J, Yang B, Zhou Q, Wu Y, Shang D, Guo Y, Song Z, Zheng Q, Xiong J (2013) Autophagy promotes hepatocellular carcinoma cell invasion through activation of epithelial-mesenchymal transition. Carcinogenesis 34:1343–1351

Li JX, Feng JM, Wang Y, Li XH, Chen XX, Su Y, Shen YY, Chen Y, Xiong B, Yang CH, Ding J, Miao ZH (2014a) The B-Raf(V600E) inhibitor dabrafenib selectively inhibits RIP3 and alleviates acetaminophen-induced liver injury. Cell Death Dis 5:e1278

Li X, Wang S, Chen Y, Liu G, Yang X (2014b) miR-22 targets the 3′ UTR of HMGB1 and inhibits the HMGB1-associated autophagy in osteosarcoma cells during chemotherapy. Tumour Biol 35(6):6021–6028

Liedtke C, Bangen JM, Freimuth J, Beraza N, Lambertz D, Cubero FJ, Hatting M, Karlmark KR, Streetz KL, Krombach GA, Tacke F, Gassler N, Riethmacher D, Trautwein C (2011) Loss of caspase-8 protects mice against inflammation-related hepatocarcinogenesis but induces non-apoptotic liver injury. Gastroenterology 141:2176–2187

Lin CW, Zhang H, Li M, Xiong X, Chen X, Dong XC, Yin XM (2013) Pharmacological promotion of autophagy alleviates steatosis and injury in alcoholic and non-alcoholic fatty liver conditions in mice. J Hepatol 58:993–999

Linkermann A, Green DR (2014) Necroptosis. N Engl J Med 370:455–465

Liu Y, Levine B (2015) Autosis and autophagic cell death: the dark side of autophagy. Cell Death Differ 22:367–376

Liu X, Kim CN, Yang J, Jemmerson R, Wang X (1996) Induction of apoptotic program in cell-free extracts: requirement for dATP and cytochrome c. Cell 86:147–157

Liu Y, Shoji-Kawata S, Sumpter RM Jr, Wei Y, Ginet V, Zhang L, Posner B, Tran KA, Green DR, Xavier RJ, Shaw SY, Clarke PG, Puyal J, Levine B (2013) Autosis is a Na+,K+-ATPase-regulated form of cell death triggered by autophagy-inducing peptides, starvation, and hypoxia-ischemia. Proc Natl Acad Sci U S A 110:20364–20371

Liu K, Huang J, Xie M, Yu Y, Zhu S, Kang R, Cao L, Tang D, Duan X (2014a) Regulates autophagy and apoptosis by targeting in the retinoblastoma cell. Autophagy 10:442–452

Liu L, Yang M, Kang R, Dai Y, Yu Y, Gao F, Wang H, Sun X, Li X, Li J, Cao L, Tang D (2014b) HMGB1-DNA complex-induced autophagy limits AIM2 inflammasome activation through RAGE. Biochem Biophys Res Commun 450:851–856

Liu K, Zhao E, Ilyas G, Lalazar G, Lin Y, Haseeb M, Tanaka KE, Czaja MJ (2015) Impaired macrophage autophagy increases the immune response in obese mice by promoting proinflammatory macrophage polarization. Autophagy 11:271–284

Livesey K, Kang R, Vernon P, Buchser W, Loughran P, Watkins SC, Zhang L, Manfredi JJ, Zeh HJ, Li L, Lotze M, Tang D (2012) p53/HMGB1 complexes regulate autophagy and apoptosis. Cancer Res 72:1996–2005

van Loo G, van Gurp M, Depuydt B, Srinivasula SM, Rodriguez I, Alnemri ES, Gevaert K, Vandekerckhove J, Declercq W, Vandenabeele P (2002) The serine protease Omi/HtrA2 is released from mitochondria during apoptosis. Omi interacts with caspase-inhibitor XIAP and induces enhanced caspase activity. Cell Death Differ 9:20–26

Lorincz T, Jemnitz K, Kardon T, Mandl J, Szarka A (2015) Ferroptosis is involved in acetaminophen induced cell death. Pathol Oncol Res 21:1115–1121

Lu B, Nakamura T, Inouye K, Li J, Tang Y, Lundback P, Valdes-Ferrer SI, Olofsson PS, Kalb T, Roth J, Zou Y, Erlandsson-Harris H, Yang H, Ting JP, Wang H, Andersson U, Antoine DJ, Chavan SS, Hotamisligil GS, Tracey KJ (2012) Novel role of PKR in inflammasome activation and HMGB1 release. Nature 488:670–674

Luedde T, Kaplowitz N, Schwabe RF (2014) Cell death and cell death responses in liver disease: mechanisms and clinical relevance. Gastroenterology 147(765–783):e764

Luger K, Mader AW, Richmond RK, Sargent DF, Richmond TJ (1997) Crystal structure of the nucleosome core particle at 2.8 a resolution. Nature 389:251–260

Madrigal-Matute J, Cuervo AM (2016) Regulation of liver metabolism by autophagy. Gastroenterology 150:328–339

Maes H, Kuchnio A, Peric A, Moens S, Nys K, De Bock K, Quaegebeur A, Schoors S, Georgiadou M, Wouters J, Vinckier S, Vankelecom H, Garmyn M, Vion AC, Radtke F, Boulanger C, Gerhardt H, Dejana E, Dewerchin M, Ghesquiere B, Annaert W, Agostinis P, Carmeliet P (2014) Tumor vessel normalization by chloroquine independent of autophagy. Cancer Cell 26:190–206

Malhi H, Gores GJ (2008) Cellular and molecular mechanisms of liver injury. Gastroenterology 134:1641–1654

Man SM, Kanneganti TD (2016) Converging roles of caspases in inflammasome activation, cell death and innate immunity. Nat Rev Immunol 16:7–21

Marino G, Niso-Santano M, Baehrecke EH, Kroemer G (2014) Self-consumption: the interplay of autophagy and apoptosis. Nat Rev Mol Cell Biol 15:81–94

Martinez-Lopez N, Singh R (2015) Autophagy and lipid droplets in the liver. Annu Rev Nutr 35:215–237

Martinon F, Mayor A, Tschopp J (2009) The inflammasomes: guardians of the body. Annu Rev Immunol 27:229–265

McIlwain DR, Berger T, Mak TW (2015) Caspase functions in cell death and disease. Cold Spring Harb Perspect Biol 7:a026716

Metzler KD, Goosmann C, Lubojemska A, Zychlinsky A, Papayannopoulos V (2014) A myeloperoxidase-containing complex regulates neutrophil elastase release and actin dynamics during NETosis. Cell Rep 8:883–896

Michalak EM, Vandenberg CJ, Delbridge AR, Wu L, Scott CL, Adams JM, Strasser A (2010) Apoptosis-promoted tumorigenesis: gamma-irradiation-induced thymic lymphomagenesis requires Puma-driven leukocyte death. Genes Dev 24:1608–1613

Mizushima N, Levine B (2010) Autophagy in mammalian development and differentiation. Nat Cell Biol 12:823–830

Mizushima N, Yoshimori T, Levine B (2010) Methods in mammalian autophagy research. Cell 140:313–326

Murphy JM, Czabotar PE, Hildebrand JM, Lucet IS, Zhang JG, Alvarez-Diaz S, Lewis R, Lalaoui N, Metcalf D, Webb AI, Young SN, Varghese LN, Tannahill GM, Hatchell EC, Majewski IJ, Okamoto T, Dobson RC, Hilton DJ, Babon JJ, Nicola NA, Strasser A, Silke J, Alexander WS (2013) The pseudokinase MLKL mediates necroptosis via a molecular switch mechanism. Immunity 39:443–453

Newton K, Dugger DL, Wickliffe KE, Kapoor N, de Almagro MC, Vucic D, Komuves L, Ferrando RE, French DM, Webster J, Roose-Girma M, Warming S, Dixit VM (2014) Activity of protein kinase RIPK3 determines whether cells die by necroptosis or apoptosis. Science 343:1357–1360

Ni HM, Bockus A, Boggess N, Jaeschke H, Ding WX (2012) Activation of autophagy protects against acetaminophen-induced hepatotoxicity. Hepatology 55:222–232

Ni HM, Woolbright BL, Williams J, Copple B, Cui W, Luyendyk JP, Jaeschke H, Ding WX (2014) Nrf2 promotes the development of fibrosis and tumorigenesis in mice with defective hepatic autophagy. J Hepatol 61:617–625

Ni HM, McGill MR, Chao X, Du K, Williams JA, Xie Y, Jaeschke H, Ding WX (2016) Removal of acetaminophen-protein adducts by autophagy protects against acetaminophen-induced liver injury in mice. J Hepatol 65:354–362

Oberhammer FA, Pavelka M, Sharma S, Tiefenbacher R, Purchio AF, Bursch W, Schulte-Hermann R (1992) Induction of apoptosis in cultured hepatocytes and in regressing liver by transforming growth factor beta 1. Proc Natl Acad Sci U S A 89:5408–5412

Park JS, Svetkauskaite D, He Q, Kim JY, Strassheim D, Ishizaka A, Abraham E (2004) Involvement of toll-like receptors 2 and 4 in cellular activation by high mobility group box 1 protein. J Biol Chem 279:7370–7377

Pasparakis M, Vandenabeele P (2015) Necroptosis and its role in inflammation. Nature 517:311–320

Petrasek J, Bala S, Csak T, Lippai D, Kodys K, Menashy V, Barrieau M, Min SY, Kurt-Jones EA, Szabo G (2012) IL-1 receptor antagonist ameliorates inflammasome-dependent alcoholic steatohepatitis in mice. J Clin Invest 122:3476–3489

Qiao J, Huang Y, Xia Y, Chu P, Yao H, Xu L, Qi K, Liu Y, Xu K, Zeng L (2015) Busulfan and cyclosphamide induce liver inflammation through NLRP3 activation in mice after hematopoietic stem cell transplantation. Sci Rep 5:17828

Radaeva S, Sun R, Jaruga B, Nguyen VT, Tian Z, Gao B (2006) Natural killer cells ameliorate liver fibrosis by killing activated stellate cells in NKG2D-dependent and tumor necrosis factor-related apoptosis-inducing ligand-dependent manners. Gastroenterology 130:435–452

Ramachandran A, McGill MR, Xie Y, Ni HM, Ding WX, Jaeschke H (2013) Receptor interacting protein kinase 3 is a critical early mediator of acetaminophen-induced hepatocyte necrosis in mice. Hepatology 58:2099–2108

Ranger AM, Zha J, Harada H, Datta SR, Danial NN, Gilmore AP, Kutok JL, Le Beau MM, Greenberg ME, Korsmeyer SJ (2003) Bad-deficient mice develop diffuse large B cell lymphoma. Proc Natl Acad Sci U S A 100:9324–9329

Rathinam VA, Fitzgerald KA (2016) Inflammasome complexes: emerging mechanisms and effector functions. Cell 165:792–800

Remijsen Q, Kuijpers TW, Wirawan E, Lippens S, Vandenabeele P, Vanden Berghe T (2011) Dying for a cause: NETosis, mechanisms behind an antimicrobial cell death modality. Cell Death Differ 18:581–588

Riley BE, Kaiser SE, Shaler TA, Ng AC, Hara T, Hipp MS, Lage K, Xavier RJ, Ryu KY, Taguchi K, Yamamoto M, Tanaka K, Mizushima N, Komatsu M, Kopito RR (2010) Ubiquitin accumulation in autophagy-deficient mice is dependent on the Nrf2-mediated stress response pathway: a potential role for protein aggregation in autophagic substrate selection. J Cell Biol 191:537–552

Roychowdhury S, McMullen MR, Pisano SG, Liu X, Nagy LE (2013) Absence of receptor interacting protein kinase 3 prevents ethanol-induced liver injury. Hepatology 57:1773–1783

Sanders YY, Liu H, Zhang X, Hecker L, Bernard K, Desai L, Liu G, Thannickal VJ (2013) Histone modifications in senescence-associated resistance to apoptosis by oxidative stress. Redox Biol 1:8–16

Schneider JL, Cuervo AM (2014) Liver autophagy: much more than just taking out the trash. Nat Rev Gastroenterol Hepatol 11:187–200

Schott C, Graab U, Cuvelier N, Hahn H, Fulda S (2015) Oncogenic RAS mutants confer resistance of RMS13 rhabdomyosarcoma cells to oxidative stress-induced ferroptotic cell death. Front Oncol 5:131

Schroder K, Tschopp J (2010) The inflammasomes. Cell 140:821–832

Seifert L, Werba G, Tiwari S, Giao Ly NN, Alothman S, Alqunaibit D, Avanzi A, Barilla R, Daley D, Greco SH, Torres-Hernandez A, Pergamo M, Ochi A, Zambirinis CP, Pansari M, Rendon M, Tippens D, Hundeyin M, Mani VR, Hajdu C, Engle D, Miller G (2016) The necrosome promotes pancreatic oncogenesis via CXCL1 and Mincle-induced immune suppression. Nature 532:245–249

Semeraro F, Ammollo CT, Morrissey JH, Dale GL, Friese P, Esmon NL, Esmon CT (2011) Extracellular histones promote thrombin generation through platelet-dependent mechanisms: involvement of platelet TLR2 and TLR4. Blood 118:1952–1961

Seok S, Fu T, Choi SE, Li Y, Zhu R, Kumar S, Sun X, Yoon G, Kang Y, Zhong W, Ma J, Kemper B, Kemper JK (2014) Transcriptional regulation of autophagy by an FXR-CREB axis. Nature 516:108–111

Shamas-Din A, Kale J, Leber B, Andrews DW (2013) Mechanisms of action of Bcl-2 family proteins. Cold Spring Harb Perspect Biol 5:a008714

Sharma A, Singh K, Almasan A (2012) Histone H2AX phosphorylation: a marker for DNA damage. Methods Mol Biol 920:613–626

Shi J, Zhao Y, Wang Y, Gao W, Ding J, Li P, Hu L, Shao F (2014) Inflammatory caspases are innate immune receptors for intracellular LPS. Nature 514:187–192

Shi J, Zhao Y, Wang K, Shi X, Wang Y, Huang H, Zhuang Y, Cai T, Wang F, Shao F (2015) Cleavage of GSDMD by inflammatory caspases determines pyroptotic cell death. Nature 526:660–665

Silke J, Rickard JA, Gerlic M (2015) The diverse role of RIP kinases in necroptosis and inflammation. Nat Immunol 16:689–697

Song JX, Lu JH, Liu LF, Chen LL, Durairajan SS, Yue Z, Zhang HQ, Li M (2014) HMGB1 is involved in autophagy inhibition caused by SNCA/alpha-synuclein overexpression: a process modulated by the natural autophagy inducer corynoxine B. Autophagy 10:144–154

Steinberg BE, Grinstein S (2007) Unconventional roles of the NADPH oxidase: signaling, ion homeostasis, and cell death. Sci STKE 2007:pe11

Strahl BD, Allis CD (2000) The language of covalent histone modifications. Nature 403:41–45

Sun X, Tang D (2014) HMGB1-dependent and -independent autophagy. Autophagy 10:1873–1876

Sun X, Tang D (2015) Hepatocyte-specific Hmgb1 deletion. Autophagy 11:1189–1191

Sun L, Wang H, Wang Z, He S, Chen S, Liao D, Wang L, Yan J, Liu W, Lei X, Wang X (2012) Mixed lineage kinase domain-like protein mediates necrosis signaling downstream of RIP3 kinase. Cell 148:213–227

Sun X, Ou Z, Xie M, Kang R, Fan Y, Niu X, Wang H, Cao L, Tang D (2015) HSPB1 as a novel regulator of ferroptotic cancer cell death. Oncogene 34:5617–5625

Sun X, Niu X, Chen R, He W, Chen KR, Tang D (2016a) Metallothionein-1G facilitates sorafenib resistance through inhibition of ferroptosis. Hepatology 64:488–500

Sun X, Ou Z, Chen R, Niu X, Chen D, Kang R, Tang D (2016b) Activation of the p62-Keap1-NRF2 pathway protects against ferroptosis in hepatocellular carcinoma cells. Hepatology 63:173–184

Szabo G, Petrasek J (2015) Inflammasome activation and function in liver disease. Nat Rev Gastroenterol Hepatol 12:387–400

Taguchi K, Fujikawa N, Komatsu M, Ishii T, Unno M, Akaike T, Motohashi H, Yamamoto M (2012) Keap1 degradation by autophagy for the maintenance of redox homeostasis. Proc Natl Acad Sci U S A 109:13561–13566

Takamura A, Komatsu M, Hara T, Sakamoto A, Kishi C, Waguri S, Eishi Y, Hino O, Tanaka K, Mizushima N (2011) Autophagy-deficient mice develop multiple liver tumors. Genes Dev 25:795–800

Takeda K, Hayakawa Y, Smyth MJ, Kayagaki N, Yamaguchi N, Kakuta S, Iwakura Y, Yagita H, Okumura K (2001) Involvement of tumor necrosis factor-related apoptosis-inducing ligand in surveillance of tumor metastasis by liver natural killer cells. Nat Med 7:94–100

Takehara T, Tatsumi T, Suzuki T, Rucker EB 3rd, Hennighausen L, Jinushi M, Miyagi T, Kanazawa Y, Hayashi N (2004) Hepatocyte-specific disruption of Bcl-xL leads to continuous hepatocyte apoptosis and liver fibrotic responses. Gastroenterology 127:1189–1197

Takeuchi O, Akira S (2010) Pattern recognition receptors and inflammation. Cell 140:805–820

Tang D, Kang R, Cheh CW, Livesey KM, Liang X, Schapiro NE, Benschop R, Sparvero LJ, Amoscato AA, Tracey KJ, Zeh HJ, Lotze MT (2010a) HMGB1 release and redox regulates autophagy and apoptosis in cancer cells. Oncogene 29:5299–5310

Tang D, Kang R, Livesey KM, Cheh CW, Farkas A, Loughran P, Hoppe G, Bianchi ME, Tracey KJ, Zeh HJ 3rd, Lotze MT (2010b) Endogenous HMGB1 regulates autophagy. J Cell Biol 190:881–892

Tang D, Kang R, Livesey KM, Kroemer G, Billiar TR, Van Houten B, Zeh HJ 3rd, Lotze MT (2011a) High-mobility group box 1 is essential for mitochondrial quality control. Cell Metab 13:701–711

Tang D, Kang R, Zeh HJ, Lotze MT (2011b) High-mobility group box 1, oxidative stress, and disease. Antioxid Redox Signal 14:1315–1335

Tang D, Lotze MT, Kang R, Zeh HJ (2011c) Apoptosis promotes early tumorigenesis. Oncogene 30:1851–1854

Tang D, Billiar TA, Lotze MT (2012a) A Janus tale of two active HMGB1 redox states. Mol Med 18:1360–1362.

Tang D, Kang R, Coyne CB, Zeh HJ, Lotze MT (2012b) PAMPs and DAMPs: signal 0s that spur autophagy and immunity. Immunol Rev 249:158–175

Tang D, Kang R, Van Houten B, Zeh HJ, Billiar TR, Lotze MT (2014) High mobility group box 1 (HMGB1) phenotypic role revealed with stress. Mol Med 20:359–362

Tian J, Avalos AM, Mao SY, Chen B, Senthil K, Wu H, Parroche P, Drabic S, Golenbock D, Sirois C, Hua J, An LL, Audoly L, La Rosa G, Bierhaus A, Naworth P, Marshak-Rothstein A, Crow MK, Fitzgerald KA, Latz E, Kiener PA, Coyle AJ (2007) Toll-like receptor 9-dependent activation by DNA-containing immune complexes is mediated by HMGB1 and RAGE. Nat Immunol 8:487–496

Tohme S, Yazdani HO, Al-Khafaji AB, Chidi AP, Loughran P, Mowen K, Wang Y, Simmons RL, Huang H, Tsung A (2016) Neutrophil extracellular traps promote the development and progression of liver metastases after surgical stress. Cancer Res 76:1367–1380

Toshima T, Shirabe K, Fukuhara T, Ikegami T, Yoshizumi T, Soejima Y, Ikeda T, Okano S, Maehara Y (2014) Suppression of autophagy during liver regeneration impairs energy charge and hepatocyte senescence in mice. Hepatology 60:290–300

Trejo-Becerril C, Perez-Cardenas E, Trevino-Cuevas H, Taja-Chayeb L, Garcia-Lopez P, Segura-Pacheco B, Chavez-Blanco A, Lizano-Soberon M, Gonzalez-Fierro A, Mariscal I, Wegman-Ostrosky T, Duenas-Gonzalez A (2003) Circulating nucleosomes and response to chemotherapy: an in vitro, in vivo and clinical study on cervical cancer patients. Int J Cancer J Int du Cancer 104:663–668

Tsung A, Sahai R, Tanaka H, Nakao A, Fink MP, Lotze MT, Yang H, Li J, Tracey KJ, Geller DA, Billiar TR (2005) The nuclear factor HMGB1 mediates hepatic injury after murine liver ischemia-reperfusion. J Exp Med 201:1135–1143

Vanaja SK, Russo AJ, Behl B, Banerjee I, Yankova M, Deshmukh SD, Rathinam VA (2016) Bacterial outer membrane vesicles mediate cytosolic localization of LPS and caspase-11 activation. Cell 165:1106–1119

Vanden Berghe T, Linkermann A, Jouan-Lanhouet S, Walczak H, Vandenabeele P (2014) Regulated necrosis: the expanding network of non-apoptotic cell death pathways. Nat Rev Mol Cell Biol 15:135–147

Venereau E, Casalgrandi M, Schiraldi M, Antoine DJ, Cattaneo A, De Marchis F, Liu J, Antonelli A, Preti A, Raeli L, Shams SS, Yang H, Varani L, Andersson U, Tracey KJ, Bachi A, Uguccioni M, Bianchi ME (2012) Mutually exclusive redox forms of HMGB1 promote cell recruitment or proinflammatory cytokine release. J Exp Med 209:1519–1528

Wallach D, Kang TB, Dillon CP, Green DR (2016) Programmed necrosis in inflammation: toward identification of the effector molecules. Science 352:aaf2154

Wang K (2014) Molecular mechanisms of hepatic apoptosis. Cell Death Dis 5:e996

Wang C, Youle RJ (2009) The role of mitochondria in apoptosis*. Annu Rev Genet 43:95–118

Wang S, Ni HM, Dorko K, Kumer SC, Schmitt TM, Nawabi A, Komatsu M, Huang H, Ding WX (2016) Increased hepatic receptor interacting protein kinase 3 expression due to impaired proteasomal functions contributes to alcohol-induced steatosis and liver injury. Oncotarget 7:17681–17698

Watanabe T, Kubota S, Nagaya M, Ozaki S, Nagafuchi H, Akashi K, Taira Y, Tsukikawa S, Oowada S, Nakano S (2005) The role of HMGB-1 on the development of necrosis during hepatic ischemia and hepatic ischemia/reperfusion injury in mice. J Surg Res 124:59–66

Weber A, Boger R, Vick B, Urbanik T, Haybaeck J, Zoller S, Teufel A, Krammer PH, Opferman JT, Galle PR, Schuchmann M, Heikenwalder M, Schulze-Bergkamen H (2010) Hepatocyte-specific deletion of the antiapoptotic protein myeloid cell leukemia-1 triggers proliferation and hepatocarcinogenesis in mice. Hepatology 51:1226–1236

Weinlich R, Oberst A, Dillon CP, Janke LJ, Milasta S, Lukens JR, Rodriguez DA, Gurung P, Savage C, Kanneganti TD, Green DR (2013) Protective roles for caspase-8 and cFLIP in adult homeostasis. Cell Rep 5:340–348

Welz PS, Wullaert A, Vlantis K, Kondylis V, Fernandez-Majada V, Ermolaeva M, Kirsch P, Sterner-Kock A, van Loo G, Pasparakis M (2011) FADD prevents RIP3-mediated epithelial cell necrosis and chronic intestinal inflammation. Nature 477:330–334

Wen Z, Liu Y, Li F, Ren F, Chen D, Li X, Wen T (2013) Circulating histones exacerbate inflammation in mice with acute liver failure. J Cell Biochem 114:2384–2391

White E (2015) The role for autophagy in cancer. J Clin Invest 125:42–46

Wong PM, Puente C, Ganley IG, Jiang X (2013) The ULK1 complex: sensing nutrient signals for autophagy activation. Autophagy 9:124–137

Wree A, Eguchi A, McGeough MD, Pena CA, Johnson CD, Canbay A, Hoffman HM, Feldstein AE (2014) NLRP3 inflammasome activation results in hepatocyte pyroptosis, liver inflammation, and fibrosis in mice. Hepatology 59:898–910

Wree A, Johnson CD, Font-Burgada J, Eguchi A, Povero D, Karin M, Feldstein AE (2015) Hepatocyte-specific Bid depletion reduces tumor development by suppressing inflammation-related compensatory proliferation. Cell Death Differ 22:1985–1994

Wu D, Ingram A, Lahti JH, Mazza B, Grenet J, Kapoor A, Liu L, Kidd VJ, Tang D (2002) Apoptotic release of histones from nucleosomes. J Biol Chem 277:12001–12008

Xiao Y, Liu H, Yu J, Zhao Z, Xiao F, Xia T, Wang C, Li K, Deng J, Guo Y, Chen S, Chen Y, Guo F (2016) MAPK1/3 regulate hepatic lipid metabolism via ATG7-dependent autophagy. Autophagy 12:592–593

Xie Y, Kang R, Sun X, Zhong M, Huang J, Klionsky DJ, Tang D (2015) Posttranslational modification of autophagy-related proteins in macroautophagy. Autophagy 11:28–45

Xie Y, Hou W, Song X, Yu Y, Huang J, Sun X, Kang R, Tang D (2016) Ferroptosis: process and function. Cell Death Differ 23:369–379

Xu WS, Parmigiani RB, Marks PA (2007) Histone deacetylase inhibitors: molecular mechanisms of action. Oncogene 26:5541–5552

Xu J, Zhang X, Pelayo R, Monestier M, Ammollo CT, Semeraro F, Taylor FB, Esmon NL, Lupu F, Esmon CT (2009) Extracellular histones are major mediators of death in sepsis. Nat Med 15:1318–1321

Xu J, Zhang X, Monestier M, Esmon NL, Esmon CT (2011) Extracellular histones are mediators of death through TLR2 and TLR4 in mouse fatal liver injury. J Immunol 187:2626–2631

Yagoda N, von Rechenberg M, Zaganjor E, Bauer AJ, Yang WS, Fridman DJ, Wolpaw AJ, Smukste I, Peltier JM, Boniface JJ, Smith R, Lessnick SL, Sahasrabudhe S, Stockwell BR (2007) RAS-RAF-MEK-dependent oxidative cell death involving voltage-dependent anion channels. Nature 447:864–868

Yan W, Chang Y, Liang X, Cardinal JS, Huang H, Thorne SH, Monga SP, Geller DA, Lotze MT, Tsung A (2012) High-mobility group box 1 activates caspase-1 and promotes hepatocellular carcinoma invasiveness and metastases. Hepatology 55:1863–1875

Yang WS, Stockwell BR (2015) Ferroptosis: death by lipid peroxidation. Trends Biol 26:165–176

Yang H, Hreggvidsdottir HS, Palmblad K, Wang H, Ochani M, Li J, Lu B, Chavan S, Rosas-Ballina M, Al-Abed Y, Akira S, Bierhaus A, Erlandsson-Harris H, Andersson U, Tracey KJ (2010a) A critical cysteine is required for HMGB1 binding to toll-like receptor 4 and activation of macrophage cytokine release. Proc Natl Acad Sci U S A 107:11942–11947

Yang L, Li P, Fu S, Calay ES, Hotamisligil GS (2010b) Defective hepatic autophagy in obesity promotes ER stress and causes insulin resistance. Cell Metab 11:467–478

Yang WS, SriRamaratnam R, Welsch ME, Shimada K, Skouta R, Viswanathan VS, Cheah JH, Clemons PA, Shamji AF, Clish CB, Brown LM, Girotti AW, Cornish VW, Schreiber SL, Stockwell BR (2014) Regulation of ferroptotic cancer cell death by GPX4. Cell 156:317–331

Yang M, Liu L, Xie M, Sun X, Yu Y, Kang R, Yang L, Zhu S, Cao L, Tang D (2015) Poly-ADP-ribosylation of HMGB1 regulates TNFSF10/TRAIL resistance through autophagy. Autophagy 11(2):214–224

Yang G, Lee HE, Lee JY (2016) A pharmacological inhibitor of NLRP3 inflammasome prevents non-alcoholic fatty liver disease in a mouse model induced by high fat diet. Sci Rep 6:24399

Yin XM, Ding WX, Gao W (2008) Autophagy in the liver. Hepatology 47:1773–1785

Yipp BG, Petri B, Salina D, Jenne CN, Scott BN, Zbytnuik LD, Pittman K, Asaduzzaman M, Wu K, Meijndert HC, Malawista SE, de Boisfleury Chevance A, Zhang K, Conly J, Kubes P (2012) Infection-induced NETosis is a dynamic process involving neutrophil multitasking in vivo. Nat Med 18:1386–1393

Yu Y, Tang D, Kang R (2015a) Oxidative stress-mediated HMGB1 biology. Front Physiol 6:93

Yu Y, Xie Y, Cao L, Yang L, Yang M, Lotze MT, Zeh HJ, Kang R, Tang D (2015b) The ferroptosis inducer erastin enhances sensitivity of acute myeloid leukemia cells to chemotherapeutic agents. Mol Cell Oncol 2:e1054549

Zeng S, Dun H, Ippagunta N, Rosario R, Zhang QY, Lefkowitch J, Yan SF, Schmidt AM, Emond JC (2009) Receptor for advanced glycation end product (RAGE)-dependent modulation of early growth response-1 in hepatic ischemia/reperfusion injury. J Hepatol 50:929–936

Zhang Y, Cheng Y, Ren X, Zhang L, Yap KL, Wu H, Patel R, Liu D, Qin ZH, Shih IM, Yang JM (2012) NAC1 modulates sensitivity of ovarian cancer cells to cisplatin by altering the HMGB1-mediated autophagic response. Oncogene 31:1055–1064

Zhou W, Yuan J (2014) Necroptosis in health and diseases. Semin Cell Dev Biol 35:14–23

Zhou RR, Zhao SS, Zou MX, Zhang P, Zhang BX, Dai XH, Li N, Liu HB, Wang H, Fan XG (2011) HMGB1 cytoplasmic translocation in patients with acute liver failure. BMC Gastroenterol 11:21

Zhu P, Duan L, Chen J, Xiong A, Xu Q, Zhang H, Zheng F, Tan Z, Gong F, Fang M (2011) Gene silencing of NALP3 protects against liver ischemia-reperfusion injury in mice. Hum Gene Ther 22:853–864

Zong WX, Thompson CB (2006) Necrotic death as a cell fate. Genes Dev 20:1–15

Chapter 8
The Critical Role of Mitochondria in Drug-Induced Liver Injury

Carl W. Decker, Joseph G. Casian, Kim Tho Nguyen, Luke A. Horton, Madhuri P. Rao, Kai H. Silkwood, and Derick Han

Abbreviations

ALT	Alanine aminotransferase
AMPK	AMP-activated protein kinase
APAP	Acetaminophen
ARE	Antioxidant response element
ASK1	Apoptosis signal-regulating kinase 1
Bcl-2	B-cell lymphoma-2
Cu3	Cullin-dependent E3 ubiquitin ligase complex
CYP450	Cytochrome P450
DILI	Drug-induced liver injury
Drp1	Dynamin-related protein 1
GCL	Glutamate cysteine ligase
GSH	Glutathione
GSK-3β	Glycogen synthase kinase-3β
GWAS	Genome-wide association studies

C.W. Decker • J.G. Casian • K.T. Nguyen • D. Han (✉)
Department of Biopharmaceutical Sciences, School of Pharmacy, Keck Graduate Institute, 535 Watson Drive, Claremont, California, CA 91711, USA
e-mail: dhan@kgi.edu

L.A. Horton
W.M. Keck Science Department, Claremont McKenna College, Claremont, CA 91711, USA

M.P. Rao
W.M. Keck Science Department, Scripps College, Claremont, CA 91711, USA

K.H. Silkwood
Department of Chemistry and Biochemistry, University of California San Diego, La Jolla, San Diego, CA 92093, USA

W.-X. Ding, X.-M. Yin (eds.), *Molecules, Systems and Signaling in Liver Injury*, Cell Death in Biology and Diseases, DOI 10.1007/978-3-319-58106-4_8

X.-M. Yin and Z. Dong (Series eds.), *Cell Death in Biology and Diseases*

HLA	Human leukocyte antigen
JNK	c-Jun *N*-terminal kinase
Keap1	Kelch-like ECH-associated protein 1
LPS	Lipopolysaccharide
Mcl-1	Induced myeloid leukemia cell differentiation protein
Mfn	Mitofusin
MLK3	Mixed-lineage kinase-3
MOMP	Mitochondrial outer membrane permeabilization
MPT	Mitochondrial permeability transition
NAPQI	*N*-Acetyl-*p*-benzo-quinoneimine
Nrf-2	Nuclear factor (erythroid-derived 2)-like-2
Opa1	Optic atrophy 1
PGC-1α	Proliferator-activated receptor gamma coactivator-1α
PKC	Protein kinase C
RIP1	Receptor-interacting serine/threonine-protein kinase 1
ROS	Reactive oxygen species
Sab	SH3 homology associated BTK-binding protein
SOD	Superoxide dismutase
TNF	Tumor necrosis factor-α

8.1 Introduction

The liver is the major organ responsible for drug detoxification and consequently the major site of drug injury (Kaplowitz 2005; Han et al. 2006a). Drug-induced liver injury (DILI) is the leading cause of black box warnings, drug withdrawals, and clinical trial failures (Kaplowitz 2005; Thames 2004; Kaplowitz 2002). DILI describes a condition where the medical intake of a drug(s) causes abnormal liver tests (i.e., increased serum ALT levels), which may lead to catastrophic consequences, including acute liver failure (Ulrich 2007). While drugs that cause liver injury are not structurally or chemically related, most hepatotoxic drugs have been shown to "stress" mitochondria (Boelsterli and Lim 2007; Han et al. 2010, 2013). In most cases, mitochondrial stress caused by hepatotoxic drugs is not sufficient to cause liver injury. However, mitochondrial stress may alter signaling pathways including adaptation/survival pathways that sensitize hepatocytes to extrinsic factors, such as inflammation and the adaptive immune system, to trigger liver injury (Han et al. 2013). This review will examine the critical role mitochondria play in mediating DILI.

8.2 Overview of DILI

DILI is categorized into two broad categories: predictable and idiosyncratic. Acetaminophen (APAP) is one of the few marketed drugs that cause a predictable, dose-dependent hepatotoxicity in animal models and in patients. APAP, due to unintentional or deliberate overdose, is the most common cause of drug-induced liver failure, accounting for ~46% of acute liver failure cases in the United States (Ostapowicz et al. 2002). Idiosyncratic DILI is the rare form of liver injury (1 in 1000–20,000) that occurs in patients taking therapeutic doses of drugs (Kaplowitz 2005; Holt and Ju 2010). Idiosyncratic DILI is not strictly dose dependent, but a dose threshold generally exists, as a minimum dosage of drug (50–100 mg daily) is required for liver injury to occur. Idiosyncratic DILI is generally not observed in animal models, as most drugs pass animal testing before being administered to patients. Idiosyncratic DILI is also generally not observed during clinical trials due to insufficient population sizes (~low thousands) (Han et al. 2010; Holt and Ju 2010). Consequently, only when a drug becomes consumed by a large number of patients does idiosyncratic DILI become unequivocally recognized. Idiosyncratic DILI accounts for ~13% of acute liver failure cases in the United States.

The underlying mechanisms of why a small subpopulation of patients develops liver injury with therapeutic drug doses, while the majority of patients do not, are complex. Idiosyncratic DILI results from the convergence of multiple risk factors, such as genetics, age, gender, diet, and infections (Ulrich 2007; Lee 2003). Recent genome-wide association studies (GWAS) have shown that patients with certain human leukocyte antigen (HLA; the human equivalent of MHC) haplotypes are susceptible to idiosyncratic DILI caused by a number of drugs, including lumiracoxib, lapatinib, ticlopidine, amoxicillin-clavulanate, flucloxacillin, and ximelagatran (Hautekeete et al. 1999; O'Donohue et al. 2000; Hirata et al. 2008; Daly et al. 2009; Singer et al. 2010; Lucena et al. 2011; Alfirevic et al. 2012). For example, the HLA-DRB1*1501 haplotype increases a patient's risk for amoxicillin-clavulanate-induced DILI, while HLA-DRB1*5701 increases the risk for flucloxacillin-induced DILI. The strong relationship between hepatotoxic drugs and HLA polymorphisms suggests that the adaptive immune system plays an important role in a large number of idiosyncratic DILI cases (Han et al. 2013). A recent study has shown that abacavir, which causes hypersensitive reactions, binds within the F pocket of the peptide-binding groove of the HLA receptor (HLA-B*57:01 haplotype) (Illing et al. 2012). Thus, patients with the HLA-B*57:01 haplotype have a greater potential for autoimmune reactions because the drug is presented on the HLA receptor to T-cells (Adam et al. 2014). Although patients with certain HLA haplotypes have increased risk for DILI, overall rates remain low for most drugs (Watkins 2009). Other extrinsic factors (infections, inflammation, diet) also play important roles in the development of idiosyncratic DILI (Ulrich 2007; Han et al. 2010).

8.3 Drugs that Cause DILI Stress Mitochondria

Mitochondria – as major generators of ATP, key regulators of apoptosis, and major sources of reactive oxygen species (ROS) – play a central role in hepatocyte survival (Han et al. 2013; Cadenas and Davies 2000). Stress or injury to mitochondria has profound effects on ATP and ROS levels and can trigger the release of pro-death

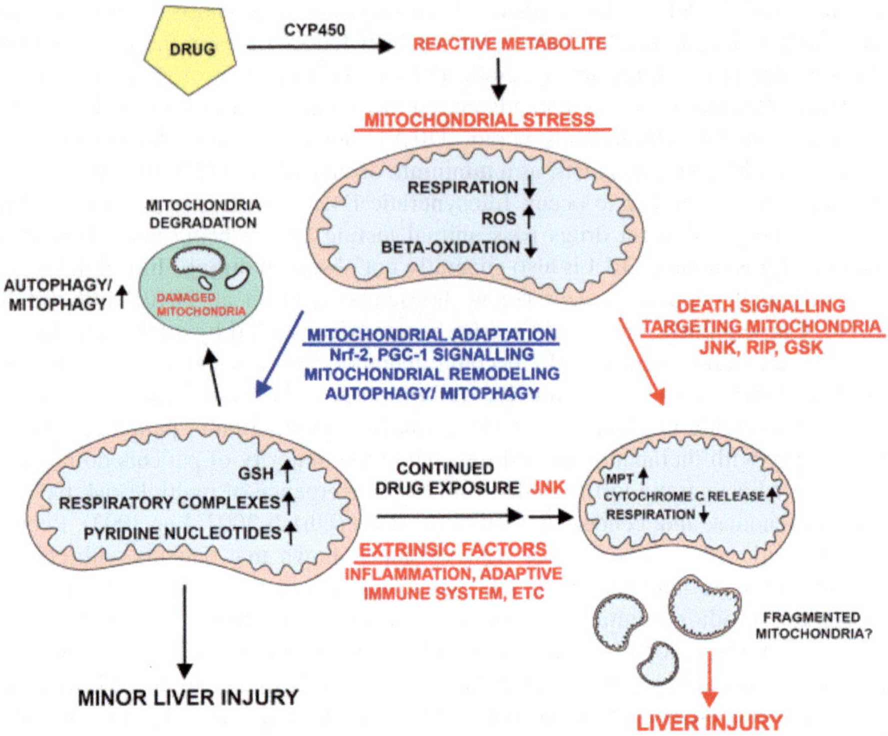

Fig. 8.1 *Critical role of mitochondria in DILI.* Hepatotoxic drugs or drug metabolites formed from metabolism by cytochrome P450 (*CYP450*) have been shown to inhibit mitochondrial respiration, disrupt beta-oxidation, increase mitochondrial ROS generation, and cause other types of mitochondrial disruptions that stress or injure mitochondria. In most cases, mitochondrial stress induced by a drug is not sufficient to cause hepatotoxicity, because adaptation pathways including autophagy/mitophagy, mitochondrial remodeling, and mitochondrial fusion-fission become activated. Mitochondrial fusion-fission could also potentially play a deleterious role by initiating mitochondrial fragmentation that promotes hepatocyte death. A small number of patients may progress to severe liver injury due to failure to adapt resulting from inflammation and other extrinsic factors. When mitochondrial and hepatocellular injury reaches a critical threshold, death signaling pathways involving JNK that target mitochondria become activated. Activated JNK binds to Sab on the outer mitochondrial membrane to trigger the mitochondrial permeability transition (*MPT*), which causes mitochondrial swelling, and release of cytochrome *c* that results in hepatocyte death and liver injury

proteins such as cytochrome *c* (Kaplowitz 2002). Mitochondria play a critical role in DILI during all stages of the pathology (Kaplowitz 2005): (1) inhibition of mitochondrial function by drugs or reactive metabolites that can initiate DILI (Han et al. 2006a), (2) adaptation pathways involving mitochondria that help most patients tolerate drugs (Thames 2004), and (3) activation of death signaling pathways involving c-Jun *N*-terminal kinase (JNK) that targets mitochondria to cause hepatocyte death and liver injury (Fig. 8.1). Most drugs that cause DILI, predictable or idiosyncratic, have been shown to stress or injure mitochondria by disrupting mitochondrial function (Boelsterli and Lim 2007; Han et al. 2013; Jones et al. 2010). Hepatotoxic drugs, or drug metabolites formed from metabolism by cytochrome P450 (CYP450), have been shown to inhibit mitochondrial respiration, disrupt beta-oxidation, deplete mitochondrial glutathione (GSH), increase mitochondrial ROS generation, and cause other mitochondrial disruptions that stress or injure mitochondria (Han et al. 2013). However, in most cases, mitochondrial stress induced by a drug is not sufficient to cause hepatotoxicity, and the majority of patients taking drugs do not develop liver injury. Most patients taking idiosyncratic hepatotoxic drugs may be "adaptors" that avoid liver injury by upregulating adaptation signaling pathways, many involving mitochondria. Mitochondrial adaptation pathways include autophagy/mitophagy, mitochondrial remodeling, and mitochondrial fusion-fission (Fig. 8.1). A small fraction of patients ("susceptibles") may progress to severe injury due to a failure of the liver to adapt because of inflammation and other extrinsic factors. When mitochondrial and hepatocellular injury reaches a critical threshold, death signaling pathways involving JNK become activated and target mitochondria (Han et al. 2013; Hanawa et al. 2008). The binding of activated JNK (phosphorylated JNK) to mitochondria triggers the mitochondrial permeability transition (MPT), which causes mitochondrial swelling and release of cytochrome *c* that results in hepatocyte death and liver injury (Han et al. 2013).

DILI has been difficult to study due to the rarity of the disease in patients and lack of animal models (Kaplowitz 2005; Boelsterli 2003). Because APAP is one of the few marketed drugs that cause a predictable, dose-dependent hepatotoxicity, most of our mechanistic knowledge on the role of mitochondria in DILI has come from work with APAP in cultured hepatocytes and in animals (Kaplowitz 2005). While APAP treatment causes very predictable liver injury, insights regarding idiosyncratic DILI have been obtained using APAP in animal models, since its hepatotoxicity is similarly affected by age, sex, diet, inflammation caused by bacteria and viruses, and genetic factors (Han et al. 2010; Maddox et al. 2010). Because the APAP animal model has revealed a great deal of information regarding the role of mitochondria in drug-induced hepatotoxicity, we will begin with an overview of the central role mitochondria play in APAP-induced liver injury.

8.4 Central Role of Mitochondria in APAP-Induced Liver Injury

Mitochondrial injury plays an essential role in APAP-induced liver injury. It has been demonstrated that serum levels of mitochondrial biomarkers (mtDNA, glutamate dehydrogenase) are predictive of patients surviving APAP-induced acute liver failure (McGill et al. 2014). Patients with greater mitochondrial injury that causes the release of mitochondrial constituents into the plasma are much more likely to suffer acute liver failure. Consequently, there has been increased research exploring mitochondria-targeted therapies, such as mitochondria-targeted antioxidants (Mito-TEMPO), and electron transport mediators, such as methylene blue, to protect against APAP-induced liver injury in animal models (Lee et al. 2015; Du et al. 2017).

8.4.1 Inhibition of Mitochondrial Bioenergetics by APAP

APAP hepatotoxicity can be directly attributed to the formation of *N*-acetyl-*p*-benzo-quinoneimine (NAPQI), a reactive metabolite generated during the metabolism of APAP by CYP450, primarily by the CYP2EI isoform (Dahlin et al. 1984; Chen et al. 1998). NAPQI is a highly reactive metabolite that forms covalent bonds with protein thiols and nonprotein thiols including glutathione (GSH), the major antioxidant in cells. At hepatotoxic doses of APAP, NAPQI causes severe depletion of GSH in the cytoplasm and mitochondria, which contain separate GSH pools (Han et al. 2006a; Kaplowitz et al. 1985). APAP treatment has also been shown to decrease levels of glutamate cysteine ligase (GCL), the rate-limiting enzyme in GSH synthesis, further depleting GSH in the liver (Shinohara et al. 2010). The depletion of mitochondrial GSH is considered the critical event in APAP hepatotoxicity, as it results in the partial inhibition of mitochondrial respiration and enhanced mitochondrial ROS generation – important in activating JNK death signaling. NAPQI may directly bind thiols in the respiratory complexes to inhibit mitochondrial respiration (Burcham and Harman 1991), with complex II being particularly sensitive to NAPQI inhibition (Lee et al. 2015). APAP treatment enhances mitochondrial ROS generation by (Kaplowitz 2005) reducing the efficacy of GSH peroxidase, which utilizes GSH to detoxify H_2O_2, and (Han et al. 2006a) by inhibiting mitochondrial respiratory complexes, which generally enhance superoxide generation (Hanawa et al. 2008; Han et al. 2003a, b). There is a critical threshold for mitochondrial GSH depletion that determines whether liver injury occurs. If the threshold is reached (>90%), extensive NAPQI inhibition of mitochondrial proteins and/or enhanced mitochondrial H_2O_2production will activate the JNK death signaling cascade that causes severe liver injury (Han et al. 2013). When the GSH threshold is not reached, adaptation pathways involving nuclear factor (erythroid-derived 2)-like-2 (Nrf-2) help mitochondrial and liver GSH levels recover, resulting in little liver injury (Copple et al. 2008). Low non-hepatotoxic doses of APAP have been shown to cause reversible mitochondrial depolarization in the liver, without hepatocyte necrosis or ALT release (Hu et al. 2016).

8.4.2 Adaptation of Mitochondria to Promote Survival in Response to APAP

Adaptive mechanisms in the liver, including mitochondrial adaptation, may be crucial in patients consuming even therapeutic doses of APAP (4 grams daily). One study reported that the majority of patients (up to 81%) receiving therapeutic doses of APAP initially develop elevated plasma ALT levels (Watkins et al. 2006). However, adaptation mechanisms appear to protect the majority of patients from APAP, and in most cases, continuous APAP intake does not result in liver injury. Mitochondrial adaptation mechanisms including enhancement of mitochondrial

GSH levels, autophagy/mitophagy, mitochondria remodeling, and alteration of mitochondrial fusion-fission have been characterized in various liver pathologies. We will next examine mitochondrial adaptation mechanisms that may be important in protecting against APAP hepatotoxicity.

8.4.2.1 Enhancement of Mitochondrial GSH Levels

Mitochondria lack enzymes necessary for GSH synthesis, and consequently GSH must be transported into the mitochondrial matrix (Kaplowitz et al. 1985). Interestingly, once GSH is oxidized to GSSG, it appears to become trapped inside the mitochondrial matrix: GSSG transporters have not been observed in mitochondria (Olafsdottir and Reed 1988; Garcia et al. 2010). We have demonstrated that feeding respiratory substrates (i.e., glutamate/malate, succinate, etc.) into the electron transport chain generates NADPH, which is essential in maintaining GSH and protein thiols in their reduced forms, as well as reversing posttranslational modifications such as S-glutathionylation and S-nitrosylation (Garcia et al. 2010; Chang et al. 2014). During times of oxidative stress, respiratory substrates are critical in keeping the mitochondrial redox environment reduced and fully functional (Garcia et al. 2010).

The transcription factor Nrf-2 plays a central role in adaptation to APAP in the liver by replenishing cytoplasmic and mitochondrial GSH levels (Copple et al. 2010). Nrf-2 is usually bound in the cytoplasm to Kelch-like ECH-associated protein 1 (Keap1), a redox-sensitive anchoring protein. Also attached to Keap1 is a cullin-dependent E3 ubiquitin ligase complex (Cu3) that facilitates the ubiquitination of Nrf-2. Keap1 binds newly synthesized Nrf-2, which results in Cu3-dependent ubiquitination and proteasomal degradation (Copple et al. 2010; Tong et al. 2007). Consequently, Nrf-2 has an extremely short half-life (10–30 min). The binding of critical thiols in Keap1 (~25 cysteine residues) by NAPQI and other electrophiles causes a conformational change that prevents Nrf-2 ubiquitination and degradation (Han et al. 2013). Therefore, the NAPQI-induced conformational change of Nrf-2 keeps Keap1 occupied without being degraded, allowing newly synthesized Nrf-2 to translocate to the nucleus and bind to the antioxidant response element (ARE), a DNA sequence in the promoter of antioxidant enzymes such as GCL (Copple et al. 2008; Tong et al. 2007). Nrf-2 upregulation of GCL is essential in replenishing mitochondrial and cytoplasmic GSH following even therapeutic doses of APAP. Even low, nontoxic doses of APAP have been shown to activate Nrf-2 and increase GSH synthesis in the liver of mice (Goldring et al. 2004).

8.4.2.2 Autophagy/Mitophagy

Autophagy is a lysosome-mediated process where intracellular proteins and organelles, including mitochondria, are degraded and recycled (Yin et al. 2008). Mitophagy is a selective form of autophagy, where damaged mitochondria are removed and degraded by lysosomes (Ding and Yin 2012). Both autophagy and mitophagy are enhanced by various stresses, including starvation. Since APAP treatment is known to cause mitochondrial dysfunction, it is not surprising that both autophagy and mitophagy occur following hepatotoxic doses of APAP. Activation of autophagy and mitophagy is believed to be protective, as damaged mitochondria are removed, especially those with enhanced ROS generation (Ni et al. 2012). Pharmacological inhibition of autophagy aggravates APAP-induced liver injury (Baulies et al. 2015), while enhancement of autophagy using rapamycin protects against APAP-induced liver injury in mice. AMP-activated protein kinase (AMPK), which is regulated by protein kinase C (PKC) (Saberi et al. 2008), is believed to promote autophagy in the liver and help protect against APAP (Saberi et al. 2014). Mitophagy has not been extensively characterized during APAP hepatotoxicity but is believed to be mediated by Parkin in the liver (Ding et al. 2012a).

8.4.2.3 Mitochondria Remodeling

The mitochondrial respiratory chain can undergo remodeling and biogenesis to increase respiratory capacity in response to injury (Han et al. 2013; Than et al. 2011). We observed that chronic alcohol feeding in mice and rats caused various types of mitochondrial remodeling, including alterations in respiratory complex proteins, increased mitochondrial NADH-NADPH levels, and increased mitochondrial respiration as adaptations to alcohol (Han et al. 2012, 2016). Mitochondrial remodeling is likely mediated by proliferator-activated receptor gamma coactivator-1α (PGC-1α), a coactivator and master regulator of mitochondria (Than et al. 2011; Han et al. 2012). Mitochondrial injury in the liver caused by the hepatotoxin CCl_4, which induces extensive lipid peroxidation, has similarly been shown to increase mitochondrial respiration and levels of respiratory complexes in the liver (Shiryaeva et al. 2008; Morimoto et al. 1988). Thus, APAP treatment, which injures mitochondria and enhances mitochondrial degradation by autophagy and mitophagy, is likely to activate PGC-1α and other mitochondrial biogenesis pathways to promote mitochondrial remodeling and biogenesis. However, to our knowledge a detailed study examining mitochondrial remodeling and biogenesis following APAP treatment has not been performed in the liver, and further studies are needed.

8.4.2.4 Alteration of Mitochondrial Fusion-Fission

Mitochondria constantly undergo fusion-fission in order to exchange mitochondrial constituents including mtDNA, respiratory complexes, mitochondria proteins, and GSH (Otera and Mihara 2011; Reddy et al. 2011). Mitochondrial fusion and fission are primarily controlled by four highly conserved GTPases: mitofusins (Mfn1, Mfn2) in mitochondrial outer membranes promote fusion; optic atrophy (Opa1) in the inner membranes promotes fusion; and dynamin-related protein 1 (Drp1) from the cytoplasm translocates to mitochondria to initiate fission (Otera and Mihara

2011; Chan 2006). Stress can alter mitochondrial fusion-fission rates and mitochondrial morphology, which can promote cell death or cell survival depending on the circumstances. Mitochondrial fission mediated by Drp1 can promote mitochondrial fragmentation and cell death (Otera and Mihara 2011; Westermann 2010). Conversely, alterations in mitochondrial fusion-fission can promote cell survival, in cases such as starvation, where mitochondrial fission decreases to produce elongated mitochondria with greater cristae surface areas that increase respiration and ATP production (Gomes et al. 2011). APAP hepatotoxicity is associated with dramatic morphological alterations in mitochondria, including both fragmentation and elongation (Ruepp et al. 2002), as well as formation of spheroid-shaped mitochondria (Ding et al. 2012b; Ni et al. 2013). The implications of these mitochondrial morphological changes caused by hepatotoxic doses of APAP remain uncertain. It has been suggested that spheroid-shaped mitochondria formation mediated by Mfn1 or Mfn2 may be a protective mechanism of inactivating damaged mitochondria (Ding et al. 2012a), but further research is needed. ROS appear to be important in triggering mitochondrial morphological changes and spheroid-shaped mitochondria formation in hepatocytes (Ding et al. 2012b). Drp1 has been observed to translocate to mitochondria very early during APAP hepatotoxicity (Dara et al. 2015), which may help explain some observations of fragmented mitochondria during APAP hepatotoxicity. However, the importance of this Drp1 translocation to mitochondria during APAP hepatotoxicity requires further examination. Overall, mitochondrial morphological changes are very prominent during APAP-induced liver injury, but whether these changes play an adaptive, detrimental, or inconsequential role in hepatotoxicity needs further exploration.

8.4.3 Death Signaling Pathways Targeting Mitochondria

8.4.3.1 JNK Activation and Translocation to Mitochondria

Despite adaptive mechanisms, if mitochondrial GSH depletion reaches a critical threshold, the death signaling pathway involving JNK becomes activated (phosphorylation). Activated JNK targets mitochondria and plays a critical role in promoting mitochondrial dysfunction and liver injury during APAP hepatotoxicity. JNK inhibition or silencing protects against APAP-induced liver injury, despite excessive mitochondrial GSH depletion and covalent binding in the liver (Hanawa et al. 2008). JNK plays an important role in stress response; however, when JNK activation is sustained, it promotes mitochondrial dysfunction and cell death (Han et al. 2010; Seki et al. 2012). Increased mitochondrial ROS generation mediated by NAPQI is believed to activate JNK during APAP-induced liver injury through a complex feed-forward mechanism. The maximal activation of JNK appears to involve a number of upstream kinases, including PKC-α, receptor-interacting serine/threonine-protein kinase 1 (RIP1), glycogen synthase kinase-3β (GSK-3β), mixed-lineage kinase-3 (MLK3), and apoptosis signal-regulating kinase 1 (ASK1)

(Shinohara et al. 2010; Saberi et al. 2014; Dara et al. 2015; Nakagawa et al. 2008; Sharma et al. 2012). Silencing any one of these kinases in mice inhibits JNK activation and translocation to mitochondria to protect against APAP hepatotoxicity. Activated JNK translocates to mitochondria and binds SH3 homology associated BTK-binding protein (Sab or SH3BP5), a scaffold protein on the outer membrane of mitochondria that contains a kinase interaction motif (Wiltshire et al. 2004; Win et al. 2014). The binding of JNK to Sab further impairs mitochondrial respiration and enhances mitochondrial ROS generation by activating a signaling cascade involving mitochondrial protein tyrosine phosphatase, nonreceptor type 6 (SHP1), and mitochondrial Src (Win et al. 2016). Once critical levels of JNK translocate to mitochondria, the mitochondrial permeability transition (MPT) is triggered, causing mitochondrial swelling and release of cytochrome *c* that results in hepatocyte death and liver injury. The exact mechanism of how JNK triggers MPT remains to be fully characterized.

APAP hepatotoxicity is an active process, since JNK signaling pathways must be activated for liver injury to occur. This is somewhat surprising, as APAP-induced liver injury occurs primarily through necrosis rather than apoptosis (Hanawa et al. 2008; Gunawan et al. 2006). Clinical studies have shown that circulating levels of full length K18 (Fl-K18), a marker of necrosis, were much higher (85%) than the cleaved K18 fragment (cK18), a marker of apoptosis (15%), in patients suffering from APAP hepatotoxicity (Antoine et al. 2012). These findings correspond with animal studies that have demonstrated that APAP hepatotoxicity occurs primarily through hepatocyte necrosis. The fact that JNK actively mediates hepatocyte death challenges the traditional view that APAP hepatotoxicity is a passive process due to overwhelming hepatocellular injury caused by NAPQI (Han et al. 2010; Han et al. 2013). APAP hepatotoxicity can therefore be viewed as a type of "programmed necrosis" (Han et al. 2013; Jones et al. 2010).

8.4.3.2 Mitochondrial Targeting by Other Kinases

JNK is not the only kinase that has been shown to translocate to mitochondria during APAP-induced liver injury. PKC-α, GSK-3β, and RIP1 have all been shown to translocate to liver mitochondria following hepatotoxic doses of APAP. However, it has been difficult to decipher whether the translocation of these kinases contributes to mitochondrial dysfunction during APAP hepatotoxicity, since silencing these kinases inhibits JNK activation and translocation to mitochondria. For example, PKC-α translocates to mitochondria and phosphorylates mitochondrial proteins. However, silencing PKC-α inhibits JNK activation and translocation during APAP hepatotoxicity, suggesting that PKC-α is important in activating JNK. In other injury models, PKC-ε translocation to mitochondria appears to play a protective role in cardiac ischemic injury by phosphorylating mitochondrial proteins such as pyruvate dehydrogenase, while PKC-δ, which inhibits pyruvate dehydrogenase, appears to promote cardiac ischemic injury (Churchill and Mochly-Rosen 2007; Kheifets and Mochly-Rosen 2007; Budas et al. 2010; Gong et al. 2012). Similarly,

GSK-3β translocation to mitochondria is believed to be important in mediating ischemia-reperfusion injury in the heart by promoting MPT (Das et al. 2008; Maurer et al. 2006; Juhaszova et al. 2004). Therefore, the translocation of other kinases – such as PKC-α, GSK-3β, and RIP1 – to mitochondria following APAP treatment may play a role in APAP hepatotoxicity, but further studies are needed to delineate their effects from JNK (Shinohara et al. 2010).

8.4.3.3 Mitochondrial Regulation by Bcl-2 Family Members

Mitochondria are the organelles where members of the B-cell lymphoma-2 (Bcl-2) family exert their apoptotic or anti-apoptotic effects. Bax, a proapoptotic member of the Bcl-2 family, has been shown to translocate to liver mitochondria during APAP hepatotoxicity, possibly through a JNK-dependent pathway (Hanawa et al. 2008). However, Bax knockout mice exhibit the same level of liver injury as control mice following APAP treatment, suggesting that Bax is not essential for mitochondrial dysfunction and hepatocyte death (Bajt et al. 2008). JNK translocation to mitochondria has been suggested to phosphorylate and inactivate Bcl-2 and Bcl-xl, two anti-apoptotic proteins on the outer membrane of mitochondria (Hanawa et al. 2008; Latchoumycandane et al. 2007). Induced myeloid leukemia cell differentiation protein (Mcl-1), an anti-apoptotic Bcl-2 family member, is also degraded in the liver during APAP hepatotoxicity (Shinohara et al. 2010). Silencing Mcl-1 in the liver causes spontaneous apoptosis in hepatocytes (Vick et al. 2009), suggesting that the loss or inactivation of anti-apoptotic Bcl-2 family members from mitochondria may contribute to hepatocyte death following APAP treatment. Taken together, many proteins (kinases, Bcl-2 family members) translocate to mitochondria, while anti-apoptotic Bcl-2 members become inactivated in mitochondria. All of these events may be contributing to mitochondrial dysfunction during APAP hepatotoxicity, but currently only JNK translocation to mitochondria has been shown to be critical in causing mitochondrial dysfunction essential for liver injury.

8.5 Role of Mitochondria in Idiosyncratic DILI

Idiosyncratic DILI likely involves mitochondrial stress, mitochondrial adaptations, and death signaling pathways targeting mitochondria as observed with APAP (Fig. 8.1). While APAP is unique in that its toxicity is dependent on severe GSH depletion, many other features of APAP hepatotoxicity are likely to occur during liver injury caused by other hepatotoxic drugs (Han et al. 2013). Thus, hepatotoxic drugs may have many general shared characteristics, as well as some unique drug-specific features (i.e., metabolism, reactive metabolites, etc.). Since idiosyncratic DILI is rare and lacks good animal models, our overall knowledge is limited and more open to speculation.

8.5.1 Inhibition of Mitochondrial Homeostasis by Idiosyncratic Hepatotoxins

Most drugs that cause idiosyncratic DILI have been shown to stress or injure mitochondria. Reactive metabolites formed during drug metabolism, rather than the drug itself, may often be disruptive to mitochondria in hepatocytes (Kaplowitz 2005; Holt and Ju 2010). Many reactive metabolites can also directly cause oxidative stress. For example, troglitazone forms a reactive metabolite that forms adducts with molecules like GSH (Alvarez-Sanchez et al. 2006). Table 8.1 shows idiosyncratic hepatotoxic drugs that disrupt mitochondrial function in hepatocytes or isolated mitochondria. A wide range of mitochondrial functions including mitochondrial electron transport, beta-oxidation, and protein synthesis can be inhibited by hepatotoxic drugs. The majority of drugs listed in Table 8.1 do not cause liver injury (i.e., increased ALT, histology) when given to healthy animals, though mitochondrial stress can sometimes be observed in the liver. For example, chronic treatment of nimesulide to mice increased cytochrome *c* release from mitochondria in the liver, but ALT levels and histology remained similar to control (Ong et al. 2006). In most cases, mitochondrial inhibition by idiosyncratic hepatotoxic drugs listed in Table 8.1 is only observed in cultured hepatocytes or with isolated liver mitochondria. Antimicrobials cause ~46% of idiosyncratic DILI cases, possibly because mitochondria are of bacterial ancestry and still share many genetic and structural similarities with bacteria. A limited amount of human data suggests that mitochondrial abnormalities occur in the liver during idiosyncratic DILI. Tolcapone-induced DILI has been observed to be associated with mitochondrial swelling in the livers of patients (Spahr et al. 2000). Electron microscopy studies have similarly demonstrated that aspirin-induced Reye's syndrome is linked with abnormal mitochondria in the liver (Iancu et al. 1977; Osterloh et al. 1989).

Large screening studies have also confirmed that inhibition of mitochondrial homeostasis occurs for the majority of idiosyncratic hepatotoxic drugs. Screening for drugs (300 tested) that cause oxidative stress and disrupt mitochondrial homeostasis (ROS, mitochondrial membrane potential, etc.) in human hepatocytes revealed that 50–60% of idiosyncratic hepatotoxic drugs caused mitochondrial perturbations, whereas non-hepatotoxic drugs were shown not to affect mitochondria (0–5%) (Xu et al. 2008). Similarly, screenings utilizing isolated mouse liver mitochondria showed that most drugs that cause idiosyncratic DILI disrupt mitochondrial function (i.e., membrane potential, respiration, cytochrome *c* release, etc.; 89% positive predictive value) (Porceddu et al. 2012). Thus, these types of mitochondrial screening studies, in combination with "omics" technologies (genomics, metabolomics, etc.) may better identify idiosyncratic hepatotoxins during drug development (Winnike et al. 2010; Daly 2012).

If a majority of hepatotoxic drugs disrupt mitochondrial homeostasis, the question remains why liver injury only occurs in a small number of patients, while the majority remains unaffected. It is clear that mitochondrial stress alone is not enough to cause liver injury in most cases, and other factors contribute to DILI. The mechanism of idiosyncratic DILI is multifactorial (genetics, age, gender, diet, infections),

Table 8.1 Drugs associated with mitochondrial dysfunction

Drugs	Model system	Effect on mitochondria
Acetaminophen	In vivo – mice liver, cultured hepatocytes	Depletion of mitochondrial GSH, inhibition of respiration, increased ROS generation
Amineptine	In vivo – mice liver	Inhibition of beta-oxidation
Amiodarone, dronedarone	Isolated liver mitochondria, cultured hepatocytes	Increased ROS generation, inhibition of respiration, inhibition of beta-oxidation
Aspirin	In vivo – mice liver, isolated liver mitochondria, cultured hepatocytes	MPT, mitochondrial uncoupling, inhibition of beta-oxidation
Benzarone, Benzbromarone	Cultured hepatocytes, isolated liver mitochondria	Inhibition of respiration, mitochondrial uncoupling
Diclofenac	Cultured hepatocytes, isolated mitochondria	Mitochondrial uncoupling, inhibition of ANT and ATPase, MPT, Bax-induced MOMP
Diflunisal	Isolated liver mitochondria	Mitochondrial uncoupling, inhibition of medium-chain acyl-CoA synthetase
Fialuridine	In vivo – woodchuck and rat liver mitochondria, HepG2 cells	Enlarged mitochondria, decreased mtDNA, reduced cristae
Isoniazid	HepG2 cells, cultured hepatocytes, isolated liver mitochondria	Inhibition of complex I, decreased membrane potential, increased ROS generation
Mefenamic acid	Isolated kidney or liver mitochondria	Mitochondrial uncoupling, inhibition of various acyl-CoA synthetase
Nefazodone	Human hepatocytes, isolated liver mitochondria, HepG2 cells	Inhibition of mitochondrial respiration (complex I and IV)
Nimesulide	Isolated liver mitochondria, rat hepatocytes	Mitochondrial uncoupling, increased ROS generation, MPT
Perhexiline	Cultured hepatocytes, isolated liver mitochondria	Mitochondrial uncoupling, inhibition of mitochondrial respiration (complex I and II), inhibition of beta-oxidation
Stavudine	Liver of patients, HepG2 cells, cultured hepatocytes	Inhibition of beta-oxidation, increased ROS generation, reduced and abnormal cristae
Tamoxifen	Isolated liver mitochondria	Inhibition of mitochondrial respiration (complex I), mitochondrial uncoupling, increased ROS generation
Tianeptine	In vivo – mice liver	Inhibition of beta-oxidation (short- and medium-chain fatty acid)
Tolcapone	Isolated liver mitochondria, in vivo – rat liver	Mitochondrial uncoupling, mitochondrial swelling
Trazodone	Human hepatocytes	Loss of mitochondrial membrane potential
Troglitazone	HepG2 cells, isolated liver mitochondria	Promotes PGC-1 degradation to decrease mitochondrial mass, decreased mitochondrial respiration, MPT
Valproate	In vivo – mice liver, cultured hepatocytes, submitochondrial particles	Inhibition of beta-oxidation, inhibition of mitochondrial pyruvate uptake, inhibition of mitochondrial respiration, depletion of mitochondrial GSH

and it is possible that the same drug could cause DILI through different mechanisms in different patients. Genetic alterations that predispose patients to oxidative stress appear to cause liver injury in a number of idiosyncratic hepatotoxins. Mice that are heterozygous for Mn-superoxide dismutase (SOD $2^{+/-}$mice) represent an animal model with greater mitochondrial oxidative stress, due to reduced levels of mitochondrial SOD responsible for clearing superoxide in the matrix. Two idiosyncratic hepatotoxic drugs (troglitazone, flutamide) were shown to cause liver injury in SOD $2^{+/-}$mice but not in wild-type mice (Ong et al. 2007; Kashimshetty et al. 2009). It must be noted that troglitazone-induced liver injury is somewhat controversial in this model (Fujimoto et al. 2009). Patients with mutations in Mn-SOD or mutations in GSH peroxidase (GPX1), responsible for detoxification of H_2O_2, are at greater risk for developing cholestatic DILI (Lucena et al. 2010). Similarly, patients, predominantly women, with mutations in glutathione S-transferase (GST) have an increased risk for developing idiosyncratic DILI regardless of the type of drug involved (Lucena et al. 2008). Genetic alterations represent only one contributing factor to idiosyncratic DILI, and the failure of adaptation pathways due to extrinsic factors, such as infections, may also be important.

8.5.2 Mitochondrial Adaptation During Idiosyncratic DILI

Clinical studies suggest that the liver readily adapts to chronic drug intake. Idiosyncratic hepatotoxic drugs often cause higher rates of mild, asymptomatic, and transient elevated serum ALT levels that disappear with continuous drug treatment (Watkins 2009; Watkins 2005). Drug screening studies in primary hepatocytes have shown that idiosyncratic toxicants (troglitazone, trovafloxacin) alter gene expression patterns much more than their non-hepatotoxic counterparts, possibly due to the mitochondrial stress induced by hepatotoxic drugs (Kier et al. 2004; Liguori et al. 2005). Many of these signaling changes in hepatocytes likely involve similar adaptation mechanisms previously characterized for APAP. Unfortunately, very little information regarding mitochondrial adaptation is available for most drugs that cause DILI due to the lack of reliable animal models. There is some evidence that many hepatotoxic drugs cause changes in mitochondrial morphology. In primary hepatocytes, chloramphenicol treatment induces the formation of large mitochondria (mega-mitochondria) associated with lower respiration (Karbowski et al. 1999). Mega-mitochondria with enhanced H_2O_2generation were observed with troglitazone treatment, but not with its non-hepatotoxic counterparts (pioglitazone, ciglitazone) (Shishido et al. 2003). Finally, trovafloxacin treatment has been shown to downregulate many mitochondrial genes, including Mfn1, suggesting that mitochondrial morphological changes occur in hepatocytes (Liguori et al. 2005). Given that many drugs that cause DILI inhibit mitochondrial respiration and/or induce oxidative stress – both of which are known to trigger mitochondrial morphological changes – these observations of mitochondrial morphological alterations are not surprising. Whether these mitochondrial morphological changes are protective, detrimental, or insignificant remains to be explored.

8.5.2.1 Failure of Adaptation Signaling Pathways Due to Drug Intake

Inflammation caused by bacteria or viruses is a clinically important risk factor for the development of idiosyncratic DILI (Ulrich 2007; Roth and Ganey 2010). Some idiosyncratic hepatotoxic drugs may promote liver injury by inhibiting signaling pathways involved in adaptation to inflammation, particularly cytokines such as tumor necrosis factor-α (TNF) released during inflammation (Han et al. 2009). TNF binding to TNF receptors in hepatocytes simultaneously activates a cell death signaling pathway and a survival/adaptation pathway. The death signaling pathway involves JNK activation and translocation to mitochondria that perpetuate a self-amplifying loop, similar to that observed during APAP hepatotoxicity (Han et al. 2009). TNF is not normally toxic to hepatocytes because TNF also activates survival/adaptation pathways involving NF-κB, which transcribes genes (GAPDD45b, XIAP, A20) and antioxidant proteins (Mn-SOD, ferritin) responsible for turning off JNK and promoting cell survival (Han et al. 2009; Wong et al. 1989; Wullaert et al. 2006). Consequently, NF-κB activation allows hepatocytes to adapt to TNF released during inflammation, and JNK activation is transient. However, oxidative stress and/or inhibition of mitochondrial respiration associated with many idiosyncratic hepatotoxins has been shown to inhibit NF-κB activation and sensitize hepatocytes to TNF-induced apoptosis (Nagai et al. 2002; Matsumaru et al. 2003; Han et al. 2006b; Lou and Kaplowitz 2007). Thus, it is not surprising that APAP and the idiosyncratic toxicant chlorpromazine have been shown to sensitize cultured hepatocytes to TNF-induced apoptosis (Gandhi et al. 2010).

The failure of hepatocytes to adapt to TNF has also been observed in vivo. Idiosyncratic hepatotoxic drugs (chlorpromazine, trovafloxacin, ranitidine) that do not normally cause liver injury in rats will induce liver injury when co-treated with a sublethal dose of lipopolysaccharide (LPS), which activates the innate immune system and promotes TNF secretion (Shaw et al. 2007; Tukov et al. 2007; Luyendyk et al. 2003). The sensitizing effects of LPS to idiosyncratic hepatotoxic drugs were shown to be mediated by TNF released by the innate immune system (Liguori et al. 2010). Therefore, drugs that induce mitochondrial stress and/or enhanced ROS generation are likely blocking adaptation signaling pathways (NF-κB) and/or helping to sustain JNK activation to sensitize hepatocytes to TNF-induced apoptosis in vivo (Han et al. 2013). With atorvastatin, a statin associated with idiosyncratic liver injury, the opposite phenomenon is observed: inflammation inhibits drug-induced adaptation pathways (Wu et al. 2016). In ApoE-/- mice, atorvastatin treatment was shown to induce Nrf-2 translocation to the nucleus and upregulate antioxidant enzymes (catalase, SOD). Inflammation induced by casein injection was observed to inhibit Nrf-2 activation induced by atorvastatin, leading to liver injury including hepatic steatosis and fibrosis. Given that both hepatotoxic drugs and inflammation are associated with oxidative stress that can inhibit adaptation signaling pathways (Nrf-2, NF-κB) and sustain JNK activation, it is easy to see how these two factors may work synergistically to promote liver injury. Surprisingly, the therapeutic potential of antioxidants to protect against idiosyncratic DILI has not been extensively explored, especially since vitamin E has been shown to protect against non-alcoholic steatosis (NASH) in patients (Sanyal et al. 2010).

The increased risk of DILI that occurs with certain HLA haplotypes suggests that the adaptive immune system is important in mediating DILI caused by some drugs. Certain haplotypes of HLA, which bind drugs or reactive metabolites, may promote the adaptive immune system to attack the liver (Holt and Ju 2010). These drug-induced autoimmune reactions may increase during inflammation, when the immune system is in a heightened state. Drug-induced mitochondrial dysfunction may also be sensitizing hepatocytes to components of the adaptive immune system, such as cytotoxic T-cells (Han et al. 2013). It is also possible that idiosyncratic hepatotoxic drugs may inhibit adaptation to TNF and other cytotoxic agents (i.e., FasL, perforin, granulysin, granzyme) released by the adaptive immune system. Clearly the role the immune system plays in idiosyncratic DILI merits further investigation.

8.5.3 Mitochondria Targeting by Death Signaling Pathways During Idiosyncratic DILI

Idiosyncratic DILI caused by drug intake and extrinsic factor(s) is likely to involve death signaling pathways, such as JNK, once substantial mitochondrial and hepatocyte injury occurs. JNK is a strong candidate to mediate DILI caused by a wide range of drugs since it plays a central role in many liver pathologies. JNK plays a critical role in hepatocellular injury caused by bile acids, fatty acids, ER stress, TRAIL-Fas, ischemia/reperfusion, and TNF-induced liver injury in vivo (Con A plus D-galactosamine, LPS plus D-galactosamine, and TNF plus D-galactosamine) (Seki et al. 2012; Schwabe 2006; Ibrahim et al. 2011; Win et al. 2011, 2014, 2015). At sufficient levels, JNK translocation to mitochondria will induce either MPT or mitochondrial outer membrane permeabilization (MOMP) to trigger hepatocyte death. Along with JNK, Bcl-2 family members including Bax and Bcl-xl are likely to determine if MPT or MOMP occurs. While we hypothesize that JNK plays a central role in mediating idiosyncratic DILI caused by many drugs, further studies are needed.

8.6 Concluding Remarks

DILI appears to involve mitochondria during all stages of the disease (Xu et al. 2008; Porceddu et al. 2012). Although idiosyncratic hepatotoxic drugs are not structurally or chemically related, most drugs that cause DILI "stress" mitochondria. In most cases, hepatocytes adapt to the drug-induced mitochondrial stress by activating adaptive signaling pathways, including mitochondrial adaptation responses such as autophagy/mitophagy, mitochondrial remodeling, and alterations in mitochondrial fusion-fission. While mitochondrial stress, such as increased mitochondrial ROS generation, may not be sufficient to trigger hepatocyte death, it may inhibit adaptation/survival signaling pathways (NF-κB, Nrf-2) needed to survive extrinsic factors such as inflammation. After a critical threshold of mitochondrial and hepatocellular injury occurs, JNK is activated and targets mitochondria to induce MPT or MOMP, leading to cell death. Thus, mitochondria are the central points where oxidative injury, pro-death proteins, and adaptation pathways converge to determine the fate of hepatocytes during DILI.

Bibliography

Adam J, Wuillemin N, Watkins S, Jamin H, Eriksson KK, Villiger P et al (2014) Abacavir induced T cell reactivity from drug naive individuals shares features of allo-immune responses. PLoS One. 9(4):e95339

Alfirevic A, Gonzalez-Galarza F, Bell C, Martinsson K, Platt V, Bretland G et al (2012) In silico analysis of HLA associations with drug-induced liver injury: use of a HLA-genotyped DNA archive from healthy volunteers. Genome Med. 4(6):51

Alvarez-Sanchez R, Montavon F, Hartung T, Pahler A (2006) Thiazolidinedione bioactivation: a comparison of the bioactivation potentials of troglitazone, rosiglitazone, and pioglitazone using stable isotope-labeled analogues and liquid chromatography tandem mass spectrometry. Chem Res Toxicol. 19(8):1106–1116

Antoine DJ, Jenkins RE, Dear JW, Williams DP, McGill MR, Sharpe MR et al (2012) Molecular forms of HMGB1 and keratin-18 as mechanistic biomarkers for mode of cell death and prognosis during clinical acetaminophen hepatotoxicity. J Hepatol. 56(5):1070–1079

Bajt ML, Farhood A, Lemasters JJ, Jaeschke H (2008) Mitochondrial bax translocation accelerates DNA fragmentation and cell necrosis in a murine model of acetaminophen hepatotoxicity. J Pharmacol Exp Ther. 324(1):8–14

Baulies A, Ribas V, Nunez S, Torres S, Alarcon-Vila C, Martinez L et al (2015) Lysosomal Cholesterol Accumulation Sensitizes To Acetaminophen Hepatotoxicity by Impairing Mitophagy. Sci Rep. 5:18017

Boelsterli UA (2003) Animal models of human disease in drug safety assessment. J Toxicol Sci. 28(3):109–121

Boelsterli UA, Lim PL (2007) Mitochondrial abnormalities--a link to idiosyncratic drug hepatotoxicity? Toxicol Appl Pharmacol. 220(1):92–107

Budas GR, Churchill EN, Disatnik MH, Sun L, Mochly-Rosen D (2010) Mitochondrial import of PKCepsilon is mediated by HSP90: a role in cardioprotection from ischaemia and reperfusion injury. Cardiovasc Res. 88(1):83–92

Burcham PC, Harman AW (1991) Acetaminophen toxicity results in site-specific mitochondrial damage in isolated mouse hepatocytes. J Biol Chem. 266(8):5049–5054

Cadenas E, Davies KJ (2000) Mitochondrial free radical generation, oxidative stress, and aging. Free Radic Biol Med. 29(3-4):222–230

Chan DC (2006) Mitochondrial fusion and fission in mammals. Annu Rev Cell Dev Biol. 22:79–99

Chang AH, Sancheti H, Garcia J, Kaplowitz N, Cadenas E, Han D (2014) Respiratory substrates regulate S-nitrosylation of mitochondrial proteins through a thiol-dependent pathway. Chem Res Toxicol. 27(5):794–804

Chen W, Koenigs LL, Thompson SJ, Peter RM, Rettie AE, Trager WF et al (1998) Oxidation of acetaminophen to its toxic quinone imine and nontoxic catechol metabolites by baculovirus-expressed and purified human cytochromes P450 2E1 and 2A6. Chem Res Toxicol. 11(4):295–301

Churchill EN, Mochly-Rosen D (2007) The roles of PKCdelta and epsilon isoenzymes in the regulation of myocardial ischaemia/reperfusion injury. Biochem Soc Trans. 35(Pt 5):1040–1042

Copple IM, Goldring CE, Jenkins RE, Chia AJ, Randle LE, Hayes JD et al (2008) The hepatotoxic metabolite of acetaminophen directly activates the Keap1-Nrf2 cell defense system. Hepatology. 48(4):1292–1301

Copple IM, Goldring CE, Kitteringham NR, Park BK (2010) The keap1-nrf2 cellular defense pathway: mechanisms of regulation and role in protection against drug-induced toxicity. Handb Exp Pharmacol. 196:233–266

Dahlin DC, Miwa GT, Lu AY, Nelson SD (1984) N-acetyl-p-benzoquinone imine: a cytochrome P-450-mediated oxidation product of acetaminophen. Proc Natl Acad Sci U S A. 81(5):1327–1331

Daly AK (2012) Using genome-wide association studies to identify genes important in serious adverse drug reactions. Annu Rev Pharmacol Toxicol. 52:21–35

Daly AK, Donaldson PT, Bhatnagar P, Shen Y, Pe'er I, Floratos A et al (2009) HLA-B*5701 genotype is a major determinant of drug-induced liver injury due to flucloxacillin. Nat Genet. 41(7):816–819

Dara L, Johnson H, Suda J, Win S, Gaarde W, Han D et al (2015) Receptor interacting protein kinase 1 mediates murine acetaminophen toxicity independent of the necrosome and not through necroptosis. Hepatology. 62(6):1847–1857

Das S, Wong R, Rajapakse N, Murphy E, Steenbergen C (2008) Glycogen Synthase Kinase 3 Inhibition Slows Mitochondrial Adenine Nucleotide Transport and Regulates Voltage-Dependent Anion Channel Phosphorylation. Circ Res. 103:983–991

Ding WX, Yin XM (2012) Mitophagy: mechanisms, pathophysiological roles, and analysis. Biol Chem. 393(7):547–564

Ding WX, Guo F, Ni HM, Bockus A, Manley S, Stolz DB et al (2012a) Parkin and mitofusins reciprocally regulate mitophagy and mitochondrial spheroid formation. J Biol Chem. 287(50):42379–42388

Ding WX, Li M, Biazik JM, Morgan DG, Guo F, Ni HM et al (2012b) Electron microscopic analysis of a spherical mitochondrial structure. J Biol Chem. 287(50):42373–42378

Du K, Farhood A, Jaeschke H (2017) Mitochondria-targeted antioxidant Mito-Tempo protects against acetaminophen hepatotoxicity. Arch Toxicol 91:761–773

Fujimoto K, Kumagai K, Ito K, Arakawa S, Ando Y, Oda S et al (2009) Sensitivity of liver injury in heterozygous Sod2 knockout mice treated with troglitazone or acetaminophen. Toxicol Pathol. 37(2):193–200

Gandhi A, Guo T, Ghose R (2010) Role of c-Jun N-terminal kinase (JNK) in regulating tumor necrosis factor-alpha (TNF-alpha) mediated increase of acetaminophen (APAP) and chlorpromazine (CPZ) toxicity in murine hepatocytes. J Toxicol Sci. 35(2):163–173

Garcia J, Han D, Sancheti H, Yap LP, Kaplowitz N, Cadenas E (2010) Regulation of mitochondrial glutathione redox status and protein glutathionylation by respiratory substrates. J Biol Chem. 285(51):39646–39654

Goldring CE, Kitteringham NR, Elsby R, Randle LE, Clement YN, Williams DP et al (2004) Activation of hepatic Nrf2 in vivo by acetaminophen in CD-1 mice. Hepatology. 39(5): 1267–1276

Gomes LC, Di Benedetto G, Scorrano L (2011) During autophagy mitochondria elongate, are spared from degradation and sustain cell viability. Nat Cell Biol. 13(5):589–598

Gong J, Hoyos B, Acin-Perez R, Vinogradov V, Shabrova E, Zhao F et al (2012) Two protein kinase C isoforms, delta and epsilon, regulate energy homeostasis in mitochondria by transmitting opposing signals to the pyruvate dehydrogenase complex. FASEB J. 26(8):3537–3549

Gunawan BK, Liu ZX, Han D, Hanawa N, Gaarde WA, Kaplowitz N (2006) c-Jun N-terminal kinase plays a major role in murine acetaminophen hepatotoxicity. Gastroenterology. 131(1):165–178

Han D, Antunes F, Canali R, Rettori D, Cadenas E (2003a) Voltage-dependent anion channels control the release of the superoxide anion from mitochondria to cytosol. J Biol Chem. 278(8):5557–5563

Han D, Canali R, Rettori D, Kaplowitz N (2003b) Effect of glutathione depletion on sites and topology of superoxide and hydrogen peroxide production in mitochondria. Mol Pharmacol. 64(5):1136–1144

Han D, Hanawa N, Saberi B, Kaplowitz N (2006a) Mechanisms of liver injury. III. Role of glutathione redox status in liver injury. Am J Physiol Gastrointest Liver Physiol. 291(1):G1–G7

Han D, Hanawa N, Saberi B, Kaplowitz N (2006b) Hydrogen peroxide and redox modulation sensitize primary mouse hepatocytes to TNF-induced apoptosis. Free Radic Biol Med. 41(4):627–639

Han D, Ybanez MD, Ahmadi S, Yeh K, Kaplowitz N (2009) Redox regulation of tumor necrosis factor signaling. Antioxid Redox Signal 11:2245–2263

Han D, Shinohara M, Ybanez MD, Saberi B, Kaplowitz N (2010) Signal transduction pathways involved in drug-induced liver injury. Handb Exp Pharmacol. 196:267–310

Han D, Ybanez MD, Johnson HS, McDonald JN, Mesropyan L, Sancheti H et al (2012) Dynamic adaptation of liver mitochondria to chronic alcohol feeding in mice: biogenesis, remodeling, and functional alterations. J Biol Chem. 287(50):42165–42179

Han D, Dara L, Win S, Than TA, Yuan L, Abbasi SQ et al (2013) Regulation of drug-induced liver injury by signal transduction pathways: critical role of mitochondria. Trends Pharmacol Sci. 34(4):243–253

Han D, Johnson HS, Rao MP, Martin G, Sancheti H, Silkwood KH et al (2016) Mitochondrial remodeling in the liver following chronic alcohol feeding to rats. Free Radic Biol Med. 102:100–110

Hanawa N, Shinohara M, Saberi B, Gaarde WA, Han D, Kaplowitz N (2008) Role of JNK translocation to mitochondria leading to inhibition of mitochondria bioenergetics in acetaminophen-induced liver injury. J Biol Chem. 283(20):13565–13577

Hautekeete ML, Horsmans Y, Van Waeyenberge C, Demanet C, Henrion J, Verbist L et al (1999) HLA association of amoxicillin-clavulanate--induced hepatitis. Gastroenterology. 117(5):1181–1186

Hirata K, Takagi H, Yamamoto M, Matsumoto T, Nishiya T, Mori K et al (2008) Ticlopidine-induced hepatotoxicity is associated with specific human leukocyte antigen genomic subtypes in Japanese patients: a preliminary case-control study. Pharmacogenomics J. 8(1):29–33

Holt M, Ju C (2010) Drug-induced liver injury. Handb Exp Pharmacol. 196:3–27

Hu J, Ramshesh VK, McGill MR, Jaeschke H, Lemasters JJ (2016) Low Dose Acetaminophen Induces Reversible Mitochondrial Dysfunction Associated with Transient c-Jun N-Terminal Kinase Activation in Mouse Liver. Toxicol Sci. 150(1):204–215

Iancu TC, Mason WH, Neustein HB (1977) Ultrastructural abnormalities of liver cells in Reye's syndrome. Hum Pathol. 8(4):421–431

Ibrahim SH, Akazawa Y, Cazanave SC, Bronk SF, Elmi NA, Werneburg NW et al (2011) Glycogen synthase kinase-3 (GSK-3) inhibition attenuates hepatocyte lipoapoptosis. J Hepatol. 54(4): 765–772

Illing PT, Vivian JP, Dudek NL, Kostenko L, Chen Z, Bharadwaj M et al (2012) Immune self-reactivity triggered by drug-modified HLA-peptide repertoire. Nature. 486(7404):554–558

Jones DP, Lemasters JJ, Han D, Boelsterli UA, Kaplowitz N (2010) Mechanisms of pathogenesis in drug hepatotoxicity putting the stress on mitochondria. Mol Interv. 10(2):98–111

Juhaszova M, Zorov DB, Kim SH, Pepe S, Fu Q, Fishbein KW et al (2004) Glycogen synthase kinase-3beta mediates convergence of protection signaling to inhibit the mitochondrial permeability transition pore. J Clin Invest. 113(11):1535–1549

Kaplowitz N (2002) Biochemical and cellular mechanisms of toxic liver injury. Semin Liver Dis. 22(2):137–144

Kaplowitz N (2005) Idiosyncratic drug hepatotoxicity. Nat Rev Drug Discov. 4(6):489–499

Kaplowitz N, Aw TY, Ookhtens M (1985) The regulation of hepatic glutathione. Annu Rev Pharmacol Toxicol. 25:715–744

Karbowski M, Kurono C, Wozniak M, Ostrowski M, Teranishi M, Nishizawa Y et al (1999) Free radical-induced megamitochondria formation and apoptosis. Free Radic Biol Med. 26(3-4):396–409

Kashimshetty R, Desai VG, Kale VM, Lee T, Moland CL, Branham WS et al (2009) Underlying mitochondrial dysfunction triggers flutamide-induced oxidative liver injury in a mouse model of idiosyncratic drug toxicity. Toxicol Appl Pharmacol. 238(2):150–159

Kheifets V, Mochly-Rosen D (2007) Insight into intra- and inter-molecular interactions of PKC: design of specific modulators of kinase function. Pharmacol Res. 55(6):467–476

Kier LD, Neft R, Tang L, Suizu R, Cook T, Onsurez K et al (2004) Applications of microarrays with toxicologically relevant genes (tox genes) for the evaluation of chemical toxicants in Sprague Dawley rats in vivo and human hepatocytes in vitro. Mutat Res. 549(1-2):101–113

Latchoumycandane C, Goh CW, Ong MM, Boelsterli UA (2007) Mitochondrial protection by the JNK inhibitor leflunomide rescues mice from acetaminophen-induced liver injury. Hepatology. 45(2):412–421

Lee WM (2003) Drug-induced hepatotoxicity. N Engl J Med. 349(5):474–485

Lee KK, Imaizumi N, Chamberland SR, Alder NN, Boelsterli UA (2015) Targeting mitochondria with methylene blue protects mice against acetaminophen-induced liver injury. Hepatology. 61(1):326–336

Liguori MJ, Anderson MG, Bukofzer S, McKim J, Pregenzer JF, Retief J et al (2005) Microarray analysis in human hepatocytes suggests a mechanism for hepatotoxicity induced by trovafloxacin. Hepatology. 41(1):177–186

Liguori MJ, Ditewig AC, Maddox JF, Luyendyk JP, Lehman-McKeeman LD, Nelson DM et al (2010) Comparison of TNFalpha to Lipopolysaccharide as an Inflammagen to Characterize the Idiosyncratic Hepatotoxicity Potential of Drugs: Trovafloxacin as an Example. Int J Mol Sci. 11(11):4697–4714

Lou H, Kaplowitz N (2007) Glutathione depletion down-regulates tumor necrosis factor alpha-induced NF-kappaB activity via IkappaB kinase-dependent and -independent mechanisms. J Biol Chem. 282(40):29470–29481

Lucena MI, Andrade RJ, Martinez C, Ulzurrun E, Garcia-Martin E, Borraz Y et al (2008) Glutathione S-transferase m1 and t1 null genotypes increase susceptibility to idiosyncratic drug-induced liver injury. Hepatology. 48(2):588–596

Lucena MI, Garcia-Martin E, Andrade RJ, Martinez C, Stephens C, Ruiz JD et al (2010) Mitochondrial superoxide dismutase and glutathione peroxidase in idiosyncratic drug-induced liver injury. Hepatology. 52(1):303–312

Lucena MI, Molokhia M, Shen Y, Urban TJ, Aithal GP, Andrade RJ et al (2011) Susceptibility to amoxicillin-clavulanate-induced liver injury is influenced by multiple HLA class I and II alleles. Gastroenterology. 141(1):338–347

Luyendyk JP, Maddox JF, Cosma GN, Ganey PE, Cockerell GL, Roth RA (2003) Ranitidine treatment during a modest inflammatory response precipitates idiosyncrasy-like liver injury in rats. J Pharmacol Exp Ther. 307(1):9–16

Maddox JF, Amuzie CJ, Li M, Newport SW, Sparkenbaugh E, Cuff CF et al (2010) Bacterial- and viral-induced inflammation increases sensitivity to acetaminophen hepatotoxicity. J Toxicol Environ Health A. 73(1):58–73

Matsumaru K, Ji C, Kaplowitz N (2003) Mechanisms for sensitization to TNF-induced apoptosis by acute glutathione depletion in murine hepatocytes. Hepatology. 37(6):1425–1434

Maurer U, Charvet C, Wagman AS, Dejardin E, Green DR (2006) Glycogen synthase kinase-3 regulates mitochondrial outer membrane permeabilization and apoptosis by destabilization of MCL-1. Mol Cell. 21(6):749–760

McGill MR, Staggs VS, Sharpe MR, Lee WM, Jaeschke H (2014) Acute Liver Failure Study G. Serum mitochondrial biomarkers and damage-associated molecular patterns are higher in acetaminophen overdose patients with poor outcome. Hepatology. 60(4):1336–1345

Morimoto T, Tanaka A, Taki Y, Noguchi M, Yokoo N, Nishihira T et al (1988) Changes in concentrations of respiratory components and cytochrome oxidase activity in mitochondria obtained from carbon tetrachloride-induced cirrhotic rat liver. Clin Sci (Lond). 74(5):485–489

Nagai H, Matsumaru K, Feng G, Kaplowitz N (2002) Reduced glutathione depletion causes necrosis and sensitization to tumor necrosis factor-alpha-induced apoptosis in cultured mouse hepatocytes. Hepatology. 36(1):55–64

Nakagawa H, Maeda S, Hikiba Y, Ohmae T, Shibata W, Yanai A et al (2008) Deletion of apoptosis signal-regulating kinase 1 attenuates acetaminophen-induced liver injury by inhibiting c-Jun N-terminal kinase activation. Gastroenterology. 135(4):1311–1321

Ni HM, Bockus A, Boggess N, Jaeschke H, Ding WX (2012) Activation of autophagy protects against acetaminophen-induced hepatotoxicity. Hepatology. 55(1):222–232

Ni HM, Williams JA, Jaeschke H, Ding WX (2013) Zonated induction of autophagy and mitochondrial spheroids limits acetaminophen-induced necrosis in the liver. Redox biology. 1(1):427–432

O'Donohue J, Oien KA, Donaldson P, Underhill J, Clare M, MacSween RN et al (2000) Co-amoxiclav jaundice: clinical and histological features and HLA class II association. Gut. 47(5):717–720

Olafsdottir K, Reed DJ (1988) Retention of oxidized glutathione by isolated rat liver mitochondria during hydroperoxide treatment. Biochim Biophys Acta. 964(3):377–382

Ong MM, Wang AS, Leow KY, Khoo YM, Boelsterli UA (2006) Nimesulide-induced hepatic mitochondrial injury in heterozygous Sod2(+/-) mice. Free Radic Biol Med. 40(3):420–429

Ong MM, Latchoumycandane C, Boelsterli UA (2007) Troglitazone-induced hepatic necrosis in an animal model of silent genetic mitochondrial abnormalities. Toxicol Sci. 97(1):205–213

Ostapowicz G, Fontana RJ, Schiodt FV, Larson A, Davern TJ, Han SH et al (2002) Results of a prospective study of acute liver failure at 17 tertiary care centers in the United States. Ann Intern Med. 137(12):947–954

Osterloh J, Cunningham W, Dixon A, Combest D (1989) Biochemical relationships between Reye's and Reye's-like metabolic and toxicological syndromes. Med Toxicol Adverse Drug Exp. 4(4):272–294

Otera H, Mihara K (2011) Molecular mechanisms and physiologic functions of mitochondrial dynamics. J Biochem. 149(3):241–251

Porceddu M, Buron N, Roussel C, Labbe G, Fromenty B, Borgne-Sanchez A (2012) Prediction of liver injury induced by chemicals in human with a multiparametric assay on isolated mouse liver mitochondria. Toxicol Sci. 129(2):332–345

Reddy PH, Reddy TP, Manczak M, Calkins MJ, Shirendeb U, Mao P (2011) Dynamin-related protein 1 and mitochondrial fragmentation in neurodegenerative diseases. Brain Res Rev. 67(1-2):103–118

Roth RA, Ganey PE (2010) Intrinsic versus idiosyncratic drug-induced hepatotoxicity--two villains or one? J Pharmacol Exp Ther. 332(3):692–697

Ruepp SU, Tonge RP, Shaw J, Wallis N, Pognan F (2002) Genomics and proteomics analysis of acetaminophen toxicity in mouse liver. Toxicol Sci. 65(1):135–150

Saberi B, Shinohara M, Ybanez MD, Hanawa N, Gaarde WA, Kaplowitz N et al (2008) Regulation of H(2)O(2)-induced necrosis by PKC and AMP-activated kinase signaling in primary cultured hepatocytes. Am J Physiol Cell Physiol. 295(1):C50–C63

Saberi B, Ybanez MD, Johnson HS, Gaarde WA, Han D, Kaplowitz N (2014) Protein kinase C (PKC) participates in acetaminophen hepatotoxicity through c-jun-N-terminal kinase (JNK)-dependent and -independent signaling pathways. Hepatology. 59:1543–1554

Sanyal AJ, Chalasani N, Kowdley KV, McCullough A, Diehl AM, Bass NM et al (2010) Pioglitazone, vitamin E, or placebo for nonalcoholic steatohepatitis. N Engl J Med. 362(18):1675–1685

Schwabe RF (2006) Cell death in the liver-all roads lead to JNK. Gastroenterology. 131(1):314–316

Seki E, Brenner DA, Karin M (2012) A liver full of JNK: signaling in regulation of cell function and disease pathogenesis, and clinical approaches. Gastroenterology. 143(2):307–320

Sharma M, Gadang V, Jaeschke A (2012) Critical role for mixed-lineage kinase 3 in acetaminophen-induced hepatotoxicity. Mol Pharmacol. 82(5):1001–1007

Shaw PJ, Hopfensperger MJ, Ganey PE, Roth RA (2007) Lipopolysaccharide and trovafloxacin coexposure in mice causes idiosyncrasy-like liver injury dependent on tumor necrosis factor-alpha. Toxicol Sci. 100(1):259–266

Shinohara M, Ybanez MD, Win S, Than TA, Jain S, Gaarde WA et al (2010) Silencing glycogen synthase kinase-3beta inhibits acetaminophen hepatotoxicity and attenuates JNK activation and loss of glutamate cysteine ligase and myeloid cell leukemia sequence 1. J Biol Chem. 285(11):8244–8255

Shiryaeva A, Baidyuk E, Arkadieva A, Okovityy S, Morozov V, Sakuta G (2008) Hepatocyte mitochondrion electron-transport chain alterations in CCl(4) and alcohol induced hepatitis in rats and their correction with simvastatin. J Bioenerg Biomembr. 40(1):27–34
Shishido S, Koga H, Harada M, Kumemura H, Hanada S, Taniguchi E et al (2003) Hydrogen peroxide overproduction in megamitochondria of troglitazone-treated human hepatocytes. Hepatology. 37(1):136–147
Singer JB, Lewitzky S, Leroy E, Yang F, Zhao X, Klickstein L et al (2010) A genome-wide study identifies HLA alleles associated with lumiracoxib-related liver injury. Nat Genet. 42(8):711–714
Spahr L, Rubbia-Brandt L, Burkhard PR, Assal F, Hadengue A (2000) Tolcapone-related fulminant hepatitis: electron microscopy shows mitochondrial alterations. Dig Dis Sci. 45(9):1881–1884
Thames G (2004) Drug-induced liver injury: what you need to know. Gastroenterol Nurs. 27(1):31–33
Than TA, Lou H, Ji C, Win S, Kaplowitz N (2011) Role of cAMP-responsive element-binding protein (CREB)-regulated transcription coactivator 3 (CRTC3) in the initiation of mitochondrial biogenesis and stress response in liver cells. J Biol Chem. 286(25):22047–22054
Tong KI, Padmanabhan B, Kobayashi A, Shang C, Hirotsu Y, Yokoyama S et al (2007) Different electrostatic potentials define ETGE and DLG motifs as hinge and latch in oxidative stress response. Mol Cell Biol 27(21):7511–7521
Tukov FF, Luyendyk JP, Ganey PE, Roth RA (2007) The role of tumor necrosis factor alpha in lipopolysaccharide/ranitidine-induced inflammatory liver injury. Toxicol Sci. 100(1):267–280
Uetrecht J (2009) Immunoallergic drug-induced liver injury in humans. Semin Liver Dis. 29(4):383–392
Ulrich RG (2007) Idiosyncratic toxicity: a convergence of risk factors. Annu Rev Med. 58:17–34
Vick B, Weber A, Urbanik T, Maass T, Teufel A, Krammer PH et al (2009) Knockout of myeloid cell leukemia-1 induces liver damage and increases apoptosis susceptibility of murine hepatocytes. Hepatology. 49(2):627–636
Watkins PB (2005) Idiosyncratic liver injury: challenges and approaches. Toxicol Pathol. 33(1):1–5
Watkins PB (2009) Biomarkers for the diagnosis and management of drug-induced liver injury. Semin Liver Dis. 29(4):393–399
Watkins PB, Kaplowitz N, Slattery JT, Colonese CR, Colucci SV, Stewart PW et al (2006) Aminotransferase elevations in healthy adults receiving 4 grams of acetaminophen daily: a randomized controlled trial. JAMA. 296(1):87–93
Westermann B (2010) Mitochondrial fusion and fission in cell life and death. Nat Rev Mol Cell Biol. 11(12):872–884
Wiltshire C, Gillespie DA, May GH (2004) Sab (SH3BP5), a novel mitochondria-localized JNK-interacting protein. Biochem Soc Trans. 32(Pt 6):1075–1077
Win S, Than TA, Han D, Petrovic LM, Kaplowitz N (2011) c-Jun N-terminal kinase (JNK)-dependent acute liver injury from acetaminophen or tumor necrosis factor (TNF) requires mitochondrial Sab protein expression in mice. J Biol Chem. 286(40):35071–35078
Win S, Than TA, Fernandez-Checa JC, Kaplowitz N (2014) JNK interaction with Sab mediates ER stress induced inhibition of mitochondrial respiration and cell death. Cell Death Dis. 5:e989
Win S, Than TA, Le BH, Garcia-Ruiz C, Fernandez-Checa JC, Kaplowitz N (2015) Sab (Sh3bp5) dependence of JNK mediated inhibition of mitochondrial respiration in palmitic acid induced hepatocyte lipotoxicity. J Hepatol. 62(6):1367–1374
Win S, Than TA, Min RW, Aghajan M, Kaplowitz N (2016) c-Jun N-terminal kinase mediates mouse liver injury through a novel Sab (SH3BP5)-dependent pathway leading to inactivation of intramitochondrial Src. Hepatology. 63(6):1987–2003
Winnike JH, Li Z, Wright FA, Macdonald JM, O'Connell TM, Watkins PB (2010) Use of pharmaco-metabonomics for early prediction of acetaminophen-induced hepatotoxicity in humans. Clin Pharmacol Ther. 88(1):45–51
Wong GH, Elwell JH, Oberley LW, Goeddel DV (1989) Manganous superoxide dismutase is essential for cellular resistance to cytotoxicity of tumor necrosis factor. Cell. 58(5):923–931

Wu W, Zhao L, Yang P, Zhou W, Li B, Moorhead JF et al (2016) Inflammatory Stress Sensitizes the Liver to Atorvastatin-Induced Injury in ApoE-/- Mice. PLoS One. 11(7):e0159512
Wullaert A, Heyninck K, Beyaert R (2006) Mechanisms of crosstalk between TNF-induced NF-kappaB and JNK activation in hepatocytes. Biochem Pharmacol. 72(9):1090–1101
Xu JJ, Henstock PV, Dunn MC, Smith AR, Chabot JR, de Graaf D (2008) Cellular imaging predictions of clinical drug-induced liver injury. Toxicol Sci. 105(1):97–105
Yin XM, Ding WX, Gao W (2008) Autophagy in the liver. Hepatology. 47(5):1773–1785

Chapter 9
Mitochondrial Damage and Mitophagy in Ischemia/Reperfusion-Induced Liver Injury

Kristina L. Go, Sooyeon Lee, Kevin E. Behrns, and Jae-Sung Kim

Abbreviations

ALCAT1	Lysocardiolipin acyltransferase 1
ALLM	Acetyl-Leu-Leu-methioninal
AMPK	Adenosine monophosphate-activated protein kinase
ANT	Adenine nucleotide translocase
ATG	Autophagy-related proteins
ATP	Adenosine triphosphate
BECN1	Beclin-1
CsA	Cyclosporine A
CVP	Central venous pressure
ER	Endoplasmic reticulum
FUNDC1	FUN14 domain-containing protein-1
HCC	Hepatocellular carcinoma
IPC	Ischemic preconditioning
I/R	Ischemia/reperfusion
JNK1	c-Jun N-terminal protein kinase 1
KAT	Lysine acetyltransferase
KDAC	Lysine deacetylase
LC3	Microtubule-associated protein 1 light chain 3
MFN	Mitofusin
MPT	Mitochondrial permeability transition
mtDNA	Mitochondrial DNA

K.L. Go • S. Lee • K.E. Behrns • J.-S. Kim (✉)
Department of Surgery, University of Florida, Gainesville, FL 32610, USA
e-mail: Jae.Kim@surgery.ufl.edu

W.-X. Ding, X.-M. Yin (eds.), *Molecules, Systems and Signaling in Liver Injury*, Cell Death in Biology and Diseases, DOI 10.1007/978-3-319-58106-4_9

X.-M. Yin and Z. Dong (Series eds.), *Cell Death in Biology and Diseases*

mTOR	Mammalian target of rapamycin
NAD^+	Oxidized nicotinamide adenine dinucleotide
NAFLD	Nonalcoholic fatty liver disease
NASH	Nonalcoholic steatohepatitis
PI	Propidium iodide
PI3KIII	Class III phosphatidylinositol 3-kinase
PINK1	PTEN-induced putative kinase protein 1
RIPC	Remote ischemic preconditioning
ROS	Reactive oxygen species
SIRT1	Sirtuin 1
TMRM	Tetramethylrhodamine methyl ester
ULK1	UNC-51-like kinase 1
VDAC	Voltage-dependent adenine nucleotide

9.1 Introduction

Chronic liver disease is a consequence of many etiologies including viral infection, congenital disorders, repeated ethanol exposure, pharmacological toxicities, and metabolic disorders. The spectrum of chronic liver disease ranges from simple steatosis to steatohepatitis which can progress to fibrosis, cirrhosis, and hepatocellular carcinoma (HCC). Chronic liver disease is currently the fifth leading cause of death worldwide. Furthermore, in comparison to other types of cancer, the incidence and rates of cancer-related deaths from HCC continue to rise in both genders (Ryerson et al. 2016). Although conservative strategies are an important component of therapy for chronic liver disease, the only strategies that can provide long-term cure for most patients are surgical resection and liver transplantation. As ischemia/reperfusion (I/R) injury inevitably occurs during these surgical treatments, patients with chronic liver disease, who have compromised liver function, are at higher risk of developing hepatic failure after surgery.

In this chapter, we review and update current information on liver I/R injury, including the unique anatomy of the liver associated with its vulnerability to I/R injury, types of I/R injury, and the central role of mitochondrial dysfunction and defective autophagy in hepatocellular death secondary to I/R. Moreover, we discuss how underlying patient comorbidities such as aging and steatosis further impact I/R injury and describe proposed methods to mitigate hepatic I/R injury.

9.2 Liver Anatomy and I/R Injury

The evolution of hepatic surgery in the last century can be attributed in part to increased understanding of internal biliary and vascular anatomy of the liver. Where hepatic trauma and liver resection were once associated with significant morbidity and mortality, techniques of vascular clamping first described by J. H. Pringle in the 1900s (Pringle 1908) and later refined in Jean-Louis Lortat-Jacob's description of a right hepatectomy in the 1950s (Lortat-Jacob et al. 1952) have shifted the landscape of liver surgery from a rare operation to a regular but complex procedure. While hepatic resection has become commonplace, operative blood loss remains an independent predictor of perioperative morbidity and mortality (Jarnagin et al. 2002); therefore, an understanding of hepatic anatomy is cardinal in improving surgical outcomes. Likewise, evaluation of gross and cellular anatomy is a vital component in the study of hepatic I/R injury.

9.2.1 Vascular Anatomy

Blood flow through the liver ranges from 1,500 to 2,000 ml/min and accounts for approximately 25% of the cardiac output. The substantial blood volume that enters the liver can be inferred by examining the organ's dual vascular inflow: the portal venous supply provides 75% of hepatic inflow, which is greater than 50% of the liver's oxygen requirements (Vollmar and Menger 2009). The hepatic artery provides 25% of hepatic inflow to meet the remaining metabolic and functional needs of the liver. Blood enters the liver through terminal portal venules and hepatic arterioles, flows through the sinusoids, and drains through a central vein to the hepatic vein and, ultimately, to the suprahepatic inferior vena cava.

Vascular anatomy of the liver is highly variable, and care must be taken to examine the vasculature preoperatively and intraoperatively to ensure the safest approach during resection. The portal vein arises behind the neck of the pancreas and is formed by the confluence of the superior mesenteric and splenic veins. From the neck of the pancreas, it ascends posteriorly to the common bile duct and hepatic artery into the hilum of the liver. Here, the portal vein bifurcates into a shorter, more vertical, right portal branch, dividing further into the right anterior and posterior branches, and a longer, more horizontal left portal branch. The left portal branch enters the umbilical fissure to supply the left liver. In a common anatomical variation, the right anterior and posterior branches may come directly from the portal vein, and the right portal branch may be absent.

Variation also exists in the hepatic arterial system. In approximately 50% of patients, the hepatic artery arises as a branch of the celiac artery, becomes the proper hepatic artery after giving off gastroduodenal and right gastric arteries, courses through the hepatoduodenal ligament, and branches into the right and left hepatic arteries. The most common variations seen include a replaced right hepatic artery, which originates from the superior mesenteric artery, and a replaced left hepatic artery, which arises from the left gastric artery.

9.2.2 Functional Surgical Anatomy

Historically, the liver was divided into the right and left lobes by the falciform ligament; however, division of the liver into sections based on either arteriobiliary segmentation (Healey and Schroy 1953) or Couinaud segmentation is now widely adopted (Couinaud 1999). Following Couinaud's nomenclature, the liver is divided into four sections based on the three main branches of the hepatic veins. Each segment drains bile via a bile duct and receives blood flow from a branch of the hepatic artery and portal vein resulting in eight segments. Controversy in nomenclature and segmentation, however, exists as to whether the caudate lobe should be divided into two segments. While Couinaud's segmentation is often used, several attempts have been made to standardize terminology used during hepatic resection including the development of Brisbane terminology (Celinski and Gamblin 2010).

9.2.3 Microcirculatory Anatomy of the Liver

Specific aspects of the liver's histologic architecture contribute to its unique functions and the pathogenesis of an array of liver diseases. The smallest functional unit of the liver is the acinus, which is derived from the liver's blood supply. The dimension of the unit is based on blood flow that originates from terminal arterioles and portal venules, passes through sinusoids, and drains from a central vein to the hepatic vein. Thus, the acinus is routinely divided into zones with relation to the portal vein (periportal zone) and the central vein (perivenous, pericentral, centrilobular) and is often subdivided based on differential expression of key enzymes.

Unidirectional blood flow through the acinus creates unique gradients of metabolites and oxygen that can impact the functions of hepatocytes as well as nonparenchymal cells. A partial pressure of oxygen in periportal blood (60–65 mmHg) is substantially greater than that in perivenous blood (30–35 mmHg) (Jungermann and Kietzmann 1996). The zonal difference in oxygen concentration plays an important role in both physiology and pathology of the liver. For example, hepatocytes near the periportal area have greater capacity for oxidative metabolism, gluconeogenesis, ureagenesis, beta-oxidation, and cholesterol synthesis, as compared to pericentral counterparts. In contrast, pericentral hepatocytes demonstrate higher rates of lipogenesis, glycolysis, and drug detoxification than periportal cells. Besides this zonal dependency in basal functions, the oxygen gradient markedly influences the onset of hepatocellular injury that is often confined to a particular zone of the liver lobule. Patients with steatosis and hemochromatosis exhibit prominent lipid peroxidation and hemosiderin accumulation in the pericentral area (MacDonald et al. 2001; Niemelä 2001). In addition, hepatocytes close to the pericentral region are the first to incur hypoxic injury due to the intralobular oxygen gradient (Jungermann and Kietzmann 2000). In contrast, after partial hepatectomy or drug-induced hepatocyte damage, regeneration of liver begins in the periportal area and extends to the pericentral zone afterward (Fabrikant 1968; Lee et al. 1998).

9.3 Hepatic I/R Injury: Clinical Context and Cellular Mechanisms

9.3.1 Types of I/R Injury

I/R injury can occur in a myriad of clinical states either secondary to low-flow states such as hypovolemia, sepsis, or cardiogenic shock, sinusoidal obstruction syndrome, Budd-Chiari syndrome, or other veno-occlusive dysfunction or as a consequence of surgical resection. Despite advances in understanding hepatic anatomy, improved surgical and anesthetic techniques, and more prudent patient selection, operative blood loss remains a major concern during hepatectomy. For patients undergoing liver resection, high operative blood loss is an independent risk factor of morbidity and mortality (Shimada et al. 1994). Intraoperative blood loss more than 1000 mL has been shown to be an independent predictor of hyperbilirubinemia (>5 mg/dL) after resection (Fukumori et al. 2011). Among patients with HCC, increased blood loss heightens morbidity and mortality (Shimada et al. 1998), and the extent of blood loss independently predicts risk recurrence and survival (Katz et al. 2009).

Several techniques exist in the current perioperative armamentarium to limit blood loss during hepatectomy. Due to the absence of valves in the hepatic veins, venous bleeding in the liver directly correlates to central venous pressure (CVP). Higher CVP results in distended veins and higher blood loss. By limiting intraoperative fluids during resection, a target CVP of less than 5 mmHg minimizes back-bleeding during parenchymal transection (Celinski and Gamblin 2010). Once resection and hemostasis are achieved, the patient may be resuscitated to euvolemia. Low CVP anesthetic technique is well tolerated by patients and minimizes intraoperative blood loss and transfusion rates (Melendez et al. 1998).

In addition to low CVP anesthesia, various vascular occlusion techniques have been developed to limit bleeding and may be further categorized as strategies to occlude hepatic inflow or maneuvers that occlude both vascular inflow and outflow. During hepatic pedicle clamping, also known as Pringle's maneuver, a vascular clamp or loop encircles the hepatoduodenal ligament containing the portal vein and hepatic artery. Disappearance of the distal pulse of the hepatic artery indicates successful occlusion. Pringle's maneuver may be done continuously or intermittently throughout the procedure. In addition to I/R, risks of clamping include splanchnic congestion and air embolus (Zhou et al. 2008), which becomes higher when Pringle's maneuver is used in conjunction with a low CVP. Selective inflow occlusion has also been described wherein inflow occlusion is applied to the branches of the hepatic artery and portal vein that feed the hemi-liver that is to be resected. Although this technique minimizes I/R injury to remnant liver and reduces splanchnic congestion, significant hilar dissection is required, and risk of blood loss to the non-clamped hemi-liver still exists (Abdalla et al. 2004).

Hepatic ischemia can be categorized into warm and cold ischemia. Whereas warm I/R is observed as a consequence of vascular occlusion and low-flow states, cold I/R is unique to liver transplant where the explanted graft is subjected to a hypothermic, ischemic, preservation phase prior to warm reperfusion after vascular reanastomosis (Papadopoulos et al. 2013). While cellular death is the common outcome from either type of I/R injury, the mechanisms of injury are distinct. Cold I/R injures sinusoidal endothelium and non-parenchymal cells (Klune and Tsung 2010), whereas hepatocytes are a major cell type experiencing warm I/R injury (Cursio et al. 2015).

9.3.1.1 Warm I/R Injury

The mechanisms underlying warm I/R injury are multifactorial and occur in a progressive manner. Soon after reperfusion, calcium overloading and reactive oxygen species (ROS) accumulation in mitochondria result in onset of the mitochondrial permeability transition (MPT), mitochondrial dysfunction, and necrotic death of hepatocytes. Continued reperfusion activates non-parenchymal cells like Kupffer cells and propagates further ROS production, leading to stimulation of chemokines and cytokines that actively recruit neutrophils (Jaeschke 1998; Klune and Tsung 2010). During the late stage of reperfusion, uncontrolled neutrophil infiltration causes irreversible inflammation (Zhai et al. 2011). Although inflammation plays an important role in late reperfusion injury, hepatocyte damage at the early stage of reperfusion is the prominent pathological event that instigates permanent hepatic injury later.

Three major biochemical changes occur during warm ischemia; (1) anoxia, (2) nutrient depletion, and (3) cytosolic acidosis. Interruption of blood flow limits oxygen delivery to mitochondria and depletes adenosine triphosphate (ATP) synthesis from the mitochondrial electron transport chain, leading to disruption of energy-dependent metabolic processes (Cursio et al. 2015). ATP-dependent channels and exchangers are likewise compromised, and dysregulation of sodium, chloride, and calcium homeostasis results (Cursio et al. 2015). Halted supply of nutrient and energy substrates resulting from impeded blood flow also potentiates ATP loss. Moreover, anoxia lowers cytosol pH due to the accumulation of lactate and hydrogen ion, limiting the efficacy of enzymes that optimally operate at a neutral pH. While prolonged tissue acidosis will eventually cause liver cell death, acidosis during acute ischemia confers cytoprotection due to blockade of MPT onset and inactivation of injurious catabolic enzymes (Kim et al. 2003b). However, damage initiated during ischemia is aggravated upon restoration of blood flow and the return of physiologic pH, a phenomenon called pH paradox (Kim et al. 2003b).

9.3.1.2 Cold I/R Injury

In contrast to warm I/R injury, endothelial and non-parenchymal cells are major contributors to reperfusion injury after cold ischemia (Klune and Tsung 2010). As such, pathological manifestations observed in cold I/R reflect aberrant functions of non-parenchymal cells, including decreased graft microcirculation, increased platelet activation, vasoconstriction, and upregulation of adhesion molecules. Of importance, these alterations further activate Kupffer cells, recruit neutrophils, and enhance production of ROS, ultimately potentiating hepatocyte death (Peralta et al. 2013). Because cold I/R is unique to orthotopic liver transplantation, while warm I/R is observed in a multitude of clinical settings, we will focus on the mechanisms, clinical implications, and therapies involved in warm I/R injury.

9.4 Mitochondrial Dysfunction After I/R

The liver plays a central role in the clearance of toxins, synthesis of hormones and proteins, vitamin storage, blood sugar regulation, bile synthesis, ketone body synthesis, and lipid homeostasis. Besides these multiple functions, the liver's regenerative capacity further demonstrates the need for hepatocytes to generate high amounts of ATP to meet energy demands for hepatic regrowth. Mitochondria, a cell's power plant, generate more than 90% of cellular ATP and are composed of two membranes with differential permeability, an outer membrane with more permissive permeability and an inner membrane, which is nearly impermeable to all solutes except those with specific carriers or exchangers. Structural integrity of mitochondrial membranes is critical for mitochondrial functions and cell survival. Hence, it is not surprising to observe that alterations in structural barrier in mitochondria often precede hepatocellular injury and death. In particular, loss of mitochondrial inner membrane permeabilization and subsequent onset of mitochondrial dysfunction are a key causal mechanism of hepatocyte death after I/R. In this section, we will describe the role of mitochondrial injury after I/R and discuss how underlying mitochondrial dysfunction is involved in aging and fatty liver disease, all of which predispose hepatocytes to further I/R injury.

9.4.1 Mitochondrial Permeability Transition (MPT)

Mitochondrial membrane integrity is prerequisite for the maintenance of proton motive force across mitochondria and can be challenged by prolonged opening of MPT pores. Though further studies are warranted to elucidate its precise molecular composition, the general consensus is that MPT pores have three major components: an adenine nucleotide translocator (ANT) on the inner mitochondrial membrane, voltage-dependent anion channels (VDAC) on the outer mitochondrial membrane, and cyclophilin D located in the mitochondrial matrix (Kim et al. 2003a). Cyclosporine A (CsA) blocks cyclophilin D, a peptidyl-prolyl *cis-trans* isomerase that catalyzes conformational changes in ANT leading to pore opening. Bongkrekic acid, a toxin generated from fermented coconut products, inhibits ANT (Halestrap et al. 2002). In addition, Mg^{2+}, acidic pH, and trifluoroperazine are known to inhibit MPT opening (Lemasters et al. 1998). Under basal conditions, transient opening and closing of these high-conductance permeability transition pores stochastically take place to temporarily create a burst of superoxide production, matrix alkalinization, and oscillation of the mitochondrial membrane potential, suggesting that mitochondrial flashes may be intimately integrated with the regulation of intramitochondrial Ca^{2+} and ROS (Wang et al. 2008; Hou et al. 2013, 2016). However, under pathological conditions like I/R, permanent opening of MPT pores ensues, resulting in either necrotic or apoptotic cell death. During ischemia, there is a profound drop in ATP with a concomitant increase in inorganic phosphate that stimulates glycolysis and causes a buildup of lactate. Although this intracellular acidosis prevents opening of high-conductance permeability transition pores of the mitochondrial inner membrane, it loads the cell with Na^+ through the Na^+/H^+ antiporter. As low ATP levels during ischemia also reduce Na^+/K^+ ATPase activity, Na^+ overloading becomes persistent and eventually reverses Na^+/Ca^{2+} antiporter which culminates in excess Ca^{2+} loading. Upon reperfusion, recovery of oxygen and energy substrates together with a compromised electron transport chain outbursts ROS and incites MPT onset in mitochondria. Thus, the combination of oxidative stress, ATP depletion, elevated inorganic phosphate, and increased matrix calcium after reperfusion permanently opens pores to initiate the onset of MPT (Qian et al. 1997; Halestrap and Richardson 2015). The instigation of MPT collapses the mitochondrial inner membrane permeability barrier, allowing solutes up to 1,500 Da to diffuse freely across the mitochondrial inner membrane (Kim et al. 2003a, 2006a, 2008a, 2012). Free diffusion into the mitochondrial matrix induces swelling of mitochondria and uncoupling of oxidative phosphorylation and depolarization of mitochondria (Go et al. 2015). In the end, the loss of mitochondrial membrane potential and collapse of pH gradient nullify mitochondrial proton motive force, causing depletion of ATP and hepatocellular death.

Near-complete depletion of ATP during prolonged ischemia makes the occurrence of apoptosis less likely than necrosis. Using confocal microscopy, the MPT, mitochondrial depolarization, and necrotic cell death after reperfusion can be visualized in live hepatocytes. Calcein, a green-fluorescing fluorophore, is a polyanionic solute of 623 Da that is impermeable to the mitochondrial inner membrane (Fig. 9.1). As a result, confocal images of calcein-loaded hepatocytes have a honeycomb appearance with numerous dark round or oval voids corresponding to individual respiring mitochondria. Co-loading with tetramethylrhodamine methyl ester (TMRM), a red-fluorescing cationic fluorophore that accumulates electrophoretically in mitochondria, confirms that virtually every dark void is a polarized mitochondrion. TMRM does not quench calcein fluorescence to cause these voids. After 4 h of simulated ischemia when isolated rodent hepatocytes are subjected to anoxic, nutrient-depleted, and acidotic conditions, mitochondria depolarize and release TMRM due to lack of oxygen, but do not become permeable to calcein, as shown by the persistence of dark voids in calcein images, indicative of lack of MPT onset during ischemia. After reperfusion (reoxygenation at pH 7.4), mitochondria begin to repolarize rapidly. The cell, however, depolarizes after 10 min (loss of TMRM). Concomitant with depolarization, calcein redistribution into some of the

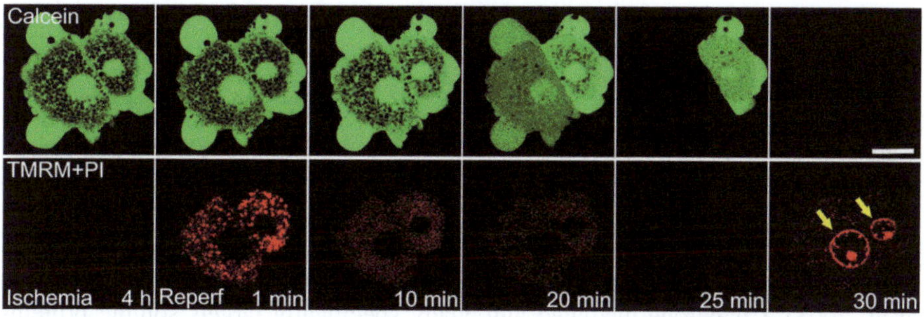

Fig. 9.1 The onset of MPT and hepatocyte death after reperfusion. Confocal images of calcein (*green*), TMRM (*red*), and propidium iodide (PI, *red*) after 4 h of simulated ischemia of rodent hepatocytes. During ischemia, calcein was excluded by mitochondria, indicative of permeability pore closure. After 1 min of reperfusion, mitochondria repolarized without calcein redistribution. After 10 min, calcein redistributed into some mitochondria due to MPT onset, and mitochondria began to depolarize, as evidenced by loss of TMRM fluorescence. Within 20 min, widespread onset of MPT occurred in nearly all mitochondria. Later, calcein fluoresce disappeared because of the loss of plasma membrane integrity, concomitant with nuclear labeling of PI (*arrows*), denoting cell death. Scale bar, 10 μm

mitochondrial matrices is evident, leading to the disappearance of the dark voids of the mitochondria (onset of the MPT). One noticeable event at this time point is that redistribution of calcein is confined to a discrete population of mitochondria, suggesting that some mitochondria are more susceptible to MPT induction. A few minutes later, all mitochondria become permeable to calcein, and cells completely lose the mitochondrial membrane potential. After 30 min, cell viability is lost, as shown by total loss of calcein and nuclear staining with propidium iodide (PI). When cells are reperfused in conditions that block the MPT, including CsA at pH 7.4 or acidic pH, mitochondria continue to repolarize after reperfusion, mitochondrial calcein voids remain intact, and cell viability is sustained (Qian et al. 1997). Hence, these observations indicate that reperfusion causes CsA-sensitive mitochondrial inner membrane permeabilization and depolarization, leading to cell death. Thus, onset of the MPT is a key mechanism underlying pH-dependent necrosis after I/R in hepatocytes. While necrosis is the dominant fate of cell death during I/R, apoptosis can occur in the liver after I/R. Studies with isolated rodent hepatocytes have demonstrated that the onset of MPT induces apoptotic cell death in the presence of glycolytic energy supply (Qian et al. 1999; Kim et al. 2003a, b, c). After MPT onset, mitochondrial swelling ruptures the mitochondrial outer membranes and releases cytochrome *c*. Normally sequestered in the mitochondrial intermembrane space, this 12 kDa protein is released into the cytosol, binds to apoptosis-inducing factor-1, and pro-caspases to form the apoptosome. As activation of caspases is an energy-dependent process, the presence of ATP following mitochondrial membrane permeabilization triggers apoptotic signaling (Lemasters 1999; Lemasters et al. 2002; Kim et al. 2003a). In sharp contrast, ATP-deficient hepatocytes undergo necrotic cell death in spite of the presence of upstream proapoptotic signals. Thus, the onset of MPT and glycolytic ATP availability govern the fate of hepatocyte death after reperfusion as to whether cells after I/R undergo necrosis or apoptosis (Kim et al. 2003b).

9.4.2 *Mitophagy: Mitochondrial Quality Control Machinery*

9.4.2.1 Autophagy

Autophagy is an evolutionarily conserved pathway that removes protein aggregates, surplus cytoplasmic contents, or dysfunctional organelles through lysosomal digestion (Baehrecke 2005). First described by Christian de Duve as a cellular "self-eating" process (Yang and Klionsky 2010), three types of autophagy exist: chaperone-mediated autophagy, microautophagy, and macroautophagy (Mizushima et al. 2008). Chaperone-mediated autophagy specifically targets soluble proteins with the KFERQ pentapeptide motif. Proteins with this motif are recognized by 70 kDa molecular chaperone heat shock protein. After interacting with lysosome-associated membrane protein 2A, cargo is delivered to the lysosome for degradation (Massey et al. 2006). Microautophagy delivers cytosolic substrates directly to the endosomal or lysosomal membrane where cellular cargos are invaginated into the lumen for degradation (Santambrogio and Cuervo 2011). Mechanistic details of microautophagy are largely unknown. Macroautophagy, herein referred to as "autophagy," is the most common form of autophagy. Two types of macroautophagy have been described: canonical, where multiple autophagy-related proteins (ATG) interact to form a cup-shaped double membrane complex called the phagophore, and the less understood noncanonical autophagy that can occur in the absence of some key ATG (Codogno et al. 2012). To date, over 30 ATG have been identified, and 18 ATG are essential to formation of the autophagosome (Vlada et al. 2015).

Five key steps occur during autophagy: initiation, nucleation, elongation, fusion, and degradation (Kim and Lee 2014). During basal conditions where autophagy occurs infrequently, mammalian target of rapamycin complex 1 (mTORC1) phosphorylates UNC-51-like kinase 1 (ULK1), the mammalian ortholog of yeast ATG1, at the Ser757 residue. When phosphorylated by mTORC1, ULK1 is dissociated from adenosine monophosphate-activated protein kinase (AMPK), which prevents autophagy initiation but promotes cellular proliferation and growth. Reduction in energy content, such as starvation or hypoxia, drives AMPK-mediated phosphorylation of Ser 317, 555, and 777 residues of ULK1 (Jung et al. 2010; Kuma and Mizushima 2010; Rabinowitz and White 2010; Kim et al. 2011; Settembre et al. 2011; Sanchez et al. 2012; Rubinsztein et al. 2012). By binding RAPTOR, phosphorylated ULK1 suppresses the kinase activity of mTORC1 (Jung et al. 2011). Additionally, activated ULK1 phosphorylates other targets including ATG13 and focal adhesion kinase family interacting protein of 200 kDa (FIP200), forming the ULK1-ATG13-FIP200 complex on the surface of the double-membrane phagophore. It has been reported that membranes of the phagophore are originated from endoplasmic reticulum (ER), Golgi, or mitochondria (Mijaljica et al. 2006; Hara et al. 2008; Hosokawa et al. 2009; Kuma and Mizushima 2010). In parallel with initiation events, further phosphorylation of Beclin-1 (BECN1) by stress-responsive c-Jun N-terminal protein kinase 1 (JNK1) liberates BECN1 from its B-cell lymphoma 2 (Bcl-2)-associated complex (Wei et al. 2008; Dai et al. 2011). Free BECN1 enables the recruitment of ATG14, vacuolar protein sorting 34 (VPS 34), and VPS15 to generate the BECN1-ATG15-VPS34-VPS15 complex (Russell et al. 2013). Other proteins such as VPS34, a class III lipid kinase, are likely involved in the modulation of autophagosomal membranes and vesicle trafficking (Backer 2008; Obara and Ohsumi 2011). The phagophore is further shaped by several other factors including UV radiation resistance-associated gene (UVRAG) (Liang et al. 2006), endophilin B-1 (BIF-1) (Takahashi et al. 2007), ATG14L (Itakura et al. 2008), and RUN domain Beclin-1-interacting and cysteine-rich domain-containing protein (RUBICON) (Matsunaga et al. 2009), corroborating the complexity of autophagy initiation.

Elongation of the autophagosome is the coordinated orchestration between ATG12 and microtubule-associated protein 1 light chain 3 (LC3). ATG12, covalently linked to ATG5 through the E2-like ubiquitin carrier protein ATG10, becomes activated by ATG7, an E1-ubiquitin-like enzyme. The ATG12-ATG5 complex ultimately interacts with ATG16L1 (Mizushima et al. 2011). Parallel to formation of the ATG12-ATG5-ATG16LI complex, ATG7 and ATG3 conjugate phosphatidylethanolamine to LC3-I to generate LC3-II, a biomarker of autophagy. While unconjugated LC3-I predominantly resides in the cytosol, LC3-II localizes to autophagosomal membranes (Ravikumar et al. 2010; Mizushima et al. 2010). ATG4B, a cysteine protease, cleaves the C-terminus of LC3 for conversion of LC3-I into LC3-II, but also possesses the ability to release the phosphatidylethanolamine group from LC3-II (Levine and Klionsky 2004). Removal of this phospholipid group and release of LC3 from the autophagosome induce fusion of the autophagosome with endosomes or lysosomes. Thus, ATG4B is essential both in recycling LC3-II and in the fusion and degradation of autophagosomal contents.

Fusion between the mature autophagosome and a lysosome generates the autolysosome (Ao et al. 2014). Components that mediate fusion include Ras-like GTPase RAB7, which tethers the autophagosome to the lysosome (Chua et al. 2011), and the UVRAG-BECN1-class III phosphatidylinositol 3-kinase (PI3KIII) and class C vacuolar protein-sorting complexes, whose interaction likely facilitates Rab7-mediated fusion (Moreau et al. 2013). Soluble N-ethylmaleimide-sensitive fusion factors also affect autolysosome formation. During degradation, catabolic enzymes residing in autolysosomal lumen including acidic proteases, lipases, nucleases, and glycosidases degrade cellular cargos into amino acids, which reenter the cell for other metabolic purposes (Weber et al. 1998; Rabinowitz and White 2010; Moreau et al. 2013).

Growing evidence indicates that autophagy may operate either selectively or nonselectively. The degradation of specific cellular constituents and organelles occurs during selective autophagy and includes pexophagy (peroxisomes) (Dunn et al. 2005), mitophagy (mitochondria) (Lemasters 2014), ribophagy (ribosomes) (Kraft et al. 2008), reticulophagy (endoplasmic reticulum) (Hamasaki et al. 2005), lipophagy (lipid droplets) (Singh et al. 2009), and ferrintinophagy (iron) (Mancias et al. 2014).

9.4.2.2 Types of Mitophagy

Mitochondria are the major source of ROS generation, and mitochondrial DNA (mtDNA) is distinctly prone to ROS-mediated injury (Shadel and Clayton 1997). Aberrant alterations in mtDNA are implicated in a broad range of human diseases due to its high susceptibility to oxidative damage and its limited DNA repair capability, as compared to nuclear DNA (Penta et al. 2001). Therefore, accumulation of abnormal mitochondria resulting from defective mitochondrial clearance causes uncontrolled ROS formation, mtDNA mutation, energetic failure, and ultimately cell death. As MPT onset and mitochondrial dysfunction are causative events attributing to hepatocellular death after I/R, timely clearance of abnormal or dysfunctional mitochondria is pivotal to hepatocellular survival after reperfusion. Accordingly, factors that impair the removal of damaged mitochondria promote I/R injury, and factors that improve clearance may prevent I/R injury. Mitophagy is a selective autophagy that sequesters and degrades damaged or abnormal mitochondria in a timely manner (Fig. 9.2). Mitophagy is also involved in normal mitochondrial turnover. Lemasters has recently proposed three distinct types of mitophagy (Lemasters 2014). Type I mitophagy follows the mechanisms similar to the canonical autophagic pathway described previously and is phosphatidylinositol-3-kinase (PI3K) dependent. In this type of mitophagy, the autolysosome initially sequesters polarized mitochondria that later become depolarized as a consequence of MPT onset or the acidic microenvironment in the autophagic vesicles (Czaja et al. 2013). Type I mitophagy is evident in response to nutrient starvation or acute ischemic events (Fig. 9.2).

In contrast, dissipation of mitochondrial membrane potential by mitochondrial uncouplers such as carbonyl cyanide *m*-chlorophenyl hydrazine or carbonyl cyanide 4-(trifluoromethoxy) phenylhydrazone initiates the PI3K-independent and PTEN-induced putative kinase protein 1 (PINK1)-dependent type II mitophagy. Accumulating evidence indicates that differential localization to the inner or outer mitochondrial membranes regulates PINK1 stability and function. In healthy mitochondria, PINK1 remains in a cleaved state due to robust serine protease activity of presenilin-associated rhomboid-like protein (PARL) in the mitochondrial

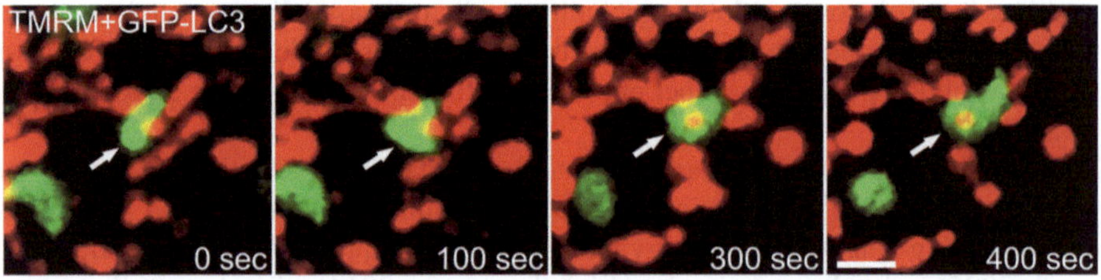

Fig. 9.2 Visualization of mitophagy. Time-lapse images of confocal microscopy with GFP-LC3 and TMRM in isolated mouse hepatocytes. Under autophagy-stimulating conditions such as nutrient depletion and starvation, hepatocytes develop mitophagy wherein green-fluorescing autophagosome completely sequesters a mitochondrion within 5 min after the phagophore formation (*arrow*). Note that the mitochondrion in the lumen of autophagosome remains polarized at 400 s. Scale bar, 2 μm

intermembrane space (Jin et al. 2010; Deas et al. 2011). As PARL activation is electrogenic, mitochondrial depolarization abrogates PARL activity, resulting in inhibition of the constitutive degradation of PINK1 and accumulation of the 63-kDa full-length form on the outer mitochondrial membrane, where it binds to BECN1 and recruits PARKIN to depolarized mitochondria. Subsequently, the E3 ubiquitin ligase activity of PARKIN ubiquitinates substrates including p62 and VDAC (Michiorri et al. 2010; Geisler et al. 2010; Okatsu et al. 2012). Transcription factor p62 has been shown to link autophagic contents to phagophores or autophagosomes (Kirkin et al. 2009). Among the factors that influence type II mitophagy, changes in membrane potential rather than pH or ATP levels appear to prompt PARKIN accumulation and the onset of type II mitophagy (Narendra et al. 2009). Additionally, mitochondrial receptor Bcl2/adenovirus EB 19 kDa interacting protein 3 (BNIP3) has been implicated in mitophagy (Feng et al. 2013). BNIP3 expression changes in response to levels of hypoxia-inducible factor 1α (HIF-1α), and increased translocation of BNIP3 to the mitochondrial outer membrane is observed in the onset of type II mitophagy (Chen et al. 1997; Bruick 2000; Bellot et al. 2009). Of importance, execution of type II mitophagy appears to follow a canonical autophagy pathway (Ray et al. 2000; Zhu et al. 2013). Other factors are likely involved in type II mitophagy. Recently, it has been demonstrated that mitochondrial depolarization activates FUN14 domain-containing protein-1 (FUNDC1), a mitochondrial outer membrane protein that accumulates at the ER-mitochondria contact sites (Liu et al. 2012; Wu et al. 2014). FUNDC1-mediated mitophagy occurs through the coordinated dephosphorylation of Ser13 by mitochondrial phosphatase PGAM family member 5 (Chen et al. 2014) and phosphorylation of Ser17 by ULK1 (Wu et al. 2014). It has been proposed that WXXL motif of BNIP3 and/or FUNDC1 binds to LC3 prior to mitophagy induction (Chen et al. 2008; Sandoval et al. 2008; Feng et al. 2013).

Type III mitophagy is another form of mitophagy where damaged and oxidized components of the mitochondria bud off as vesicles and transit into multivesicular bodies (Soubannier et al. 2012). Generation of mitochondria-derived vesicles increases as an early response to oxidative stress. Type III mitophagy has been proposed to complement mitophagy as delivery to the lysosomes through this pathway is independent of mitochondrial membrane potential, ATG5 and LC3. In addition, silencing of dynamin-1-like protein (DRP1), a mitochondrial fission protein enhancing mitophagy (Twig et al. 2008), does not affect vesicle formation and transport. Hence, type III mitophagy may be distinct from DRP1-dependent mitophagy. It is, however, unknown whether type III mitophagy relies on PI3K activation or not.

9.4.2.3 Role of Mitophagy in Hepatic I/R Injury

MPT onset and mitochondrial failure after reperfusion amass toxic mitochondrial by-products and precipitate I/R-induced hepatocyte death, insinuating that reperfused hepatocytes might have defective or insufficient mitophagy. Biochemical, genetic, and imaging approaches have indeed revealed that reperfusion of ischemic

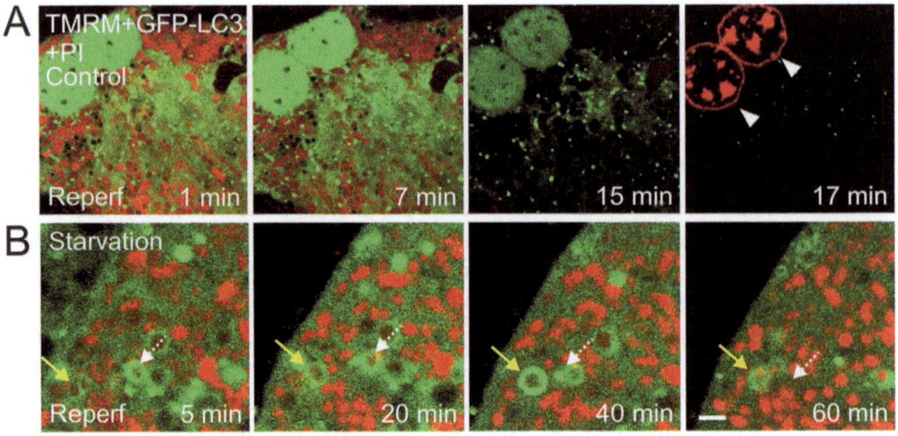

Fig. 9.3 Impaired autophagy after I/R. Visualization of autophagy and mitophagy after reperfusion. Hepatocytes labeled with GFP-LC3 and TMRM were subjected to 4 h of simulated ischemia. Within 7 min of reperfusion, mitochondria repolarized. Although some green punctate autophagosomes were visible, they were unable to enwrap abnormal mitochondria (*top panels*). The cell lost mitochondrial membrane potential and cell viability after 17 min, as indicated by PI labeling in the nuclei (*arrow heads*). In stark contrast, when hepatocytes were exposed to starvation prior to ischemia, cells executed mitophagy, as shown by green-fluorescing ring structure surrounding a mitochondrion (*dotted white arrow*). More importantly, the cell sustained mitochondrial membrane potential and viability after reperfusion (*bottom panels*). Solid yellow arrow displays an initiation of new mitophagy. Scale bar, 2 μm

hepatocytes impairs autophagy and mitophagy (Fig. 9.3). Autophagic impairment after I/R stems from two factors: energy insufficiency and loss of key autophagy proteins. Autophagy is a highly energy-dependent process. While nutrient depletion and starvation are powerful stimuli for autophagy induction in normal livers, the formation of autophagic vesicles during ischemia is nonetheless halted due to ATP depletion. Lack of functional autophagy can be assessed biochemically by monitoring the changes in LC3-II, a biomarker of autophagy, before and after administration of lysosomal inhibitors like bafilomycin or chloroquine. Since blockade of autolysosomal degradation of autophagic cargos by these agents accumulates autophagosomes, the increase in LC3-II after lysosomal inhibition reflects the extent of autophagic flux. Similarly, fluorescence imaging of GFP-tagged LC3 can visualize autophagosomes that uniquely display punctate pattern of green fluorescence. Hepatocytes after 4 h of ischemia do not exhibit chloroquine-mediated increase in LC3-II, indicative of lack of autophagic flux. However, cell death under this condition is negligible since naturally occurring tissue acidosis prevents the onset of MPT and mitochondrial dysfunction. During the early stage of reperfusion, mitochondria transiently repolarize and exclude calcein, denoting that MPT pores remain closed (Fig. 9.1). At this time, as hepatocytes temporarily recover ATP synthesis, cells are able to execute autophagy and sustain viability. Nevertheless, the energy needed to eliminate swollen and damaged mitochondria during ischemia and early reperfusion begins to outweigh the capacity to generate adequate levels of

ATP. During the late stage of reperfusion, such imbalance becomes exacerbated, autophagy eventually ceases, and hepatocytes increasingly accumulate abnormal or dysfunctional mitochondria that release greater levels of Ca^{2+} and ROS to neighboring mitochondria. Widespread onset of MPT occurs, and necrotic cell death ensues thereafter (Kim et al. 2008b, 2013; Wang et al. 2011b). Consistent with this observation, it has been demonstrated that MPT onset initially affects a subpopulation of mitochondria prior to extensive MPT, supporting the presence of nonuniform pool of mitochondria (Pacher and Hajnoczky 2001). The toxic metabolites and by-products from a discrete subpopulation of mitochondria disperse damage to healthy adjacent mitochondria before systemic mitochondrial dysfunction develops. Thus, functional mitophagy in order to sequester and clear a mitochondrial subpopulation that is vulnerable to I/R injury may be pivotal to sustain both mitochondrial bioenergetics and cell survival.

Apart from energy insufficiency after reperfusion, depletion of key autophagy proteins is another factor contributing to autophagy failure after I/R. During ischemia, cytosolic Ca^{2+} increases progressively without affecting mitochondrial Ca^{2+} due to ATP depletion and collapse of Na^{+} and K^{+} gradients that inhibit ATP-driven Ca^{2+} pumps and secondary ion exchangers (Kim et al. 2012). Moreover, mitochondrial depolarization during anoxia blocks mitochondrial Ca^{2+} exchanger that is modulated by mitochondrial membrane potential (Gunter and Pfeiffer 1990). Cytosolic increase in Ca^{2+} stimulates calpains that are known to maintain substantial activity even under acidic pH (Maddock et al. 2005), which in turn degrades ATG7 and BECN1, two key autophagy proteins involved in autophagosome formation (Wang et al. 2011b; Kim et al. 2008b, 2013). After 4 h of ischemia in isolated hepatocytes, ATG7 and BECN1 decrease to 8.5% and 31.9% of normal values, respectively (Kim et al. 2008a). During reperfusion, mitochondrial Ca^{2+} increases substantially (Kim et al. 2012), calpains become further activated, and levels of both ATG7 and BECN1 decline by 90%. The importance of calcium and calpains in autophagy failure in reperfused hepatocytes is also substantiated by findings that pharmacological inhibitors like acetyl-Leu-Leu-methioninal (ALLM) or carbamazepine considerably suppress the loss of both ATG7 and BECN1 and promote autophagy and cell survival after reperfusion (Kim et al. 2008a, 2013). Of importance, overexpression of either ATG7 or BECN1 or calpain inhibitors all prevent the onset of MPT and cell death, indicating that calpain-mediated defective mitophagy precedes the onset of MPT and cell death. Taken together, with the combination of ATP depletion and ATG loss, the formation of autophagic vesicles is impeded, and autophagy becomes dysfunctional after I/R.

9.4.2.4 Role of Sirtuin 1 (SIRT1) in Autophagy After I/R

Posttranslational modification of proteins by the covalent addition of an acetyl group to ε-lysine residues is referred to as acetylation. This covalent modification neutralizes the positive charge of lysine, affecting the total charge balance of target proteins and altering the steric environment of interaction sites. Acetylation/

deacetylation is involved in a variety of cellular processes such as DNA binding, enzyme activity, and the stability and localization of target proteins. Lysine acetyltransferases (KATs) and deacetylases (KDACs) are two major enzymes to support the balance between acetylation and deacetylation. Four subtypes of KDACs exist including class III KDACs, also known as sirtuins. Among seven mammalian sirtuin isoforms, SIRT1 requires oxidized nicotinamide adenine dinucleotide (NAD^+) as a cofactor (Finnin et al. 2001; Min et al. 2001). SIRT1 resides in both the cytosol and nucleus (Tanno et al. 2007) and regulates a myriad of cellular events such as energy metabolism (Rodgers et al. 2005; Gerhart-Hines et al. 2007; Purushotham et al. 2009), mitochondrial biogenesis (Lagouge et al. 2006; Aquilano et al. 2010), and autophagy (Lee et al. 2008; Hariharan et al. 2010; Jang et al. 2012; Fang et al. 2014). As implied by its nuclear location, SIRT1-mediated deacetylation is a central chemical modification in regulation of various gene transcriptions. Some examples encompass peroxisome proliferator-activated receptor gamma coactivator-1 alpha (PGC-1α) (Rodgers et al. 2005), CREB-regulated transcription coactivator 2 (Liu et al. 2008), forkhead transcription factors (FOXO) (Park et al. 2010), fibroblast growth factor (FGF21) (Li et al. 2014), and signal transducer and activator of transcription (STAT3) (Nie et al. 2009). There exists growing evidence that acetylation/deacetylation impacts autophagy at the transcriptional, translational, and posttranslational level. Inhibition of ATG7 acetylation with spermidine increases autophagy (Kuma and Mizushima 2010; Sanchez et al. 2012). Conversely, overexpression of acetyltransferase p300 inhibits autophagy by increasing acetylation of ATG5, ATG7, ATG8, and ATG12 and its interaction with ATG7 (Lee and Finkel 2009). It has also been suggested that SIRT1 can interact directly with ATG5, ATG7, and LC3 during nutrient insufficiency (Lee et al. 2008). The importance of acetylation/deacetylation in autophagy is further supported by recent studies demonstrating that acetylated BECN1 suppresses autophagosome maturation and endocytic trafficking (Sun et al. 2015) and that the acetylation status of LC3 governs its distribution between the nucleus and cytoplasm (Huang et al. 2015). While deacetylation of ATG, in general, promotes autophagy, acetylation can also augment autophagy. Using cell lines, Lin et al. have demonstrated that acetylated ULK1 drives autophagy initiation under the condition of serum deprivation (Lin et al. 2012). Interestingly, when cells are exposed to glucose insufficiency, a condition stimulating autophagy, ULK1 becomes phosphorylated instead, implicating the presence of an intimate cross talk between acetylation/deacetylation and phosphorylation in autophagy regulation.

SIRT1 is critical in hepatocyte survival after I/R. We have recently reported that I/R depletes the liver of SIRT1, which in turn contributes to mechanistic chains of defective autophagy, mitochondrial dysfunction, and hepatocyte injury (Biel et al. 2016). In this study, human liver biopsies collected after 15 min of inflow occlusion during hepatectomy decreases SIRT1 expression by 30%. Similar to patient findings, both murine liver homogenates obtained from an in vivo model of I/R and isolated hepatocytes subjected to a simulated in vitro I/R show a significant reduction in SIRT1, corroborating the sensitivity of SIRT1 to I/R. SIRT1 depletion is likely, at least in part, mediated by calpain activation secondary to Ca^{2+} overloading

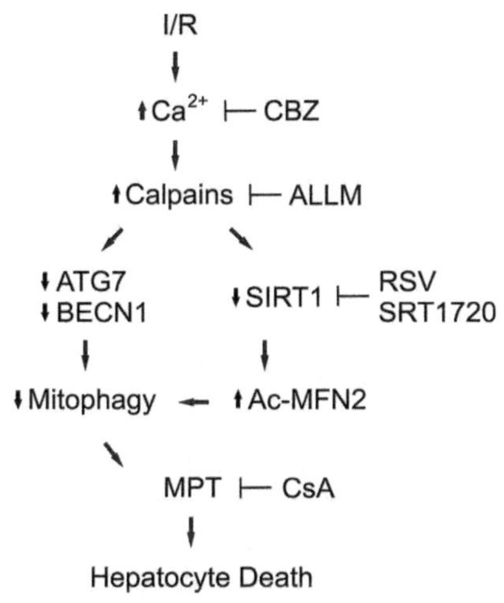

Fig. 9.4 Scheme of I/R-induced autophagy impairment and cell death. After reperfusion, hepatocytes are overloaded with calcium, which in turn stimulates calpains. These enzymes subsequently hydrolyze ATG7 and BECN1, causing defective autophagy. Since impaired autophagy fails to clear dysfunctional or abnormal mitochondria, hepatocytes accumulate ROS and calcium, toxic precursors inducing the MPT, and ultimately lose viability. At the same time, activated calpains degrade SIRT1, leading to accumulation of acetylated MFN2 (Ac-MFN2) and suppression of mitophagy. Suppression of calcium elevation, inhibition of calpains, activation of SIRT1, or blockade of the MPT all attenuate MPT-dependent hepatocyte death after reperfusion. (Abbreviations: *ALLM* acetyl-Leu-Leu-methioninal, *CBZ* carbamazepine, *CsA* cyclosporine A, *MPT* mitochondrial permeability transition, *RSV* resveratrol)

during I/R since pharmacological inhibition of calpains considerably suppresses the loss of SIRT1 as well as autophagy defects and necrotic cell death. Genetic overexpression of SIRT1 or pharmacological activation with resveratrol or SRT1720 also maintains mitochondrial structural integrity, enhances mitophagy, and prevents MPT onset. Conversely, hepatocytes from SIRT1 conditional knockout mice are vulnerable to I/R injury, as evidenced by impaired autophagic flux, rapid MPT onset, and increased cell death (Fig. 9.4). Collectively, SIRT1 depletion is an important causal mechanism to heighten reperfusion injury.

About 35% of all mitochondrial proteins are acetylated (Anderson and Hirschey 2012), and 24% of acetylated mitochondrial proteins are involved in energy homeostasis during fasting/starvation (Kim et al. 2006b). This suggests that the acetylation status in mitochondria may be a core element to command mitochondrial function and bioenergetics. In fact, hyperacetylation of mitochondrial proteins has been reported in obesity (Hirschey et al. 2011) and chronic alcohol consumption (Fritz et al. 2012), both of which are closely associated with mitochondrial malfunction. We have recently proposed a mechanism as to how cytosolic SIRT1 modulates mitochondrial function (Biel et al. 2016). Biochemical and genetic approaches reveal that cytosolic SIRT1 interacts with and deacetylates a mitochondrial outer membrane protein, mitofusin (MFN). Because either knockdown or deletion mutants of MFN2 abolish SIRT1-mediated autophagy and cytoprotection against I/R injury, MFN2 is necessary for SIRT1-dependent autophagy. Furthermore, as no changes are observed in the acetylation status of MFN1 after SIRT1 overexpression, MFN2, but not MFN1, is a substrate for SIRT1. Therefore, the acetylation status of this mitochondrial outer membrane protein may be an important molecular switch to facilitate the cross talk between the cytoplasm and mitochondria in the liver.

9.5 Mitochondrial Injury in Other Liver Diseases

The prevalence of chronic liver disease continues to rise. Although viral hepatitis and alcohol-related liver disease are among the most common causes of chronic liver disease, the growing obesity epidemic, a rising population of the elderly, and the impact of subacute or chronic use of medications all contribute to liver injury. It is also intuitively expected that the response of patients having liver diseases to I/R will differ greatly from healthy individuals.

9.5.1 Mitochondrial Dysfunction in Aging

In the United States, the population of individuals older than 65 is anticipated to rise to 89 million people by 2050 (Cellier et al. 2016). The growing geriatric population affects multiple aspects of liver disease. Aged individuals are more vulnerable to acute liver injury and are at increased risk to develop hepatic fibrosis and cirrhosis that can proceed to HCC; moreover, older age is a poor prognostic factor in chronic liver disease of many etiologies. Despite these findings, hepatic resection for the elderly with HCC has become increasingly accepted as a main stay of treatment (Kim et al. 2015a) and may provide a more curative outcome when compared to their young counterparts (Mizuguchi et al. 2015). Concurrently, older donor livers are more susceptible to ischemic injury during and following transplantation. Thus, understanding how aging impacts hepatocellular function under basal conditions and after I/R remains paramount.

Ultrasound studies suggest that liver volume decreases by 20–40% with age and that compared to patients younger than 40, patients older than 65 have a 35% decrease in blood volume through the liver. Moreover, functional studies likewise allude to a decrease in the mass of functional liver cells with age. In rodent models, older rats had less regenerative capacity after hepatectomy than their younger counterparts (Tsukamoto et al. 1993).

From a cellular standpoint, both hepatocytes and non-parenchymal cells are equally affected by aging. Sinusoidal epithelial cells become thicker and lose their fenestrations to acquire morphology akin to small capillaries (Huang et al. 2009; Kim et al. 2015b). In combination with an accumulation of oxidative stress and decreased oxygen flow, pseudo-capillarization in aged sinusoids diminishes the efficiency of filtration and clearance of multiple types of drugs (Le Couteur et al. 2008; Kim et al. 2015a). Evaluation of the hepatic immune cells also suggests that inflammatory responses and antitumor responses are altered with age (Kawabata et al. 2008). Observations of patient biopsies and samples from rodent models have indicated that with advancing age, hepatocytes undergo multiple intracellular alterations. Lipofuscin, an autofluorescent pigment from oxidized proteins and lipids that accumulate in lysosomes, is associated with enhanced free radical formation and a failure to clear waste products (Le Couteur et al. 1999, 2008). Lipid droplets and vesicles are abundant in aged hepatocytes, and a marked decline is evident in smooth ER surface area in aged livers which may mirror reduced hepatic microsomal protein concentrations and enzymatic activity (Zhang and Cuervo 2008; Hoare et al. 2010). However, overall function of aged livers remains intact with advancing age.

From a structural perspective, the mitochondria from aged murine livers appear more swollen and have fewer cristae, shorter cristae, and reduced density (Zhang and Cuervo 2008). Functionally, the mitochondria from senescence-accelerated murine models demonstrate more uncoupling of oxidative phosphorylation and less efficient energy production, suggesting that aged mitochondria may not be as functionally efficient as young mitochondria (Zhang and Cuervo 2008; Khraiwesh et al. 2014). These observations are well aligned with the mitochondrial free radical hypothesis of aging. First described by Harman in the 1950s, it posits that mtDNA is more susceptible to damage due to an absence of histones and their location in the free radical-enriched inner mitochondrial membrane. DNA damage in mitochondria further creates a vicious cycle of free radical propagation that ultimately harms cellular macromolecules and other organelles (Nakahara et al. 1998; Kolosova et al. 2001; Okatani et al. 2002). However, it is worth noting the controversy of this hypothesis. In particular, studies in *C. elegans* refute free radical hypothesis due to unexpected observations that life span is not shortened in *C. elegans* mutants that lack superoxide dismutase (Harman 1956, 1972, 1981) and that mitochondrial ROS can paradoxically increase longevity in *C. elegans* (Van Raamsdonk and Hekimi 2012). In spite of the controversy, solid lines of evidence, however, exist that mitochondrial dysfunction plays a central role in aging. The mitochondrial genome contains multiple copies within each mitochondrion. It encodes 13 subunits of the electron transport chain, two rRNAs, and 22 tRNAs and also possesses a noncoding regulatory region (Yang and Hekimi 2010; Yee et al. 2014). Given that specific mutations occur with aging especially in the sixth segment of a noncoding region of mtDNA (Fraga et al. 1990; Hamilton et al. 2001; Cavallini et al. 2007) and large deletion mutations are frequently found in aged mtDNA (Michikawa et al. 1999), erroneous replication or ineffective repair of mtDNA may also contribute to aging (Trifunovic et al. 2004). Accumulation of damaged protein and decreased membrane fluidity are likewise contributing factors to age-dependent decline in cellular dysfunction. As young cells efficiently eliminate surplus or unnecessary cellular constituents with autophagy, age-mediated increase in abnormal proteins and lipids may manifest decreased capacity of cellular clearance with aging. Zhang and Cuervo have shown that autophagic capacity in the liver subsides in old rats (Zhang and Cuervo 2008). Other work also supports the importance of autophagy in protein quality control with age. In human kidney cell lines and fibroblasts, aged cells favor autophagy over proteasomal degradation to turnover polyubiquitinated proteins (Gamerdinger et al. 2009). Of interest, although basal levels of autophagy are relatively unchanged in aged cells, they are unable to display a robust autophagic response to additional stresses, inferring that autophagic adaptation declines with advancing age (Donati et al. 2001, 2008). Consistent with these studies, decreased autophagic capacity has been shown to be attributable to age-mediated sensitivity to I/R injury (Wang et al. 2011). In 3-month- and 26-month-old mouse livers, the number of polarized mitochondria and their mitochondrial membrane potential are indistinguishable between two age groups. Moreover, basal levels of autophagic flux are comparable between young and old livers, and hepatocytes from both young and aged mice have a robust autophagic flux under mildly stressed conditions such as normoxia and starvation, supporting that alterations in hepatic parenchymal structure and basal function are minimal with age. However, when aged livers are subjected to additional stress such as short I/R, a condition that young livers tolerate well, they rapidly develop autophagy failure, MPT onset, and cell death. Immunoblotting, genetic, and imaging approaches in both isolated hepatocytes and in vivo livers show that calpain activation after short I/R degrades ATG4B, a key autophagy protein that activates and recycles LC3, in aged livers. As a corollary, aged livers lose autophagic responsiveness to reperfusion and increasingly accumulate dysfunctional mitochondria that provokes widespread onset of MPT and ultimately cell death. Overexpression of ATG4B mitigates this age-dependent vulnerability to reperfusion injury. Hence, despite seemingly minimal impact of aging per se on hepatocellular basal autophagy, aged livers are unable to elicit a proper autophagic response to I/R.

9.5.2 *Mitochondrial Dysfunction in the Fatty Liver*

The rising obesity epidemic is quickly making nonalcoholic fatty liver disease (NAFLD) the most common cause of chronic liver disease. NAFLD's clinical spectrum ranges from simple hepatic steatosis, where lipids account for greater than 5% of the wet weight of the liver, to hepatic inflammation seen in nonalcoholic steatohepatitis (NASH), to cirrhosis, and to HCC (Starley et al. 2010). About 29% of patients with severe NASH will develop cirrhosis and HCC within 10 years of diagnosis (Argo and Caldwell 2009); thus, understanding the underlying mechanisms of NAFLD development and implications of I/R is a growing priority.

The progression from simple steatosis to NASH has been described through a "two-hit" hypothesis by Day and colleagues (Day and James 1998). Excess hepatic triglycerides and cholesterol sensitize the liver to a second insult, hallmarked by oxidative stress, inflammation, and the induction of insulin resistance (Matsuzawa-Nagata et al. 2008). A meticulous review of the mechanisms underlying the evolution from steatosis to NASH is beyond the scope of this chapter; however, the interplay between mitochondrial dysfunction, autophagy, and I/R injury both at the bench and in the clinical setting warrants further examination.

Fat over-intake can alter mitochondrial structure and function. In comparison to rats fed with a high-fish oil diet, those fed with a high-lard diet clearly demonstrate increased mitochondrial fission, as evidenced by decreased MFN2 (Wang et al. 2015) and increased DRP1 and mitochondrial fission 1 protein (FIS1) expression (Lionetti et al. 2014). Composition of mitochondrial membranes is likewise changed with surplus fat consumption, in part by increased cardiolipin content (Aoun et al. 2012a, b). Cardiolipin, a unique phospholipid localized in the mitochondrial inner membrane, plays a pivotal role in the maintenance of membrane fluidity (Unsay et al. 2013), ion trafficking (Subramani et al. 2016), protein translocation (Gebert et al. 2009), mitochondrial bioenergetics (Beyer and Nuscher 1996), and mitochondrial-dependent apoptosis (Lutter et al. 2001). Recently, relocation of cardiolipin from the inner to the outer membrane has been proposed to mediate fusion between mitochondria and lysosomes, an integral part of mitophagy (Chu et al. 2013). In parallel with altered mitochondrial structure, excess lipid intake can adversely affect mitochondrial function, including fatty acid oxidation, expression of uncoupling protein 2 (UCP-2) (Jiang et al. 2008; Hodson et al. 2010), and proton leak across the inner mitochondrial membrane (Rolfe et al. 1994). Liver mitochondria isolated from high-fat-fed rats exhibit reduced mitochondrial respiration and increased ROS production (Kathirvel et al. 2010; Vial et al. 2011). mtDNA mutation rates also increase in high-fat diet-fed mice, compared to their normal chow counterparts (Yang et al. 2010)

Multiple murine models have been developed to demonstrate that autophagy is negatively impacted with steatosis (Yang et al. 2010). Steatotic livers manifest a profound ER stress and decreased levels of BECN1, ATG5, and ATG7 (Yang et al. 2010). Excessive lipids can be, however, removed by lipophagy (Singh et al. 2009). Onset of lipophagy requires the interaction between ancient ubiquitous protein 1 (AUP1) and E2 ubiquitin conjugase G2 (UBE2G2) at the lipid droplet surface (Spandl et al. 2011), which induces clustering of lipid droplets (Lohmann et al. 2013). Conversion of LC3-I to LC3-II in a RAB7-dependent manner occurs on the surfaces of lipid droplet aggregates or small, individual lipid droplets prior to autophagosome engulfment (Singh et al. 2009; Schroeder et al. 2015). RAB7, a small GTPase, appears to be a key mediator in lipophagy because its depletion voids lipophagic efficiency and modifies the morphology of multivesicular bodies, lysosomes, and autophagosomes (Schroeder et al. 2015). Autophagy inhibition via 3-methyladenine, knockdown of ATG5, and liver-specific ATG7 knockout all cause the aberrant accumulation of lipids, further emphasizing an integral role of lipophagy in hepatic lipid metabolism (Koga et al. 2010).

Dysfunctional mitophagy is also associated with the pathogenesis of NAFLD. The hepatic expression of lysocardiolipin acyltransferase 1 (ALCAT1), an acyl CoA-dependent enzyme catalyzing the remodeling of cardiolipin in obesity, diabetes, and cardiovascular disease, is significantly upregulated by high-fat diet and leptin deficiency. Wang et al. have shown that elevated levels of ALCAT1 observed in animal models of NAFLD are mechanistically linked to autophagic arrest, oxidative stress, mitochondrial dysfunction, and insulin resistance in hepatocytes, which can be reversed by specific knockout of ALCAT1 (Wang et al. 2015). ALCAT1-null mice improve mitochondrial respiration and autophagy in the presence of high-fat diet.

Mitochondrial dysfunction secondary to excess lipid exacerbates hepatic I/R injury (Selzner et al. 2000). ATP deficiency resulting from increased expression of UCP-2 in fatty livers makes mitochondria more vulnerable to I/R injury (Evans et al. 2008). Decreased glutathione and thioredoxin levels are common in steatotic hepatocytes, reflecting reduced antioxidant capacity (Zhang et al. 2014). From a surgical perspective, the presence of steatosis has been associated with increased morbidity and mortality during hepatectomy, including longer operative times, increased intraoperative blood loss, and increased blood transfusions (Behrns et al. 1998; Veteläinen et al. 2007; de Meijer et al. 2010). In addition, much literature has discussed the impact of steatotic donor livers in the context of transplantation. Macrosteatosis has been known to reduce graft survival 1 year after transplant (Spitzer et al. 2010) and increase the likelihood of primary non-function and initial poor function after transplant (Ploeg et al. 1993).

9.6 Strategies to Reduce I/R Injury

Although preclinical animal studies have offered numerous therapeutic strategies to mitigate hepatic I/R injury, promising outcomes have not been achieved in the clinical setting. This could be largely due to incomplete understanding of multifactorial nature of I/R injury. In this section, we discuss cellular and surgical approaches that may attenuate hepatocellular injury.

9.6.1 Pharmacological Approaches

MPT onset is a principal step toward hepatocyte necrosis after warm I/R, and strategies to prevent the MPT have been intensively evaluated (Fig. 9.4). CsA is the most well-studied inhibitor of the MPT and has been shown to suppress hepatic I/R injury. However, this cyclophilin D blocker has a narrow therapeutic window at nanomolar concentration ranges. In addition to its nephrotoxic effects, this drug inhibits calcineurin, a Ca^{2+}- and calmodulin-dependent phosphatase, resulting in immunosuppression. NIM811, a non-immunosuppressive CsA derivative, mitigates hepatic I/R injury and improves regeneration after hepatectomy in rodent models (Waldmeier et al. 2002; Zhong et al. 2007, 2008; Rehman et al. 2011), but has yet to be tested clinically. TRO40303, which binds to the cholesterol site of the mitochondrial benzodiazepine receptor to block the MPT, did not show noticeable benefit after cardiac ischemia in clinical trials (Atar et al. 2015), and studies in liver have not been pursued. Importantly, more fundamental issues on CsA-based therapy are that despite its capacity to prevent mitochondrial pore opening, this agent does little to mitigate upstream events of the MPT such as ROS and Ca^{2+} accumulation. Therefore, cells could continuously amass ROS and Ca^{2+} in the presence of MPT blockers. Additionally, hepatocytes can develop CsA-insensitive MPT when mitochondria encounter high dose of MPT inducers (He and Lemasters 2002).

After reperfusion, accumulation of calcium and ROS stimulates calpains, which ultimately impair the autophagic machinery and causes mitochondrial inner membrane permeabilization. Use of steroids, Ca^{2+} chelators, and protease inhibitors has been proposed to ease reperfusion injury (Uchida et al. 1994; Yamashita et al. 2001; Gurusamy et al. 2010; Pan et al. 2012; Akhtar et al. 2013). Other agents include hemin and calcium chelators (Yun et al. 2014; Go et al. 2015), glutathione (Suyavaran et al. 2015), and *N*-acetylcysteine (Fukuzawa et al. 1995; Nakano et al. 1995). Attempts to enhance autophagy either through calpain inhibition or sustaining autophagic machinery have also been evaluated in the preclinical setting. Carbamazepine, an FDA-approved anticonvulsant drug, confers cytoprotection after I/R in mouse livers by preventing calpain activation and depletion of ATG7 and BECN1 (Kim et al. 2013). ALLM, a membrane permeable peptide that blocks calpains and cathepsins, promotes autophagy after I/R by preventing cytosolic SIRT1 degradation (Biel et al. 2016). Other SIRT1 activators such as SRT1720 and resveratrol could impart a protective effect after I/R through autophagy enhancement (Biel et al. 2016). In addition to SIRT1, autophagy inducers such as cisplatin or rapamycin have been demonstrated to attenuate hepatocellular necrosis after I/R (Cardinal et al. 2009a, b; Evankovich et al. 2012; Zhu et al. 2015).

9.6.2 Surgical Approaches

Ischemic preconditioning (IPC) is a procedure whereby repeated short episodes of occlusion and release of the portal vein and hepatic artery mitigate the risk of I/R injury (Murry et al. 1990). Mechanisms underlying IPC cytoprotection are multifactorial and affect both hepatocytes and non-parenchymal liver cells. Mitochondrial integrity is better maintained, and cell death is reduced by IPC (Yadav et al. 1999; Clavien et al. 2003; Selzner et al. 2003b; Lee and Lee 2006; Ko et al. 2013). Autophagy may also be enhanced by IPC in a heme oxygenase 1-dependent manner (Liu et al. 2014). However, clinical benefits of IPC remain controversial, and there exists a disparity between IPC protocols and the heterogeneity of patients evaluated within and among studies (Clavien et al. 2000; Ye et al. 2014).

Similar to IPC, intermittent clamping utilizes shorter periods of occlusion and reperfusion (Selzner et al. 2003a). A prospective randomized study comparing prolonged Pringle to intermittent clamping demonstrates that intermittent clamping reduces peak transaminase levels and is associated with lower incidence of postoperative liver failure (Belghiti et al. 1999). Petrowsky et al. have reported that either IPC or intermittent clamping confers a comparable protection as measured by peak transaminase levels, duration of stay in the intensive care unit, total hospital length of stay, and complication rate, although IPC presents somewhat lower intraoperative blood loss and shorter transection times (Petrowsky et al. 2006). Murine models further interrogating IPC and intermittent clamping suggest that beyond 75 min of continuous occlusion, intermittent clamping confers better cytoprotection than IPC (Rüdiger et al. 2002). Other studies with animal models have indicated that IPC and intermittent clamping provide equivalent protection in steatotic (Saidi et al. 2007) and cirrhotic livers (Jang et al. 2008). In contrast, intermittent clamping has been proposed as a better strategy to restore bile flow and reduce hepatocellular injury in aged livers (Schiesser et al. 2009). It is, however, unknown how intermittent clamping confers such benefits.

Whereas IPC and intermittent clamping primes the liver by direct occlusion and release of blood flow into the liver, remote ischemic preconditioning (RIPC) inflicts brief episodes of I/R to distant tissues or organs prior to liver resection (Gho et al. 1996). Beneficial outcomes of RIPC through occlusion of limb or intestine have been reported in the liver (Abu-Amara et al. 2011; Kageyama et al. 2015). The mechanisms by which RIPC provides protection against hepatic I/R injury remain to be elucidated, but heme oxygenase 1 may be an important player (Lai et al. 2006, 2007; Jia et al. 2015; Kageyama et al. 2015). Future studies are warranted to evaluate whether these promising preclinical results can be implemented in the clinical setting.

9.7 Concluding Remarks and Future Perspectives

As aging, obesity, diabetes, cancer, and metabolic syndrome prevail, a number of patients with acute and chronic liver disease continue to rise. It is anticipated that hepatic diseases will be a major socioeconomic burden in the future. Although liver resection and transplantation are still the most effective lifesaving interventions for patients with advanced liver disease, I/R injury is inevitable during these surgical treatments. To date, strategies to improve liver function after I/R have not been successful mostly due to an incomplete understanding of the pathogenesis of I/R injury. Dysfunctional autophagy and subsequent onset of mitochondrial malfunction are critical events contributing to reperfusion injury in the liver. Mitophagy selectively targets and removes damaged mitochondria laden with ROS, Ca^{2+}, and oxidative stress. Despite its therapeutic potentials for ameliorating hepatic I/R injury, our understanding of mitophagy remains limited. In addition, visible gaps are apparent between animal models and clinical settings. Thus, it is imperative to develop improved animal models that accurately reflect the patient population such as obese or aged patients. Future studies on mechanistic insights into mitophagy and its translational applications could lead to development of viable treatments for patients.

Acknowledgments This work was supported in part by US National Institutes of Health National Institute of Diabetes and Digestive and Kidney Diseases grant DK079879 and DK090115 (J-S Kim) and National Institute on Aging AG028740 (J-S Kim).

References

Abdalla EK, Noun R, Belghiti J (2004) Hepatic vascular occlusion: which technique? Surg Clin North Am 84(2):563–585

Abu-Amara M, Yang SY, Seifalian AM, Fuller B, Davidson BR (2011) Remote ischemic preconditioning by hindlimb occlusion prevents liver ischemic/reperfusion injury. Ann Surg 254(1):178–180

Akhtar MZ, Henderson T, Sutherland A, Vogel T, Friend PJ (2013) Novel approaches to preventing ischemia-reperfusion injury during liver transplantation. Transplant Proc 45(6):2083–2092

Anderson KA, Hirschey MD (2012) Mitochondrial protein acetylation regulates metabolism. Essays Biochem 52:23–35

Ao X, Zou L, Wu Y (2014) Regulation of autophagy by the Rab GTPase network. Cell Death Differ 21(3):348–358

Aoun M, Feillet-Coudray C, Fouret G, Chabi B, Crouzier D, Ferreri C, Chatgilialoglu C, Wrutniak-Cabello C, Cristol JP, Carbonneau MA, Coudray C (2012a) Rat liver mitochondrial membrane characteristics and mitochondrial functions are more profoundly altered by dietary lipid quantity than by dietary lipid quality: effect of different nutritional lipid patterns. Br J Nutr 107(5):647–659

Aoun M, Fouret G, Michel F, Bonafos B, Ramos J, Cristol JP, Carbonneau MA, Coudray C, Feillet-Coudray C (2012b) Dietary fatty acids modulate liver mitochondrial cardiolipin content and its fatty acid composition in rats with non alcoholic fatty liver disease. J Bioenerg Biomembr 44(4):439–452

Aquilano K, Vigilanza P, Baldelli S, Pagliei B, Rotilio G, Ciriolo MR (2010) Peroxisome proliferator-activated receptor gamma co-activator 1alpha (PGC-1alpha) and sirtuin 1 (SIRT1) reside in mitochondria: possible direct function in mitochondrial biogenesis. J Biol Chem 285(28):21590–21599

Argo CK, Caldwell SH (2009) Epidemiology and natural history of non-alcoholic steatohepatitis. Clin Liver Dis 13(4):511–531

Atar D, Arheden H, Berdeaux A, Bonnet JL, Carlsson M, Clemmensen P, Cuvier V, Danchin N, Dubois-Randé JL, Engblom H, Erlinge D, Firat H, Halvorsen S, Hansen HS, Hauke W, Heiberg E, Koul S, Larsen AI, Le Corvoisier P, Nordrehaug JE, Paganelli F, Pruss RM, Rousseau H, Schaller S, Sonou G, Tuseth V, Veys J, Vicaut E, Jensen SE (2015) Effect of intravenous TRO40303 as an adjunct to primary percutaneous coronary intervention for acute ST-elevation myocardial infarction: MITOCARE study results. Eur Heart J 36(2):112–119

Backer JM (2008) The regulation and function of class III PI3Ks: novel roles for Vps34. Biochem J 410(1):1–17

Baehrecke EH (2005) Autophagy: dual roles in life and death? Nat Rev Mol Cell Biol 6(6):505–510

Behrns KE, Tsiotos GG, DeSouza NF, Krishna MK, Ludwig J, Nagorney DM (1998) Hepatic steatosis as a potential risk factor for major hepatic resection. J Gastrointest Surg 2(3):292–298

Belghiti J, Noun R, Malafosse R, Jagot P, Sauvanet A, Pierangeli F, Marty J, Farges O (1999) Continuous versus intermittent portal triad clamping for liver resection: a controlled study. Ann Surg 229(3):369–375

Bellot G, Garcia-Medina R, Gounon P, Chiche J, Roux D, Pouyssegur J, Mazure NM (2009) Hypoxia-induced autophagy is mediated through hypoxia-inducible factor induction of BNIP3 and BNIP3L via their BH3 domains. Mol Cell Biol 29(10):2570–2581

Beyer K, Nuscher B (1996) Specific cardiolipin binding interferes with labeling of sulfhydryl residues in the adenosine diphosphate/adenosine triphosphate carrier protein from beef heart mitochondria. Biochemistry 35(49):15784–15790

Biel TG, Lee S, Flores-Toro JA, Dean JW, Go KL, Lee MH, Law BK, Law ME, Dunn WA Jr, Zendejas I, Behrns KE, Kim JS (2016) Sirtuin 1 suppresses mitochondrial dysfunction of ischemic mouse livers in a mitofusin 2-dependent manner. Cell Death Differ 23(2):279–290

Bruick RK (2000) Expression of the gene encoding the proapoptotic Nip3 protein is induced by hypoxia. Proc Natl Acad Sci U S A 97(16):9082–9087

Cardinal J, Pan P, Dhupar R, Ross M, Nakao A, Lotze M, Billiar T, Geller D, Tsung A (2009a) Cisplatin prevents high mobility group box 1 release and is protective in a murine model of hepatic ischemia/reperfusion injury. Hepatology 50(2):565–574
Cardinal J, Pan P, Tsung A (2009b) Protective role of cisplatin in ischemic liver injury through induction of autophagy. Autophagy 5(8):1211–1212
Cavallini G, Donati A, Taddei M, Bergamini E (2007) Evidence for selective mitochondrial autophagy and failure in aging. Autophagy 3(1):26–27
Celinski SA, Gamblin TC (2010) Hepatic resection nomenclature and techniques. Surg Clin North Am 90(4):737–748
Cellier L, Tamareille S, Kalakech H, Guillou S, Lenaers G, Prunier F, Mirebeau-Prunier D (2016) Remote ischemic conditioning influences mitochondrial dynamics. Shock 45(2):192–197
Chen G, Ray R, Dubik D, Shi L, Cizeau J, Bleackley RC, Saxena S, Gietz RD, Greenberg AH (1997) The E1B 19K/Bcl-2-binding protein Nip3 is a dimeric mitochondrial protein that activates apoptosis. J Exp Med 186(12):1975–1983
Chen M, Sandoval H, Wang J (2008) Selective mitochondrial autophagy during erythroid maturation. Autophagy 4(7):926–928
Chen G, Han Z, Feng D, Chen Y, Chen L, Wu H, Huang L, Zhou C, Cai X, Fu C, Duan L, Wang X, Liu L, Liu X, Shen Y, Zhu Y, Chen Q (2014) A regulatory signaling loop comprising the PGAM5 phosphatase and CK2 controls receptor-mediated mitophagy. Mol Cell 54(3):362–377
Chu CT, Ji J, Dagda RK, Jiang JF, Tyurina YY, Kapralov AA, Tyurin VA, Yanamala N, Shrivastava IH, Mohammadyani D, Qiang Wang KZ, Zhu J, Klein-Seetharaman J, Balasubramanian K, Amoscato AA, Borisenko G, Huang Z, Gusdon AM, Cheikhi A, Steer EK, Wang R, Baty C, Watkins S, Bahar I, Bayır H, Kagan VE (2013) Cardiolipin externalization to the outer mitochondrial membrane acts as an elimination signal for mitophagy in neuronal cells. Nat Cell Biol 15(10):1197–1205
Chua CE, Gan BQ, Tang BL (2011) Involvement of members of the Rab family and related small GTPases in autophagosome formation and maturation. Cell Mol Life Sci 68(20):3349–3358
Clavien PA, Yadav S, Sindram D, Bentley RC (2000) Protective effects of ischemic preconditioning for liver resection performed under inflow occlusion in humans. Ann Surg 232(2):155–162
Clavien PA, Selzner M, Rüdiger HA, Graf R, Kadry Z, Rousson V, Jochum W (2003) A prospective randomized study in 100 consecutive patients undergoing major liver resection with versus without ischemic preconditioning. Ann Surg 238(6):843–850
Codogno P, Mehrpour M, Proikas-Cezanne T (2012) Canonical and non-canonical autophagy: variations on a common theme of self-eating? Nat Rev Mol Cell Biol 13(1):7–12
Couinaud C (1999) Liver anatomy: portal (and suprahepatic) or biliary segmentation. Dig Surg 16(6):459–467
Cursio R, Colosetti P, Gugenheim J (2015) Autophagy and liver ischemia-reperfusion injury. Biomed Res Int 2015:417590
Czaja MJ, Ding WX, Donohue TM, Friedman SL, Kim J-S, Komatsu M, Lemasters JJ, Lemoine A, Lin JD, Ou JH, Perlmutter DH, Randall G, Ray RB, Tsung A, Yin XM (2013) Functions of autophagy in normal and diseased liver. Autophagy 9(8):1131–1158
Dai DF, Johnson SC, Villarin JJ, Chin MT, Nieves-Cintrón M, Chen T, Marcinek DJ, Dorn GW, Kang YJ, Prolla TA, Santana LF, Rabinovitch PS (2011) Mitochondrial oxidative stress mediates angiotensin II-induced cardiac hypertrophy and Galphaq overexpression-induced heart failure. Circ Res 108(7):837–846
Day CP, James OF (1998) Steatohepatitis: a tale of two "hits"? Gastroenterology 114(4):842–845
Deas E, Plun-Favreau H, Gandhi S, Desmond H, Kjaer S, Loh SH, Renton AE, Harvey RJ, Whitworth AJ, Martins LM, Abramov AY, Wood NW (2011) PINK1 cleavage at position A103 by the mitochondrial protease PARL. Hum Mol Genet 20(5):867–879
Donati A, Cavallini G, Paradiso C, Vittorini S, Pollera M, Gori Z, Bergamini E (2001) Age-related changes in the regulation of autophagic proteolysis in rat isolated hepatocytes. J Gerontol A Biol Sci Med Sci 56(7):B288–B293

Donati A, Recchia G, Cavallini G, Bergamini E (2008) Effect of aging and anti-aging caloric restriction on the endocrine regulation of rat liver autophagy. J Gerontol A Biol Sci Med Sci 63(6):550–555

Dunn WA, Cregg JM, Kiel JA, van der Klei IJ, Oku M, Sakai Y, Sibirny AA, Stasyk OV, Veenhuis M (2005) Pexophagy: the selective autophagy of peroxisomes. Autophagy 1(2):75–83

Evankovich J, Zhang R, Cardinal JS, Zhang L, Chen J, Huang H, Beer-Stolz D, Billiar TR, Rosengart MR, Tsung A (2012) Calcium/calmodulin-dependent protein kinase IV limits organ damage in hepatic ischemia-reperfusion injury through induction of autophagy. Am J Physiol Gastrointest Liver Physiol 303(2):G189–G198

Evans ZP, Ellett JD, Schmidt MG, Schnellmann RG, Chavin KD (2008) Mitochondrial uncoupling protein-2 mediates steatotic liver injury following ischemia/reperfusion. J Biol Chem 283(13):8573–8579

Fabrikant JI (1968) The kinetics of cellular proliferation in regenerating liver. J Cell Biol 36(3):551–565

Fang EF, Scheibye-Knudsen M, Brace LE, Kassahun H, SenGupta T, Nilsen H, Mitchell JR, Croteau DL, Bohr VA (2014) Defective mitophagy in XPA via PARP-1 hyperactivation and NAD(+)/SIRT1 reduction. Cell 157(4):882–896

Feng D, Liu L, Zhu Y, Chen Q (2013) Molecular signaling toward mitophagy and its physiological significance. Exp Cell Res 319(12):1697–1705

Finnin MS, Donigian JR, Pavletich NP (2001) Structure of the histone deacetylase SIRT2. Nat Struct Biol 8(7):621–625

Fraga CG, Shigenaga MK, Park JW, Degan P, Ames BN (1990) Oxidative damage to DNA during aging: 8-hydroxy-2′-deoxyguanosine in rat organ DNA and urine. Proc Natl Acad Sci U S A 87(12):4533–4537

Fritz KS, Galligan JJ, Hirschey MD, Verdin E, Petersen DR (2012) Mitochondrial acetylome analysis in a mouse model of alcohol-induced liver injury utilizing SIRT3 knockout mice. J Proteome Res 11(3):1633–1643

Fukumori D, Kamiyama T, Nakanishi K, Yokoo H, Tahara M, Kamachi H, Matsushita M, Todo S (2011) Analysis of risk factors for hyperbilirubinemia occurring after hepatectomy. J Hepatobiliary Pancreat Sci 18(5):724–730

Fukuzawa K, Emre S, Senyuz O, Acarli K, Schwartz ME, Miller CM (1995) N-acetylcysteine ameliorates reperfusion injury after warm hepatic ischemia. Transplantation 59(1):6–9

Gamerdinger M, Hajieva P, Kaya AM, Wolfrum U, Hartl FU, Behl C (2009) Protein quality control during aging involves recruitment of the macroautophagy pathway by BAG3. EMBO J 28(7):889–901

Gebert N, Joshi AS, Kutik S, Becker T, McKenzie M, Guan XL, Mooga VP, Stroud DA, Kulkarni G, Wenk MR, Rehling P, Meisinger C, Ryan MT, Wiedemann N, Greenberg ML, Pfanner N (2009) Mitochondrial cardiolipin involved in outer-membrane protein biogenesis: implications for Barth syndrome. Curr Biol 19(24):2133–2139

Geisler S, Holmstrom KM, Skujat D, Fiesel FC, Rothfuss OC, Kahle PJ, Springer W (2010) PINK1/Parkin-mediated mitophagy is dependent on VDAC1 and p62/SQSTM1. Nat Cell Biol 12(2):119–131

Gerhart-Hines Z, Rodgers JT, Bare O, Lerin C, Kim SH, Mostoslavsky R, Alt FW, Wu Z, Puigserver P (2007) Metabolic control of muscle mitochondrial function and fatty acid oxidation through SIRT1/PGC-1alpha. EMBO J 26(7):1913–1923

Gho BC, Schoemaker RG, van den Doel MA, Duncker DJ, Verdouw PD (1996) Myocardial protection by brief ischemia in noncardiac tissue. Circulation 94(9):2193–2200

Go KL, Lee S, Zendejas I, Behrns KE, Kim J-S (2015) Mitochondrial dysfunction and autophagy in hepatic ischemia/reperfusion injury. Biomed Res Int 2015:183469

Gunter TE, Pfeiffer DR (1990) Mechanisms by which mitochondria transport calcium. Am J Phys 258(5 Pt 1):C755–C786

Gurusamy KS, Gonzalez HD, Davidson BR (2010) Current protective strategies in liver surgery. World J Gastroenterol 16(48):6098–6103

Halestrap AP, Richardson AP (2015) The mitochondrial permeability transition: a current perspective on its identity and role in ischaemia/reperfusion injury. J Mol Cell Cardiol 78:129–141
Halestrap AP, McStay GP, Clarke SJ (2002) The permeability transition pore complex: another view. Biochimie 84(2–3):153–166
Hamasaki M, Noda T, Baba M, Ohsumi Y (2005) Starvation triggers the delivery of the endoplasmic reticulum to the vacuole via autophagy in yeast. Traffic 6(1):56–65
Hamilton ML, Van Remmen H, Drake JA, Yang H, Guo ZM, Kewitt K, Walter CA, Richardson A (2001) Does oxidative damage to DNA increase with age? Proc Natl Acad Sci U S A 98(18):10469–10474
Hara T, Takamura A, Kishi C, Iemura S, Natsume T, Guan JL, Mizushima N (2008) FIP200, a ULK-interacting protein, is required for autophagosome formation in mammalian cells. J Cell Biol 181(3):497–510
Hariharan N, Maejima Y, Nakae J, Paik J, Depinho RA, Sadoshima J (2010) Deacetylation of FoxO by Sirt1 plays an essential role in mediating starvation-induced autophagy in cardiac myocytes. Circ Res 107(12):1470–1482
Harman D (1956) Aging: a theory based on free radical and radiation chemistry. J Gerontol 11(3):298–300
Harman D (1972) The biologic clock: the mitochondria? J Am Geriatr Soc 20(4):145–147
Harman D (1981) The aging process. Proc Natl Acad Sci U S A 78(11):7124–7128
He L, Lemasters JJ (2002) Regulated and unregulated mitochondrial permeability transition pores: a new paradigm of pore structure and function? FEBS Lett 512(1–3):1–7
Healey JE, Schroy PC (1953) Anatomy of the biliary ducts within the human liver; analysis of the prevailing pattern of branchings and the major variations of the biliary ducts. AMA Arch Surg 66(5):599–616
Hirschey MD, Shimazu T, Jing E, Grueter CA, Collins AM, Aouizerat B, Stančáková A, Goetzman E, Lam MM, Schwer B, Stevens RD, Muehlbauer MJ, Kakar S, Bass NM, Kuusisto J, Laakso M, Alt FW, Newgard CB, Farese RV, Kahn CR, Verdin E (2011) SIRT3 deficiency and mitochondrial protein hyperacetylation accelerate the development of the metabolic syndrome. Mol Cell 44(2):177–190
Hoare M, Das T, Alexander G (2010) Ageing, telomeres, senescence, and liver injury. J Hepatol 53(5):950–961
Hodson L, McQuaid SE, Humphreys SM, Milne R, Fielding BA, Frayn KN, Karpe F (2010) Greater dietary fat oxidation in obese compared with lean men: an adaptive mechanism to prevent liver fat accumulation? Am J Physiol Endocrinol Metab 299(4):E584–E592
Hosokawa N, Hara T, Kaizuka T, Kishi C, Takamura A, Miura Y, Iemura S, Natsume T, Takehana K, Yamada N, Guan JL, Oshiro N, Mizushima N (2009) Nutrient-dependent mTORC1 association with the ULK1-Atg13-FIP200 complex required for autophagy. Mol Biol Cell 20(7):1981–1991
Hou T, Zhang X, Xu J, Jian C, Huang Z, Ye T, Hu K, Zheng M, Gao F, Wang X, Cheng H (2013) Synergistic triggering of superoxide flashes by mitochondrial Ca2+ uniport and basal reactive oxygen species elevation. J Biol Chem 288(7):4602–4612
Hou T, Jian C, Xu J, Huang AY, Xi J, Hu K, Wei L, Cheng H, Wang X (2016) Identification of EFHD1 as a novel Ca(2+) sensor for mitoflash activation. Cell Calcium 59(5):262–270
Huang J, Li BK, Chen GH, Li JQ, Zhang YQ, Li GH, Yuan YF (2009) Long-term outcomes and prognostic factors of elderly patients with hepatocellular carcinoma undergoing hepatectomy. J Gastrointest Surg 13(9):1627–1635
Huang R, Xu Y, Wan W, Shou X, Qian J, You Z, Liu B, Chang C, Zhou T, Lippincott-Schwartz J, Liu W (2015) Deacetylation of nuclear LC3 drives autophagy initiation under starvation. Mol Cell 57(3):456–466
Itakura E, Kishi C, Inoue K, Mizushima N (2008) Beclin 1 forms two distinct phosphatidylinositol 3-kinase complexes with mammalian Atg14 and UVRAG. Mol Biol Cell 19(12):5360–5372
Jaeschke H (1998) Mechanisms of reperfusion injury after warm ischemia of the liver. J Hepato-Biliary-Pancreat Surg 5(4):402–408

Jang JH, Kang KJ, Kang Y, Lee IS, Graf R, Clavien PA (2008) Ischemic preconditioning and intermittent clamping confer protection against ischemic injury in the cirrhotic mouse liver. Liver Transpl 14(7):980–988

Jang SY, Kang HT, Hwang ES (2012) Nicotinamide-induced mitophagy: event mediated by high NAD+/NADH ratio and SIRT1 protein activation. J Biol Chem 287(23):19304–19314

Jarnagin WR, Gonen M, Fong Y, DeMatteo RP, Ben-Porat L, Little S, Corvera C, Weber S, Blumgart LH (2002) Improvement in perioperative outcome after hepatic resection: analysis of 1,803 consecutive cases over the past decade. Ann Surg 236(4):397–406; discussion 406–397

Jia J, Li J, Jiang L, Zhang J, Chen S, Wang L, Zhou Y, Xie H, Zhou L, Zheng S (2015) Protective effect of remote limb ischemic perconditioning on the liver grafts of rats with a novel model. PLoS One 10(3):e0121972

Jiang Y, Zhang H, Dong LY, Wang D, An W (2008) Increased hepatic UCP2 expression in rats with nonalcoholic steatohepatitis is associated with upregulation of Sp1 binding to its motif within the proximal promoter region. J Cell Biochem 105(1):277–289

Jin SM, Lazarou M, Wang C, Kane LA, Narendra DP, Youle RJ (2010) Mitochondrial membrane potential regulates PINK1 import and proteolytic destabilization by PARL. J Cell Biol 191(5):933–942

Jung CH, Ro SH, Cao J, Otto NM, Kim DH (2010) mTOR regulation of autophagy. FEBS Lett 584(7):1287–1295

Jung CH, Seo M, Otto NM, Kim DH (2011) ULK1 inhibits the kinase activity of mTORC1 and cell proliferation. Autophagy 7(10):1212–1221

Jungermann K, Kietzmann T (1996) Zonation of parenchymal and nonparenchymal metabolism in liver. Annu Rev Nutr 16:179–203

Jungermann K, Kietzmann T (2000) Oxygen: modulator of metabolic zonation and disease of the liver. Hepatology 31(2):255–260

Kageyama S, Hata K, Tanaka H, Hirao H, Kubota T, Okamura Y, Iwaisako K, Takada Y, Uemoto S (2015) Intestinal ischemic preconditioning ameliorates hepatic ischemia/reperfusion injury in rats: role of heme oxygenase 1 in the second window of protection. Liver Transpl 21(1):112–122

Kathirvel E, Chen P, Morgan K, French SW, Morgan TR (2010) Oxidative stress and regulation of anti-oxidant enzymes in cytochrome P4502E1 transgenic mouse model of non-alcoholic fatty liver. J Gastroenterol Hepatol 25(6):1136–1143

Katz SC, Shia J, Liau KH, Gonen M, Ruo L, Jarnagin WR, Fong Y, D'Angelica MI, Blumgart LH, Dematteo RP (2009) Operative blood loss independently predicts recurrence and survival after resection of hepatocellular carcinoma. Ann Surg 249(4):617–623

Kawabata T, Kinoshita M, Inatsu A, Habu Y, Nakashima H, Shinomiya N, Seki S (2008) Functional alterations of liver innate immunity of mice with aging in response to CpG-oligodeoxynucleotide. Hepatology 48(5):1586–1597

Khraiwesh H, López-Domínguez JA, Fernández del Río L, Gutierrez-Casado E, López-Lluch G, Navas P, de Cabo R, Ramsey JJ, Burón MI, Villalba JM, González-Reyes JA (2014) Mitochondrial ultrastructure and markers of dynamics in hepatocytes from aged, calorie restricted mice fed with different dietary fats. Exp Gerontol 56:77–88

Kim KH, Lee MS (2014) Autophagy – a key player in cellular and body metabolism. Nat Rev Endocrinol 10(6):322–337

Kim J-S, He L, Lemasters JJ (2003a) Mitochondrial permeability transition: a common pathway to necrosis and apoptosis. Biochem Biophys Res Commun 304(3):463–470

Kim J-S, He L, Qian T, Lemasters JJ (2003b) Role of the mitochondrial permeability transition in apoptotic and necrotic death after ischemia/reperfusion injury to hepatocytes. Curr Mol Med 3(6):527–535

Kim J-S, Qian T, Lemasters JJ (2003c) Mitochondrial permeability transition in the switch from necrotic to apoptotic cell death in ischemic rat hepatocytes. Gastroenterology 124(2):494–503

Kim J-S, Jin Y, Lemasters JJ (2006a) Reactive oxygen species, but not Ca2+ overloading, trigger pH- and mitochondrial permeability transition-dependent death of adult rat myocytes after ischemia-reperfusion. Am J Physiol Heart Circ Physiol 290(5):H2024–H2034

Kim SC, Sprung R, Chen Y, Xu Y, Ball H, Pei J, Cheng T, Kho Y, Xiao H, Xiao L, Grishin NV, White M, Yang XJ, Zhao Y (2006b) Substrate and functional diversity of lysine acetylation revealed by a proteomics survey. Mol Cell 23(4):607–618

Kim J-S, Nitta T, Mohuczy D, O'Malley KA, Moldawer LL, Dunn WA, Behrns KE (2008a) Impaired autophagy: a mechanism of mitochondrial dysfunction in anoxic rat hepatocytes. Hepatology 47(5):1725–1736

Kim J-S, Nitta T, Mohuczy D, O'Malley KA, Moldawer LL, Dunn WA Jr, Behrns KE (2008b) Impaired autophagy: a mechanism of mitochondrial dysfunction in anoxic rat hepatocytes. Hepatology 47:1725–1736

Kim J, Kundu M, Viollet B, Guan KL (2011) AMPK and mTOR regulate autophagy through direct phosphorylation of Ulk1. Nat Cell Biol 13(2):132–141

Kim J-S, Wang JH, Lemasters JJ (2012) Mitochondrial permeability transition in rat hepatocytes after anoxia/reoxygenation: role of Ca^{2+}-dependent mitochondrial formation of reactive oxygen species. Am J Physiol Gastrointest Liver Physiol 302(7):G723–G731

Kim J-S, Wang JH, Biel TG, Kim DS, Flores-Toro JA, Vijayvargiya R, Zendejas I, Behrns KE (2013) Carbamazepine suppresses calpain-mediated autophagy impairment after ischemia/reperfusion in mouse livers. Toxicol Appl Pharmacol 273(3):600–610

Kim IH, Kisseleva T, Brenner DA (2015a) Aging and liver disease. Curr Opin Gastroenterol 31(3):184–191

Kim JM, Cho BI, Kwon CH, Joh JW, Park JB, Lee JH, Kim SJ, Paik SW, Park CK (2015b) Hepatectomy is a reasonable option for older patients with hepatocellular carcinoma. Am J Surg 209(2):391–397

Kirkin V, McEwan DG, Novak I, Dikic I (2009) A role for ubiquitin in selective autophagy. Mol Cell 34(3):259–269

Klune JR, Tsung A (2010) Molecular biology of liver ischemia/reperfusion injury: established mechanisms and recent advancements. Surg Clin North Am 90(4):665–677

Ko JS, Gwak MS, Kim GS, Shin YH, Ryu S, Kim JS, Kim SJ, Kim ST (2013) The protective effect of ischemic preconditioning against hepatic ischemic-reperfusion injury under isoflurane anesthesia in rats. Transplant Proc 45(5):1704–1707

Koga H, Kaushik S, Cuervo AM (2010) Altered lipid content inhibits autophagic vesicular fusion. FASEB J 24(8):3052–3065

Kolosova NG, Aidagulova SV, Nepomnyashchikh GI, Shabalina IG, Shalbueva NI (2001) Dynamics of structural and functional changes in hepatocyte mitochondria of senescence-accelerated OXYS rats. Bull Exp Biol Med 132(2):814–819

Kraft C, Deplazes A, Sohrmann M, Peter M (2008) Mature ribosomes are selectively degraded upon starvation by an autophagy pathway requiring the Ubp3p/Bre5p ubiquitin protease. Nat Cell Biol 10(5):602–610

Kuma A, Mizushima N (2010) Physiological role of autophagy as an intracellular recycling system: with an emphasis on nutrient metabolism. Semin Cell Dev Biol 21(7):683–690

Lagouge M, Argmann C, Gerhart-Hines Z, Meziane H, Lerin C, Daussin F, Messadeq N, Milne J, Lambert P, Elliott P, Geny B, Laakso M, Puigserver P, Auwerx J (2006) Resveratrol improves mitochondrial function and protects against metabolic disease by activating SIRT1 and PGC-1alpha. Cell 127(6):1109–1122

Lai IR, Chang KJ, Chen CF, Tsai HW (2006) Transient limb ischemia induces remote preconditioning in liver among rats: the protective role of heme oxygenase-1. Transplantation 81(9):1311–1317

Lai IR, Chang KJ, Chen CF (2007) The mechanism of hepatic heme oxygenase-1 expression by limb remote ischemic preconditioning. Transplantation 83(3):364

Le Couteur DG, Hickey HM, Harvey PJ, McLean AJ (1999) Oxidative injury reproduces age-related impairment of oxygen-dependent drug metabolism. Pharmacol Toxicol 85(5):230–232

Le Couteur DG, Warren A, Cogger VC, Smedsrød B, Sørensen KK, De Cabo R, Fraser R, McCuskey RS (2008) Old age and the hepatic sinusoid. Anat Rec (Hoboken) 291(6):672–683

Lee IH, Finkel T (2009) Regulation of autophagy by the p300 acetyltransferase. J Biol Chem 284(10):6322–6328

Lee WY, Lee SM (2006) Synergistic protective effect of ischemic preconditioning and allopurinol on ischemia/reperfusion injury in rat liver. Biochem Biophys Res Commun 349(3):1087–1093

Lee VM, Cameron RG, Archer MC (1998) Zonal location of compensatory hepatocyte proliferation following chemically induced hepatotoxicity in rats and humans. Toxicol Pathol 26(5):621–627

Lee IH, Cao L, Mostoslavsky R, Lombard DB, Liu J, Bruns NE, Tsokos M, Alt FW, Finkel T (2008) A role for the NAD-dependent deacetylase Sirt1 in the regulation of autophagy. Proc Natl Acad Sci U S A 105(9):3374–3379

Lemasters JJ (1999) V. Necrapoptosis and the mitochondrial permeability transition: shared pathways to necrosis and apoptosis. Am J Phys 276(1 Pt 1):G1–G6

Lemasters JJ (2014) Variants of mitochondrial autophagy: types 1 and 2 mitophagy and micromitophagy (type 3). Redox Biol 2:749–754

Lemasters JJ, Nieminen AL, Qian T, Trost LC, Elmore SP, Nishimura Y, Crowe RA, Cascio WE, Bradham CA, Brenner DA, Herman B (1998) The mitochondrial permeability transition in cell death: a common mechanism in necrosis, apoptosis and autophagy. Biochim Biophys Acta 1366(1–2):177–196

Lemasters JJ, Qian T, He L, Kim J-S, Elmore SP, Cascio WE, Brenner DA (2002) Role of mitochondrial inner membrane permeabilization in necrotic cell death, apoptosis, and autophagy. Antioxid Redox Signal 4(5):769–781

Levine B, Klionsky DJ (2004) Development by self-digestion: molecular mechanisms and biological functions of autophagy. Dev Cell 6(4):463–477

Li Y, Wong K, Giles A, Jiang J, Lee JW, Adams AC, Kharitonenkov A, Yang Q, Gao B, Guarente L, Zang M (2014) Hepatic SIRT1 attenuates hepatic steatosis and controls energy balance in mice by inducing fibroblast growth factor 21. Gastroenterology 146(2):539–549

Liang C, Feng P, Ku B, Dotan I, Canaani D, Oh BH, Jung JU (2006) Autophagic and tumour suppressor activity of a novel Beclin1-binding protein UVRAG. Nat Cell Biol 8(7):688–699

Lin SY, Li TY, Liu Q, Zhang C, Li X, Chen Y, Zhang SM, Lian G, Ruan K, Wang Z, Zhang CS, Chien KY, Wu J, Li Q, Han J, Lin SC (2012) GSK3-TIP60-ULK1 signaling pathway links growth factor deprivation to autophagy. Science 336(6080):477–481

Lionetti L, Mollica MP, Donizzetti I, Gifuni G, Sica R, Pignalosa A, Cavaliere G, Gaita M, De Filippo C, Zorzano A, Putti R (2014) High-lard and high-fish-oil diets differ in their effects on function and dynamic behaviour of rat hepatic mitochondria. PLoS One 9(3):e92753

Liu Y, Dentin R, Chen D, Hedrick S, Ravnskjaer K, Schenk S, Milne J, Meyers DJ, Cole P, Yates J III, Olefsky J, Guarente L, Montminy M (2008) A fasting inducible switch modulates gluconeogenesis via activator/coactivator exchange. Nature 456(7219):269–273

Liu L, Feng D, Chen G, Chen M, Zheng Q, Song P, Ma Q, Zhu C, Wang R, Qi W, Huang L, Xue P, Li B, Wang X, Jin H, Wang J, Yang F, Liu P, Zhu Y, Sui S, Chen Q (2012) Mitochondrial outer-membrane protein FUNDC1 mediates hypoxia-induced mitophagy in mammalian cells. Nat Cell Biol 14(2):177–185

Liu A, Fang H, Wei W, Dirsch O, Dahmen U (2014) Ischemic preconditioning protects against liver ischemia/reperfusion injury via heme oxygenase-1-mediated autophagy. Crit Care Med 42(12):762–771

Lohmann D, Spandl J, Stevanovic A, Schoene M, Philippou-Massier J, Thiele C (2013) Monoubiquitination of ancient ubiquitous protein 1 promotes lipid droplet clustering. PLoS One 8(9):e72453

Lortat-Jacob JL, Robert HG, Henry C (1952) Excision of the right lobe of the liver for a malignant secondary tumor. Arch Mal Appar Dig Mal Nutr 41(6):662–667

Lutter M, Perkins GA, Wang X (2001) The pro-apoptotic Bcl-2 family member tBid localizes to mitochondrial contact sites. BMC Cell Biol 2:22

MacDonald GA, Bridle KR, Ward PJ, Walker NI, Houglum K, George DK, Smith JL, Powell LW, Crawford DH, Ramm GA (2001) Lipid peroxidation in hepatic steatosis in humans is associ-

ated with hepatic fibrosis and occurs predominately in acinar zone 3. J Gastroenterol Hepatol 16(6):599–606
Maddock KR, Huff-Lonergan E, Rowe LJ, Lonergan SM (2005) Effect of pH and ionic strength on mu- and m-calpain inhibition by calpastatin. J Anim Sci 83(6):1370–1376
Mancias JD, Wang X, Gygi SP, Harper JW, Kimmelman AC (2014) Quantitative proteomics identifies NCOA4 as the cargo receptor mediating ferritinophagy. Nature 509(7498):105–109
Massey AC, Zhang C, Cuervo AM (2006) Chaperone-mediated autophagy in aging and disease. Curr Top Dev Biol 73:205–235
Matsunaga K, Saitoh T, Tabata K, Omori H, Satoh T, Kurotori N, Maejima I, Shirahama-Noda K, Ichimura T, Isobe T, Akira S, Noda T, Yoshimori T (2009) Two Beclin 1-binding proteins, Atg14L and Rubicon, reciprocally regulate autophagy at different stages. Nat Cell Biol 11(4):385–396
Matsuzawa-Nagata N, Takamura T, Ando H, Nakamura S, Kurita S, Misu H, Ota T, Yokoyama M, Honda M, Miyamoto K, Kaneko S (2008) Increased oxidative stress precedes the onset of high-fat diet-induced insulin resistance and obesity. Metabolism 57(8):1071–1077
de Meijer VE, Kalish BT, Puder M, Ijzermans JN (2010) Systematic review and meta-analysis of steatosis as a risk factor in major hepatic resection. Br J Surg 97(9):1331–1339
Melendez JA, Arslan V, Fischer ME, Wuest D, Jarnagin WR, Fong Y, Blumgart LH (1998) Perioperative outcomes of major hepatic resections under low central venous pressure anesthesia: blood loss, blood transfusion, and the risk of postoperative renal dysfunction. J Am Coll Surg 187(6):620–625
Michikawa Y, Mazzucchelli F, Bresolin N, Scarlato G, Attardi G (1999) Aging-dependent large accumulation of point mutations in the human mtDNA control region for replication. Science 286(5440):774–779
Michiorri S, Gelmetti V, Giarda E, Lombardi F, Romano F, Marongiu R, Nerini-Molteni S, Sale P, Vago R, Arena G, Torosantucci L, Cassina L, Russo MA, Dallapiccola B, Valente EM, Casari G (2010) The Parkinson-associated protein PINK1 interacts with Beclin1 and promotes autophagy. Cell Death Differ 17(6):962–974
Mijaljica D, Prescott M, Devenish RJ (2006) Endoplasmic reticulum and Golgi complex: contributions to, and turnover by, autophagy. Traffic 7(12):1590–1595
Min J, Landry J, Sternglanz R, Xu RM (2001) Crystal structure of a SIR2 homolog-NAD complex. Cell 105(2):269–279
Mizuguchi T, Kawamoto M, Meguro M, Okita K, Ota S, Ishii M, Ueki T, Nishidate T, Kimura Y, Furuhata T, Hirata K (2015) Impact of aging on morbidity and mortality after liver resection: a systematic review and meta-analysis. Surg Today 45(3):259–270
Mizushima N, Levine B, Cuervo AM, Klionsky DJ (2008) Autophagy fights disease through cellular self-digestion. Nature 451(7182):1069–1075
Mizushima N, Yoshimori T, Levine B (2010) Methods in mammalian autophagy research. Cell 140(3):313–326
Mizushima N, Yoshimori T, Ohsumi Y (2011) The role of Atg proteins in autophagosome formation. Annu Rev Cell Dev Biol 27:107–132
Moreau K, Renna M, Rubinsztein DC (2013) Connections between SNAREs and autophagy. Trends Biochem Sci 38(2):57–63
Murry CE, Richard VJ, Reimer KA, Jennings RB (1990) Ischemic preconditioning slows energy metabolism and delays ultrastructural damage during a sustained ischemic episode. Circ Res 66(4):913–931
Nakahara H, Kanno T, Inai Y, Utsumi K, Hiramatsu M, Mori A, Packer L (1998) Mitochondrial dysfunction in the senescence accelerated mouse (SAM). Free Radic Biol Med 24(1):85–92
Nakano H, Boudjema K, Alexandre E, Imbs P, Chenard MP, Wolf P, Cinqualbre J, Jaeck D (1995) Protective effects of N-acetylcysteine on hypothermic ischemia-reperfusion injury of rat liver. Hepatology 22(2):539–545
Narendra D, Tanaka A, Suen DF, Youle RJ (2009) Parkin-induced mitophagy in the pathogenesis of Parkinson disease. Autophagy 5(5):706–708

Nie Y, Erion DM, Yuan Z, Dietrich M, Shulman GI, Horvath TL, Gao Q (2009) STAT3 inhibition of gluconeogenesis is downregulated by SirT1. Nat Cell Biol 11(4):492–500

Niemelä O (2001) Distribution of ethanol-induced protein adducts in vivo: relationship to tissue injury. Free Radic Biol Med 31(12):1533–1538

Obara K, Ohsumi Y (2011) Atg14: a key player in orchestrating autophagy. Int J Cell Biol 2011:713435

Okatani Y, Wakatsuki A, Reiter RJ, Miyahara Y (2002) Hepatic mitochondrial dysfunction in senescence-accelerated mice: correction by long-term, orally administered physiological levels of melatonin. J Pineal Res 33(3):127–133

Okatsu K, Oka T, Iguchi M, Imamura K, Kosako H, Tani N, Kimura M, Go E, Koyano F, Funayama M, Shiba-Fukushima K, Sato S, Shimizu H, Fukunaga Y, Taniguchi H, Komatsu M, Hattori N, Mihara K, Tanaka K, Matsuda N (2012) PINK1 autophosphorylation upon membrane potential dissipation is essential for Parkin recruitment to damaged mitochondria. Nat Commun 3:1016

Pacher P, Hajnoczky G (2001) Propagation of the apoptotic signal by mitochondrial waves. EMBO J 20(15):4107–4121

Pan LJ, Zhang ZC, Zhang ZY, Wang WJ, Xu Y, Zhang ZM (2012) Effects and mechanisms of store-operated calcium channel blockade on hepatic ischemia-reperfusion injury in rats. World J Gastroenterol 18(4):356–367

Papadopoulos D, Siempis T, Theodorakou E, Tsoulfas G (2013) Hepatic ischemia and reperfusion injury and trauma: current concepts. Arch Trauma Res 2(2):63–70

Park JM, Kim TH, Bae JS, Kim MY, Kim KS, Ahn YH (2010) Role of resveratrol in FOXO1-mediated gluconeogenic gene expression in the liver. Biochem Biophys Res Commun 403(3–4):329–334

Penta JS, Johnson FM, Wachsman JT, Copeland WC (2001) Mitochondrial DNA in human malignancy. Mutat Res 488(2):119–133

Peralta C, Jiménez-Castro MB, Gracia-Sancho J (2013) Hepatic ischemia and reperfusion injury: effects on the liver sinusoidal milieu. J Hepatol 59(5):1094–1106

Petrowsky H, McCormack L, Trujillo M, Selzner M, Jochum W, Clavien PA (2006) A prospective, randomized, controlled trial comparing intermittent portal triad clamping versus ischemic preconditioning with continuous clamping for major liver resection. Ann Surg 244(6):921–928

Ploeg RJ, D'Alessandro AM, Knechtle SJ, Stegall MD, Pirsch JD, Hoffmann RM, Sasaki T, Sollinger HW, Belzer FO, Kalayoglu M (1993) Risk factors for primary dysfunction after liver transplantation – a multivariate analysis. Transplantation 55(4):807–813

Pringle JH (1908) V. Notes on the arrest of hepatic hemorrhage due to trauma. Ann Surg 48(4):541–549

Purushotham A, Schug TT, Xu Q, Surapureddi S, Guo X, Li X (2009) Hepatocyte-specific deletion of SIRT1 alters fatty acid metabolism and results in hepatic steatosis and inflammation. Cell Metab 9(4):327–338

Qian T, Nieminen AL, Herman B, Lemasters JJ (1997) Mitochondrial permeability transition in pH-dependent reperfusion injury to rat hepatocytes. Am J Phys 273(6 Pt 1):C1783–C1792

Qian T, Herman B, Lemasters JJ (1999) The mitochondrial permeability transition mediates both necrotic and apoptotic death of hepatocytes exposed to Br-A23187. Toxicol Appl Pharmacol 154(2):117–125

Rabinowitz JD, White E (2010) Autophagy and metabolism. Science 330(6009):1344–1348

Ravikumar B, Sarkar S, Davies JE, Futter M, Garcia-Arencibia M, Green-Thompson ZW, Jimenez-Sanchez M, Korolchuk VI, Lichtenberg M, Luo S, Massey DC, Menzies FM, Moreau K, Narayanan U, Renna M, Siddiqi FH, Underwood BR, Winslow AR, Rubinsztein DC (2010) Regulation of mammalian autophagy in physiology and pathophysiology. Physiol Rev 90(4):1383–1435

Ray R, Chen G, Vande VC, Cizeau J, Park JH, Reed JC, Gietz RD, Greenberg AH (2000) BNIP3 heterodimerizes with Bcl-2/Bcl-X(L) and induces cell death independent of a Bcl-2 homology 3 (BH3) domain at both mitochondrial and nonmitochondrial sites. J Biol Chem 275(2):1439–1448

Rehman H, Sun J, Shi Y, Ramshesh VK, Liu Q, Currin RT, Lemasters JJ, Zhong Z (2011) NIM811 prevents mitochondrial dysfunction, attenuates liver injury, and stimulates liver regeneration after massive hepatectomy. Transplantation 91(4):406–412

Rodgers JT, Lerin C, Haas W, Gygi SP, Spiegelman BM, Puigserver P (2005) Nutrient control of glucose homeostasis through a complex of PGC-1alpha and SIRT1. Nature 434(7029):113–118

Rolfe DF, Hulbert AJ, Brand MD (1994) Characteristics of mitochondrial proton leak and control of oxidative phosphorylation in the major oxygen-consuming tissues of the rat. Biochim Biophys Acta 1188(3):405–416

Rubinsztein DC, Codogno P, Levine B (2012) Autophagy modulation as a potential therapeutic target for diverse diseases. Nat Rev Drug Discov 11(9):709–730

Rüdiger HA, Kang KJ, Sindram D, Riehle HM, Clavien PA (2002) Comparison of ischemic preconditioning and intermittent and continuous inflow occlusion in the murine liver. Ann Surg 235(3):400–407

Russell RC, Tian Y, Yuan H, Park HW, Chang YY, Kim J, Kim H, Neufeld TP, Dillin A, Guan KL (2013) ULK1 induces autophagy by phosphorylating Beclin-1 and activating VPS34 lipid kinase. Nat Cell Biol 15(7):741–750

Ryerson AB, Eheman CR, Altekruse SF, Ward JW, Jemal A, Sherman RL, Henley SJ, Holtzman D, Lake A, Noone AM, Anderson RN, Ma J, Ly KN, Cronin KA, Penberthy L, Kohler BA (2016) Annual report to the nation on the status of cancer, 1975-2012, featuring the increasing incidence of liver cancer. Cancer 122(9):1312–1337

Saidi RF, Chang J, Brooks S, Nalbantoglu I, Adsay V, Jacobs MJ (2007) Ischemic preconditioning and intermittent clamping increase the tolerance of fatty liver to hepatic ischemia-reperfusion injury in the rat. Transplant Proc 39(10):3010–3014

Sanchez AM, Csibi A, Raibon A, Cornille K, Gay S, Bernardi H, Candau R (2012) AMPK promotes skeletal muscle autophagy through activation of forkhead FoxO3a and interaction with Ulk1. J Cell Biochem 113(2):695–710

Sandoval H, Thiagarajan P, Dasgupta SK, Schumacher A, Prchal JT, Chen M, Wang J (2008) Essential role for Nix in autophagic maturation of erythroid cells. Nature 454(7201):232–235

Santambrogio L, Cuervo AM (2011) Chasing the elusive mammalian microautophagy. Autophagy 7(6):652–654

Schiesser M, Wittert A, Nieuwenhuijs VB, Morphett A, Padbury RT, Barritt GJ (2009) Intermittent ischemia but not ischemic preconditioning is effective in restoring bile flow after ischemia reperfusion injury in the livers of aged rats. J Surg Res 152(1):61–68

Schroeder B, Schulze RJ, Weller SG, Sletten AC, Casey CA, McNiven MA (2015) The small GTPase Rab7 as a central regulator of hepatocellular lipophagy. Hepatology 61(6):1896–1907

Selzner M, RuDiger HA, Sindram D, Madden J, Clavien PA (2000) Mechanisms of ischemic injury are different in the steatotic and normal rat liver. Hepatology 2000(32):1280–1288

Selzner N, Rudiger H, Graf R, Clavien PA (2003a) Protective strategies against ischemic injury of the liver. Gastroenterology 125(3):917–936

Selzner N, Selzner M, Jochum W, Clavien PA (2003b) Ischemic preconditioning protects the steatotic mouse liver against reperfusion injury: an ATP dependent mechanism. J Hepatol 39(1):55–61

Settembre C, Di Malta C, Polito VA, Garcia Arencibia M, Vetrini F, Erdin S, Erdin SU, Huynh T, Medina D, Colella P, Sardiello M, Rubinsztein DC, Ballabio A (2011) TFEB links autophagy to lysosomal biogenesis. Science 332(6036):1429–1433

Shadel GS, Clayton DA (1997) Mitochondrial DNA maintenance in vertebrates. Annu Rev Biochem 66:409–435

Shimada M, Matsumata T, Akazawa K, Kamakura T, Itasaka H, Sugimachi K, Nose Y (1994) Estimation of risk of major complications after hepatic resection. Am J Surg 167(4):399–403

Shimada M, Takenaka K, Fujiwara Y, Gion T, Shirabe K, Yanaga K, Sugimachi K (1998) Risk factors linked to postoperative morbidity in patients with hepatocellular carcinoma. Br J Surg 85(2):195–198

Singh R, Kaushik S, Wang Y, Xiang Y, Novak I, Komatsu M, Tanaka K, Cuervo AM, Czaja MJ (2009) Autophagy regulates lipid metabolism. Nature 458(7242):1131–1135

Soubannier V, McLelland GL, Zunino R, Braschi E, Rippstein P, Fon EA, McBride HM (2012) A vesicular transport pathway shuttles cargo from mitochondria to lysosomes. Curr Biol 22(2):135–141

Spandl J, Lohmann D, Kuerschner L, Moessinger C, Thiele C (2011) Ancient ubiquitous protein 1 (AUP1) localizes to lipid droplets and binds the E2 ubiquitin conjugase G2 (Ube2g2) via its G2 binding region. J Biol Chem 286(7):5599–5606

Spitzer AL, Lao OB, Dick AA, Bakthavatsalam R, Halldorson JB, Yeh MM, Upton MP, Reyes JD, Perkins JD (2010) The biopsied donor liver: incorporating macrosteatosis into high-risk donor assessment. Liver Transpl 16(7):874–884

Starley BQ, Calcagno CJ, Harrison SA (2010) Nonalcoholic fatty liver disease and hepatocellular carcinoma: a weighty connection. Hepatology 51(5):1820–1832

Subramani S, Perdreau-Dahl H, Morth JP (2016) The magnesium transporter A is activated by cardiolipin and is highly sensitive to free magnesium in vitro. Elife:5. doi:10.7554/eLife.11407

Sun T, Li X, Zhang P, Chen WD, Zhang HL, Li DD, Deng R, Qian XJ, Jiao L, Ji J, Li YT, Wu RY, Yu Y, Feng GK, Zhu XF (2015) Acetylation of Beclin 1 inhibits autophagosome maturation and promotes tumour growth. Nat Commun 6:7215

Suyavaran A, Ramamurthy C, Mareeswaran R, Subastri A, Lokeswara Rao P, Thirunavukkarasu C (2015) TNF-α suppression by glutathione preconditioning attenuates hepatic ischemia reperfusion injury in young and aged rats. Inflamm Res 64(1):71–81

Takahashi Y, Coppola D, Matsushita N, Cualing HD, Sun M, Sato Y, Liang C, Jung JU, Cheng JQ, Mulé JJ, Pledger WJ, Wang HG (2007) Bif-1 interacts with Beclin 1 through UVRAG and regulates autophagy and tumorigenesis. Nat Cell Biol 9(10):1142–1151

Tanno M, Sakamoto J, Miura T, Shimamoto K, Horio Y (2007) Nucleocytoplasmic shuttling of the NAD+–dependent histone deacetylase SIRT1. J Biol Chem 282(9):6823–6832

Trifunovic A, Wredenberg A, Falkenberg M, Spelbrink JN, Rovio AT, Bruder CE, Bohlooly-Y M, Gidlöf S, Oldfors A, Wibom R, Törnell J, Jacobs HT, Larsson NG (2004) Premature ageing in mice expressing defective mitochondrial DNA polymerase. Nature 429(6990):417–423

Tsukamoto I, Nakata R, Kojo S (1993) Effect of ageing on rat liver regeneration after partial hepatectomy. Biochem Mol Biol Int 30(4):773–778

Twig G, Elorza A, Molina AJ, Mohamed H, Wikstrom JD, Walzer G, Stiles L, Haigh SE, Katz S, Las G, Alroy J, Wu M, Py BF, Yuan J, Deeney JT, Corkey BE, Shirihai OS (2008) Fission and selective fusion govern mitochondrial segregation and elimination by autophagy. EMBO J 27(2):433–446

Uchida M, Takemoto Y, Nagasue N, Dhar DK, Kohno H, Nakamura T (1994) Effect of verapamil on hepatic reperfusion injury after prolonged ischemia in pigs. J Hepatol 21(2):217–223

Unsay JD, Cosentino K, Subburaj Y, García-Sáez AJ (2013) Cardiolipin effects on membrane structure and dynamics. Langmuir 29(51):15878–15887

Van Raamsdonk JM, Hekimi S (2012) Superoxide dismutase is dispensable for normal animal lifespan. Proc Natl Acad Sci U S A 109(15):5785–5790

Veteläinen R, van Vliet A, Gouma DJ, van Gulik TM (2007) Steatosis as a risk factor in liver surgery. Ann Surg 245(1):20–30

Vial G, Dubouchaud H, Couturier K, Cottet-Rousselle C, Taleux N, Athias A, Galinier A, Casteilla L, Leverve XM (2011) Effects of a high-fat diet on energy metabolism and ROS production in rat liver. J Hepatol 54(2):348–356

Vlada CA, Kim J-S, Behrns KE (2015) Autophagy: self-preservation through cannibalism of proteins and organelles. Surgery 157(1):1–5

Vollmar B, Menger MD (2009) The hepatic microcirculation: mechanistic contributions and therapeutic targets in liver injury and repair. Physiol Rev 89(4):1269–1339

Waldmeier PC, Feldtrauer JJ, Qian T, Lemasters JJ (2002) Inhibition of the mitochondrial permeability transition by the nonimmunosuppressive cyclosporin derivative NIM811. Mol Pharmacol 62(1):22–29

Wang W, Fang H, Groom L, Cheng A, Zhang W, Liu J, Wang X, Li K, Han P, Zheng M, Yin J, Mattson MP, Kao JP, Lakatta EG, Sheu SS, Ouyang K, Chen J, Dirksen RT, Cheng H (2008) Superoxide flashes in single mitochondria. Cell 134(2):279–290

Wang JH, Ahn IS, Fischer TD, Byeon JI, Dunn WA Jr, Behrns KE, Leeuwenburgh C, Kim J-S (2011) Autophagy suppresses age-dependent ischemia and reperfusion injury in livers of mice. Gastroenterology 141(6):2188–2199

Wang L, Liu X, Nie J, Zhang J, Kimball SR, Zhang H, Zhang WJ, Jefferson LS, Cheng Z, Ji Q, Shi Y (2015) ALCAT1 controls mitochondrial etiology of fatty liver diseases, linking defective mitophagy to steatosis. Hepatology 61(2):486–496

Weber T, Zemelman BV, McNew JA, Westermann B, Gmachl M, Parlati F, Söllner TH, Rothman JE (1998) SNAREpins: minimal machinery for membrane fusion. Cell 92(6):759–772

Wei Y, Pattingre S, Sinha S, Bassik M, Levine B (2008) JNK1-mediated phosphorylation of Bcl-2 regulates starvation-induced autophagy. Mol Cell 30(6):678–688

Wu W, Tian W, Hu Z, Chen G, Huang L, Li W, Zhang X, Xue P, Zhou C, Liu L, Zhu Y, Li L, Zhang L, Sui S, Zhao B, Feng D (2014) ULK1 translocates to mitochondria and phosphorylates FUNDC1 to regulate mitophagy. EMBO Rep 15(5):566–575

Yadav SS, Sindram D, Perry DK, Clavien PA (1999) Ischemic preconditioning protects the mouse liver by inhibition of apoptosis through a caspase-dependent pathway. Hepatology 30(5):1223–1231

Yamashita Y, Shimada M, Hamatsu T, Rikimaru T, Tanaka S, Shirabe K, Sugimachi K (2001) Effects of preoperative steroid administration on surgical stress in hepatic resection: prospective randomized trial. Arch Surg 136(3):328–333

Yang W, Hekimi S (2010) A mitochondrial superoxide signal triggers increased longevity in *Caenorhabditis elegans*. PLoS Biol 8(12):e1000556. doi:10.1371/journal.pbio.1000556

Yang Z, Klionsky DJ (2010) Eaten alive: a history of macroautophagy. Nat Cell Biol 12(9):814–822

Yang L, Li P, Fu S, Calay ES, Hotamisligil GS (2010) Defective hepatic autophagy in obesity promotes ER stress and causes insulin resistance. Cell Metab 11(6):467–478

Ye B, Zhao H, Hou H, Wang G, Liu F, Zhao Y, Zhang Z, Xie K, Zhu L, Geng X (2014) Ischemic preconditioning provides no additive clinical value in liver resection of cirrhotic and non-cirrhotic patients under portal triad clamping: a prospective randomized controlled trial. Clin Res Hepatol Gastroenterol 38(4):467–474

Yee C, Yang W, Hekimi S (2014) The intrinsic apoptosis pathway mediates the pro-longevity response to mitochondrial ROS in *C. elegans*. Cell 157(4):897–909

Yun N, Cho HI, Lee SM (2014) Impaired autophagy contributes to hepatocellular damage during ischemia/reperfusion: heme oxygenase-1 as a possible regulator. Free Radic Biol Med 68:168–177

Zhai Y, Busuttil RW, Kupiec-Weglinski JW (2011) Liver ischemia and reperfusion injury: new insights into mechanisms of innate-adaptive immune-mediated tissue inflammation. Am J Transplant 11(8):1563–1569

Zhang C, Cuervo AM (2008) Restoration of chaperone-mediated autophagy in aging liver improves cellular maintenance and hepatic function. Nat Med 14(9):959–965

Zhang C, Huang C, Tian Y, Li X (2014) Polyol pathway exacerbated ischemia/reperfusion-induced injury in steatotic liver. Oxid Med Cell Longev 2014:963629

Zhong Z, Theruvath TP, Currin RT, Waldmeier PC, Lemasters JJ (2007) NIM811, a mitochondrial permeability transition inhibitor, prevents mitochondrial depolarization in small-for-size rat liver grafts. Am J Transplant 7(5):1103–1111

Zhong Z, Ramshesh VK, Rehman H, Currin RT, Sridharan V, Theruvath TP, Kim I, Wright GL, Lemasters JJ (2008) Activation of the oxygen-sensing signal cascade prevents mitochondrial injury after mouse liver ischemia-reperfusion. Am J Physiol Gastrointest Liver Physiol 295(4):G823–G832

Zhou W, Li A, Pan Z, Fu S, Yang Y, Tang L, Hou Z, Wu M (2008) Selective hepatic vascular exclusion and Pringle maneuver: a comparative study in liver resection. Eur J Surg Oncol 34(1):49–54

Zhu Y, Massen S, Terenzio M, Lang V, Chen-Lindner S, Eils R, Novak I, Dikic I, Hamacher-Brady A, Brady NR (2013) Modulation of serines 17 and 24 in the LC3-interacting region of Bnip3 determines pro-survival mitophagy versus apoptosis. J Biol Chem 288(2):1099–1113

Zhu J, Lu T, Yue S, Shen X, Gao F, Busuttil RW, Kupiec-Weglinski JW, Xia Q, Zhai Y (2015) Rapamycin protection of livers from ischemia and reperfusion injury is dependent on both autophagy induction and mammalian target of rapamycin complex 2-Akt activation. Transplantation 99(1):48–55

Chapter 10
Role of Endoplasmic Reticulum Stress in Hepatic Injury

Cheng Ji, Neil Kaplowitz, and Hui Han

Abbreviations

AA	Amino acid
ABCA1	ATP-binding cassette transporter member 1, also known as the cholesterol efflux regulatory protein (CERP)
ACC	Acetyl-CoA carboxylase
ACL	ATP citrate lyase
ADH	Alcohol dehydrogenase
AERR	Alcohol-induced ER stress response
AKT	Also known as protein kinase B (PKB), is a serine–/threonine-specific protein kinase
ALD	Alcohol-induced liver disease
APAP	Acetaminophen
Apo-B/E	Apolipoprotein B/E
ARα	Androgen receptor-α
ASK1	Apoptosis signal-regulating kinase 1
ASMase	Acidic sphingomyelinase
ATF	Activating transcription factor
BAX	bcl-2-like protein 4
BHMT	Betaine-homocysteine methyltransferase
BiP	Also known as *GRP78*, glucose-regulated protein 78
C/EBP	CCAAT-enhancer-binding protein
CaMKII	Calmodulin-dependent protein kinase II

C. Ji (✉) • N. Kaplowitz • H. Han
Department of Medicine, Keck School of Medicine of USC,
University of Southern California, Los Angeles, CA, USA
e-mail: chengji@usc.edu

W.-X. Ding, X.-M. Yin (eds.), *Molecules, Systems and Signaling in Liver Injury*,
Cell Death in Biology and Diseases, DOI 10.1007/978-3-319-58106-4_10

X.-M. Yin and Z. Dong (Series eds.), *Cell Death in Biology and Diseases*

CBS	Cystathionine β-synthase
CDIPT	CDP-diacylglycerol-inositol 3-phosphatidyltransferase
CHOP	C/EBP homologous protein
COX	Cyclooxygenases
CREB	Cyclic AMP response element-binding protein
Creld2	Cysteine rich with EGF-like domains 2
CYP2E1	Cytochrome P4502E1
DGAT2	Diacylglycerol acyltransferase-2
DR5	Death receptor 5
EDEM	ER degradation-enhancing mannosidase-like protein
eIF2α	Eukaryotic initiation factor 2α
ER	Endoplasmic reticulum
ERAD	ER-associated degradation
ERK	Extracellular signal-regulated kinases
ERO1	ER oxidase 1
ERα	Estrogen receptor-α
FAS	Fatty acid synthase
GADD34	Growth arrest and DNA damage-inducible protein 34
GGH	Ground glass hepatocyte
GSH	Glutathione
GSK3β	Glycogen synthase kinase-3β
HBV	Hepatitis B
HCA	Hepatocellular adenomas
HCC	Hepatocellular carcinoma
HCV	Hepatitis C
HDL	High-density lipoprotein
HepG2	A perpetual cell line derived from human liver carcinoma cells
HERP	Homocysteine-induced ER protein
HFD	High-fat diet
HHcy	Hyperhomocysteinemia
HIV PIs	Anti-HIV protease inhibitors
HNF1	Hepatocyte nuclear factor 1 α
Huh7	A cell line derived from hepatic cellular carcinoma cells
IP10	Interferon gamma-inducible protein 10
IRE1	Inositol-requiring enzyme 1
JNK	c-Jun N-terminal kinases
LDLT	Living donor liver transplantation
LPS	Lipopolysaccharide
MCD	Methionine-choline deficient
MDDC	Monocyte-derived dendritic cells
MEOS	Microsomal ethanol oxidizing system
NADH	Nicotinamide adenine dinucleotide
NAFLD	Nonalcoholic fatty liver disease
NASH	Nonalcoholic steatohepatitis
NF-κB	Nuclear factor κB
NNRTI	Non-nucleoside reverse transcriptase inhibitors

NRF2	NF-E2-related factor 2
NRTI	Nucleoside/nucleotide reverse transcriptase inhibitors
NSAIDs	Nonsteroidal anti-inflammatory drugs
PERK	PKR-like ER-localized eIF2α kinase
PGC1	PPAR gamma co-activator 1
PP2Ac	Protein phosphatase 2Ac
PPAR	Peroxisome proliferator-activated receptor
PTEN	Phosphatase and tensin homolog
RIDD	Regulated IRE1-dependent mRNA decay
ROS	Reactive oxygen species
S1P	Site 1 protease
S2P	Site 2 protease
SAH	S-adenosylhomocysteine
SAM	S-Adenosylmethionine
SCD	Stearoyl-CoA desaturase
SERCA	Sarco/ER calcium ATPase
SHP	Small heterodimer partner
SREBP	Sterol regulatory element-binding protein
STAT3	Signal transducer and activator of transcription 3
sXbp1	Unconventionally spliced Xbp1
TRAF2	TNF receptor-associated factor 2
TRAIL	TNF-related apoptosis-inducing ligand
TRB3	Tribbles 3
TUDCA	Tauroursodeoxycholic acid
UDCA	Ursodeoxycholic acid
UPR	The unfolded protein response
VEGF	Vascular endothelial growth factor
VLDL	Very low-density lipoprotein
XBP1	X-box binding protein 1

10.1 Introduction

The endoplasmic reticulum (ER) is a type of organelle in the cells of eukaryotes. The ER was first discovered in 1943 (Claude 1943), and its fundamental structure and function in protein synthesis were identified 2 years later (Porter et al. 1945). There are two types of ER, rough ER and smooth ER. The rough ER is an extensive membranous network of cisterns, branched tubules, and flattened sacs with the cytosolic face studded with ribosomes that are the sites of protein synthesis. The smooth ER lacks ribosomes and functions in synthesis of lipid and steroid hormones and drug detoxification. The rough ER consists of oxidizing conditions and a high calcium concentration, which are necessary for the formation of disulfide bonds and proper protein folding (Gaut and Hendershot 1993; Bulleid 2012). During protein synthesis, nascent proteins are translated into the ER lumen where they are co-translationally and/or posttranslationally modified with oligosaccharyl residues for recognition by ER chaperones and subsequent extracellular secretion or intracellular distribution to other organelles (Ogata et al. 2006). Since its discovery, the ER has been shown to be essential in maintaining cellular protein homeostasis through integrating various intracellular and extracellular signals including growth, differentiation, inflammatory, and metabolic signals. Accumulation of misfolded/unfolded proteins will stress the ER and trigger the unfolded protein response (UPR), a unique set of molecular responses that are conserved among various eukaryotic organisms (Walter and Ron 2011; Mori 2015). The ER stress response contributes to development and progression of many human disorders, including but not limited to neurodegeneration, atherosclerosis, diabetes, infectious and inflammatory disease, and cancer (Ozcan and Tabas 2012; Wang and Kaufman 2012). The liver is particularly vulnerable to the ER stress as it consists of cells with abundant ER of microsomal structures, which assume synthesis of a large amount of secretory and membrane proteins (Palade 1995; Palade and Siekevitz 1956). The ER stress contributes to a spectrum of liver injuries, which are reviewed in this chapter.

10.2 Molecular Mechanisms of ER Stress Response

The ER stress is a state in which protein folding slows and unfolded or malfolded proteins accumulate in the ER (Mori 2000; Patil and Walter 2001; Schröder and Kaufman 2005; Ji and Kaplowitz 2006; Ron and Walter 2007; Ron and Hubbard 2013), which results from perturbations of ER homeostasis such as inhibition of glycosylation, alterations of ER redox environment, calcium depletion, lipid overloading, or viral infection. The ER stress activates the UPR, which constitutes a series of ER-to-nucleus signaling pathways that are mediated by three ER-resident transmembrane sensor proteins: inositol-requiring protein 1 (IRE1), dsRNA-activated protein kinase R PKR-like ER-localized eIF2α kinase (PERK), and activating transcription factor 6 (ATF6) (Fig. 10.1). The three sensors are activated upon dissociation from their inhibitory binding with the chaperone GRP78/BiP. IRE1, which has kinase and endoribonuclease activities, is activated by trans-autophosphorylation. The RNAse activity of IRE1 under low-activation conditions participates a process known as regulated IRE1-dependent mRNA decay (RIDD), which degrades RNA and reduces protein synthesis and loading in the ER (Maurel et al. 2014). The activated IRE1 processes mRNA of the transcription factor X-box binding protein 1 (XBP1) via unconventional splicing to form transcriptionally active spliced XBP1 (sXBP1). sXBP1, often known as a UPR marker, activates a number of UPR target genes, including chaperone genes that are responsible for restoring ER folding capacity and genes of the ER-associated degradation (ERAD) pathway that are important for cell survival (Hampton 2002). The second sensor PERK phosphorylates the eukaryotic initiation factor 2α (eIF2α) subunit, leading to an inhibition of the initiation of translation and a global attenuation in protein translation. Phosphorylation of eIF2α selectively activates activating transcription factor 4 (ATF4), which regulates genes for ER

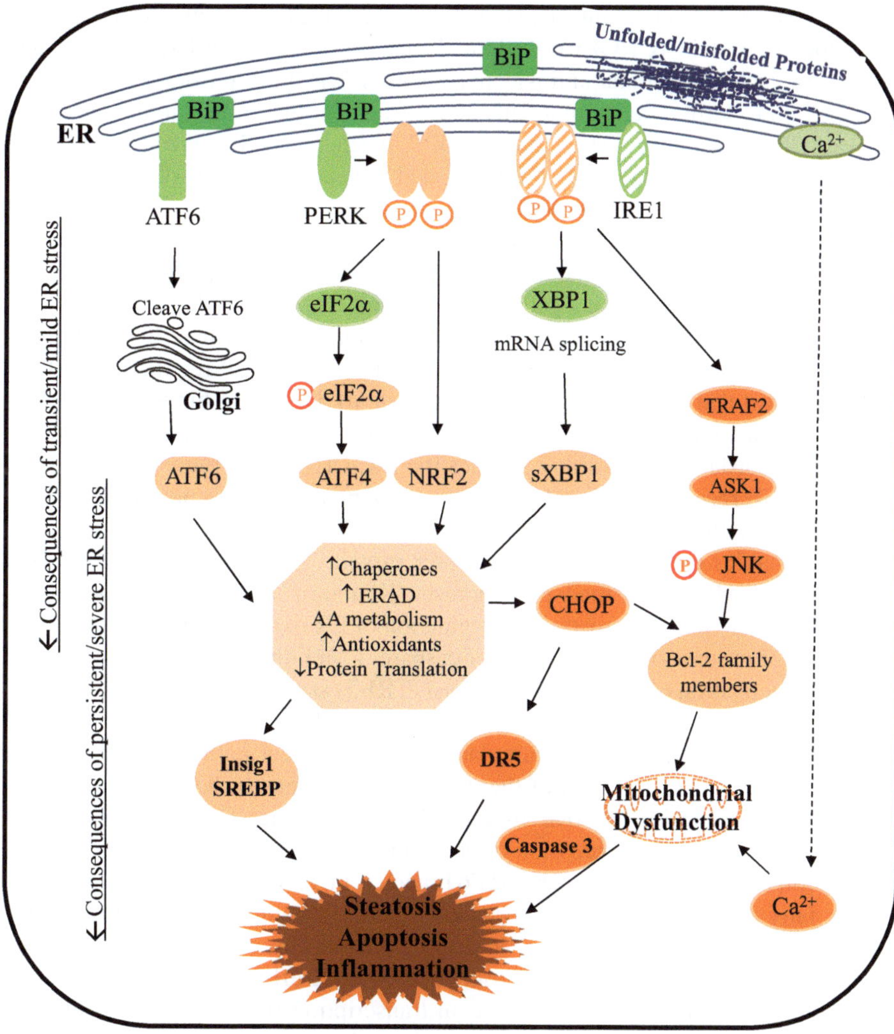

Fig. 10.1 Canonical unfolded protein response (UPR) and consequences. Accumulation of unfolded/misfolded proteins activates the three ER sensors, ATF6, PERK, and IRE1. ATF6 is translocated from the ER to the Golgi apparatus where it is activated by proteolysis. Activated PERK phosphorylates either eIF2, which enables translation of ATF4 or NRF2. Active IRE1 is a kinase as well as an endonuclease that regulates the unconventional splicing of the transcription factor XBP1 into sXBP1. Cleaved ATF6, ATF4, NRF2, and sXBP1 are all nuclear transcription factors that upregulate the expression of UPR target genes/proteins which relates to ER chaperones, translational attenuation, the ERAD pathway, antioxidant responses, and amino acid metabolism resulting in the restoration of the homeostasis. However, prolonged and severe ER stress modulates Insig1/SREBP, CHOP/DR5, JNK/Bcl-2 family proteins, cytosolic Ca^{2+} concentration, and mitochondrial function, leading to steatosis, apoptosis, and inflammation

chaperones, ERAD, and amino acid metabolism, and the transcription factor (C/EBP homologous protein (CHOP) (Oyadomari and Mori 2004). PERK also activates NRF2 that regulates cellular antioxidant responses. The third sensor ATF6 is cleaved in the Golgi to form a transcriptionally active fragment that traffics to the nucleus to activate UPR target genes. Therefore, the UPR in general results in reduced synthesis of proteins, increased capacity of folding, and increased turnover of unfolded proteins, which lead to restoration of ER homeostasis.

However, under prolonged or severe ER stress, the UPR provokes a complex network of interacting and parallel responses contributing to pathological consequences such as apoptosis, inflammation, and fat accumulation (Fig. 10.1) (Hotamisligil 2010; Ozcan and Tabas 2012; Wang and Kaufman 2012). The ER stress-induced apoptosis is mediated by a few factors. CHOP regulates GADD34 (growth arrest and DNA damage-inducible protein). GADD34 binds protein phosphatase-1 and enhances eIF2α dephosphorylation, leading to premature restoration of protein translation that further stresses the ER. CHOP also regulates expression of the TRAIL receptor DR5 and pro- and antiapoptotic Bcl-2 family proteins Bim, Bax, and Bcl-2 that modulate cell death (Tabas and Ron 2011). Sustained activation of IRE1 enables recruitment of the adaptor protein TRAF2 and activates ASK1, JNK, and NF-κB, which mediate apoptosis (Kitamura 2011). In addition, disturbance of ER calcium homeostasis, upregulation of ER oxidase 1 (ERO1) by CHOP, activation of caspase-12 (in animals) or caspase-4 (in humans), phosphorylation of AKT, and activation of GSK3β by tribbles 3 (TRB3) are other mechanisms underlying ER stress-induced inflammation and apoptosis (Ron and Hubbard 2013; Kim et al. 2015). Excessive lipid accumulation is also a main pathological feature of prolonged ER stress, and each of the three ER stress sensors has direct molecular effects on lipogenesis. The IRE1α-XBP1 branch regulates C/EBPα and C/EBPβ that directly control the expression of genes involved in de novo fatty acid biosynthesis (Rutkowski et al. 2008). The ATF6 branch is involved in phospholipid biosynthesis, fatty acid oxidation, and lipoprotein secretion (Zeng et al. 2004; Oyadomari et al. 2008). Severe or prolonged ER stress increases expression of eIF2α-specific phosphatase GADD34, which compromises the PERK-eIF2α signaling and reduced expression of C/EBPα/β and downstream transcription factor PPARγ resulting in interference with hepatic lipogenesis (Bobrovnikova-Marjon et al. 2008). The PERK-eIF2α signaling also controls activation of SREBP that regulates de novo lipid synthesis (Ji et al. 2006).

10.3 ER Stress in Alcohol-Induced Liver Disease

Alcohol is mainly metabolized in the liver. Partial role of ER in alcohol metabolism was initially realized decades ago when NADH from the hepatic oxidation of ethanol to acetaldehyde by alcohol dehydrogenase (ADH) was found to support also microsomal ethanol oxidations (Cinti et al. 1973; Lieber 1987). The inducible microsomal ethanol oxidizing system (MEOS) is associated with proliferation of

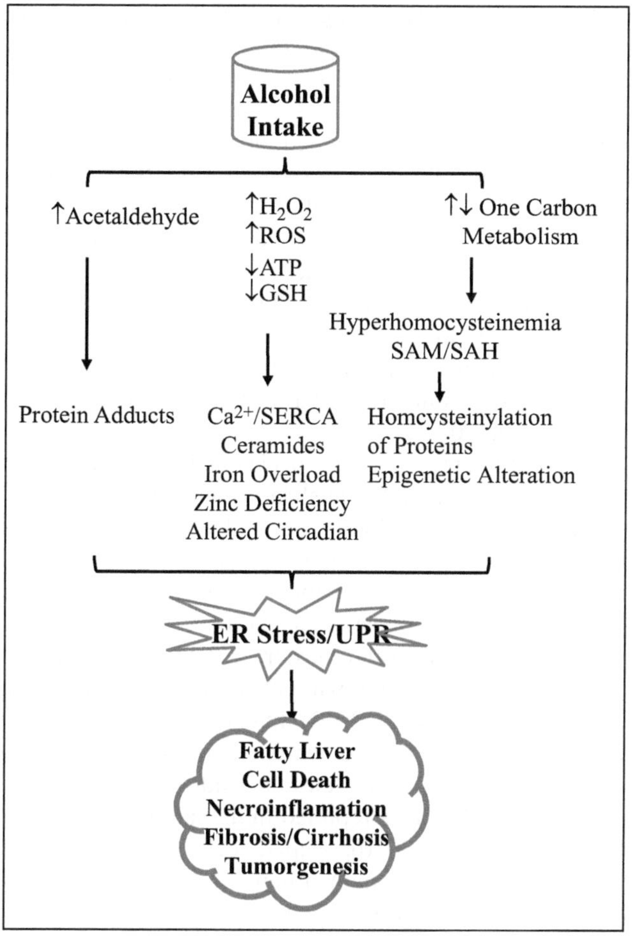

Fig. 10.2 Multiple factors involved in alcohol-induced ER stress and liver injury. Alcohol, alcohol metabolites, and alterations of cellular redox and one carbon metabolism modify proteins that disturb ER protein homeostasis and cause ER stress response resulting in liver injury

the ER and a concomitant induction of cytochrome P4502E1 (CYP2E1) in rats and humans. Free radical release as a consequence of CYP2E1 function in the ER and subsequent oxidative stress and lipid peroxidation generally contribute to ALD (Cederbaum et al. 2009). However, alcohol-induced ER stress response (AERR) was not recognized until recently (Fig. 10.2). Molecular evidence for an impaired UPR was first found in the intragastric alcohol-fed mice using microarray gene expression profiling (Ji and Kaplowitz 2003). The alterations of selected ER stress markers were associated with severe steatosis, scattered apoptosis, and necroinflammatory foci. Moderate upregulation of expression of SREBP-1c and SREBP-2 and their responsive genes was detected by immunoblotting (Ji and Kaplowitz 2003; Ji et al. 2006). SREBP-1c knockout mice were protected against triglyceride accumulation (Ji et al. 2006). Knocking out CHOP resulted in minimal alcohol-induced apoptosis in mouse liver (Ji et al. 2005). In a mouse model with moderate obesity and alcohol infusion, there are synergistic effects of accentuated ER and mitochondrial stress and nitrosative stress on hepatic necroinflammation and steatohepatitis (Xu et al. 2011). In micropigs fed alcohol, liver steatosis and apoptosis were shown to be accompanied by increased CYP2E1, GRP78, SREBP-1c, CYP2E1, and caspase-12 (Esfandiari et al. 2005), which were correlated with elevated transcripts of lipogenic genes such as fatty acid synthase (FAS), acetyl-CoA carboxylase (ACC), and stearoyl-CoA desaturase (SCD). Further, alcohol consumption increases gut-derived lipopolysaccharide (LPS), which is linked to impair UPR and advanced hepatic injury. Full UPR as indicated by activation of IRE1α, ATF-6, and eIF2α was detected in cirrhotic livers of rats challenged with LPS, whereas only eIF2α was activated in the basal state in the cirrhotic rats without the LPS challenge (Tazi et al. 2007).

Besides rodents and micropigs, other species including humans have ER stress response to alcohol consumption. Zebrafish larvae metabolize alcohol that is simply added to the water (Passeri et al. 2009). Upon alcohol challenge, zebrafish larvae developed signs of acute ALD, including hepatomegaly and steatosis accompanied by changes in the expression of genes for SREBP-mediated hepatic lipogenesis. Further, the ER stress response appeared much robust in zebrafish deficient in CDP-diacylglycerol-inositol 3-phosphatidyltransferase (CDIPT) (Thakur et al. 2011). In the species, *Caenorhabditis elegans* without a liver for alcohol digestion, little AERR has been detected (Ient et al. 2012). In baboon fed alcohol orally, upregulation of calpain 2 and calpain p94 and downregulation of eIF2α were detected (Seth and Leo 2003). The most clinically relevant studies regarding AERR are from human cells and patients. AERR is reported in human monocyte-derived dendritic cells (MDDC) (Boukli et al. 2010), HepG2 cells expressing human CYP2E1 (Magne et al. 2011), and primary human hepatocytes (Kao et al. 2012). Striking upregulation of multiple ER stress signaling molecules was detected in human patients with ALD (Longato et al. 2012; Ramirez et al. 2013).

A major cause for AERR is alcohol-induced hyperhomocysteinemia that is often seen in rodents and humans (Barak et al. 1987; Hultberg et al. 1993; Blasco et al. 2005; Lutz 2008). A few lines of molecular evidence support the homocysteine mechanism. First, betaine is a methyl donor for re-methylation of homocysteine to methionine by betaine-homocysteine methyltransferase (BHMT). Simultaneous betaine feeding or transgenic expression of BHMT in alcohol-fed mice decreased elevated homocysteine and abrogated AERR (Ji and Kaplowitz 2003; Ji et al. 2004, 2008). Second, an intragastric infusion with both high fat and alcohol induced moderate obesity and much severe ALD (Xu et al. 2011), which resulted from synergistic effects of an accentuated ER stress by elevated homocysteine in combination with mitochondrial stress and adiponectin resistance. Third, rats with alcohol infusion expressed BHMT normally and did not develop a significant hyperhomocysteinemia (Shinohara et al. 2010). Consequently, the alcohol-infused rats had a minimal ER stress response and were more resistant to ALD. Fourth, in a study with 14 inbred mouse strains with alcohol infusion, profound differences in ALD were observed among the strains in spite of consistently high levels of urine alcohol (Tsuchiya et al. 2012). ER stress-related genes were induced only in strains with the most liver injury, which were closely associated with expression patterns of methionine metabolism-related genes and plasma homocysteine levels. Thus, abnormal protein modifications by excessive homocysteine are likely responsible for the alcoholic ER stress and UPR in alcohol-infused mice that lack a sufficient upregulation of BHMT.

Multiple factors can cause AERR during ALD. Factors other than homocysteine include direct alcohol metabolites, epigenetic changes, genetic predisposition, iron overload, zinc deficiency, circadian clock, ceramide, and disrupted Ca^{2+} homeostasis (Fig. 10.2). First, alcohol metabolites, e.g., acetaldehyde, or oxidants were found contributing to AERR and development of fatty liver in a study examining effects of a selective inhibition of CYP2E1 on the development of ALD in alcohol-infused rats (Ronis et al. 2010). Second, epigenetic and genetic alterations are related to AERR. Severe steatohepatitis and upregulations of hepatic ER stress signals were associated with decreased SAM/SAH ratio in alcohol-infused mice deficient in heterozygous cystathionine β-synthase (CBS) (Esfandiari et al. 2010). The low SAM/SAH ratio was correlated with decrease levels of suppressor chromatin marker trimethylated histone H3 lysine-9 (3meH3K9) in the promoter regions of ER stress markers. Such epigenetic mechanism for AERR also occurs in human alcoholics. Epigenetic DNA hypermethylation of the promoter of the ER stress-responsive HERP gene has been reported to downregulate its mRNA expression in patients with alcohol dependence (Bleich et al. 2006; Han et al. 2013). With respect to effects of genetic defect on AERR, a robust ER stress response was observed in the mice with a liver-specific deletion of BiP and fed moderate doses of alcohol (Ji et al. 2011), which was accompanied by much aggravated hepatic steatosis and fibrosis. This suggests that the liver BiP deletion represents a genetic predisposition that unmasks a distinct mechanism by which alcohol induces ER stress, one that normally is largely obscured by compensatory changes in normal animals or presumably in the majority of human population who have low-to-moderate drinking. Third, evidence for a causal effect of iron overload on AERR was reported in mice deficient in the hemochromatosis gene ($Hfe^{-/-}$). Co-feeding ad libitum with alcohol and a high-fat diet (HFD) was found to lead to increased XBP1 splicing, activation of IRE1α and PERK, and CHOP protein expression, which were associated with profound steatohepatitis and fibrosis (Tan et al. 2013). In contrast to iron overload, alcohol induces zinc deficiency in alcoholics, which links to ER stress and cell death damages. In Wistar rats fed chronic alcohol, it was found recently that increased expression of hepatic p-eIF2α, ATF4, and CHOP and activation of caspase-3 were associated with increased cell death and reduced hepatocyte ER zinc levels. The alcohol-induced ER stress and cell death could be inhibited by zinc supplement but not by antioxidant treatment (Sun et al. 2015). Fourth, circadian clock has recently been found to be associated with AERR and fatty liver injury (Hiroyuki et al. 2014). Alcohol disturbs the nuclear receptor small heterodimer partner (SHP)-mediated circadian clock and thus impairs oscillations of ER stress response and lipid accumulation. Fifth, alterations of lipid composition of the cellular membranes by alcohol can be a direct cause for AERR. One of the common phospholipids of membrane bilayers is sphingomyelin. Acid sphingomyelinase (ASMase) hydrolyzes sphingomyelin into ceramides, which, in addition to their critical structural function in membrane bilayers, regulate apoptosis, cellular stress response, inflammation, and metabolism (Marí et al. 2014). Alcohol feeding increases ASMase expression and activity in both experimental models and patients with acute alcoholic hepatitis (Fernandez et al. 2013). Particularly, incubation of HepG2 cells with exogenous ASMase disrupts ER Ca^{2+} homeostasis. Since SERCA contributes to Ca^{2+} homeostasis in the ER (Pinton et al. 2001), it is likely that ASMase activation and subsequent ceramide production disrupt physical properties of the ER membrane, which modulates SERCA activities. Thus, SERCA activities might be a key element in AERR and liver injury. Results from other earlier studies also support the role of SERCA in AERR. In mice, alcohol exposure aggravates the inhibitory effects on SERCA and Ca^{2+} homeostasis by some anti-HIV drugs (Kao et al. 2012). In a model for dendritic regression of Purkinje neurons from the cerebellar cortex of ethanol-fed rats, dilation of the extensive smooth ER and altered SERCA activities were shown to precede activation of ER-resident caspase-12 (Dlugos 2014).

In brief, alcohol-induced hepatic ER stress occurs in the livers of several species including mouse, rat, minipigs, zebrafish, and humans, which can be caused by multiple factors and contributes significantly to ALD.

10.4 ER Stress in NAFLD/NASH

Lipid accumulation in the absence of alcohol consumption characterizes nonalcoholic fatty liver disease (NAFLD). When members of the lipid family such as free fatty acids and free cholesterol become excessive, they can cause apoptosis of hepatocytes, which advances NAFLD to severe nonalcoholic steatohepatitis (NASH). NASH is a risk factor for downstream hepatic inflammation and various degrees of fibrosis, cirrhosis, and hepatocellular carcinoma (HCC). NAFLD and NASH become increasingly common worldwide. It is a recent research focus to understand the molecular mechanisms responsible for the development of NAFLD and progression of NAFLD to NASH. The ER stress is likely an important initial factor contributing to the hepatic lipid accumulation, and lipotoxicity is central to the subsequent development of steatohepatitis. The role of non-proteasomal ER pathways in lipid metabolism was initially noticed when the ER stress-inducing agent, tunicamycin, inhibited N-linked glycosylation and interfered with the production of apolipoprotein B-100 (apoB-100) in HepG2 cells (Liao and Chan 2001) and another ER stress-inducing agent, thapsigargin (TG), activated the ER-resident caspase-12 and impaired calcium homeostasis in Huh7 cells, which was inhibited by tauroursodeoxycholic acid (TUDCA) (Xie et al. 2002). The role of ER stress in lipid accumulation was further established when alcohol-induced fatty liver disease was reported to be associated with ER stress and UPR in vivo animal models (Ji and Kaplowitz 2003) and the ER was reported to be the site of cholesterol-induced cytotoxicity

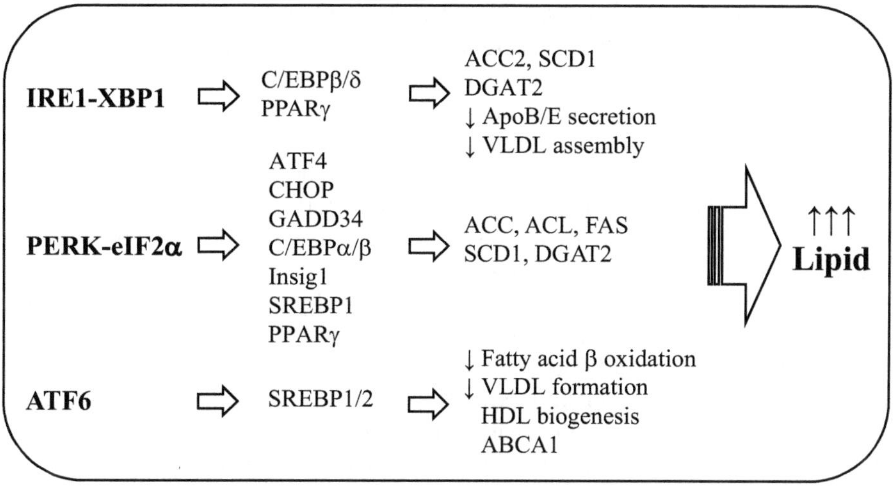

Fig. 10.3 The UPR and hepatosteatosis. Each of the three UPR branches is directly involved in regulating transcription factors and/or enzymes for lipogenesis resulting in fat accumulation in the liver under ER stress conditions

(Feng et al. 2003). Following these earlier reports, the induction of ER stress was first described in the livers of genetic and diet-induced models of NAFLD/NASH (Ozcan et al. 2004). Moreover, chemical chaperones inhibited ER stress as well as hepatic lipid accumulation in a murine model of hepatic steatosis (ob/ob mice) (Ozcan et al. 2006). These findings have since been confirmed in several other animal models with obesity or without obesity (e.g., a methionine-choline-deficient (MCD) diet-induced hepatic steatosis) (Wang et al. 2006a, b; Rahman et al. 2007; Yang et al. 2007; Sreejayan et al. 2008). Of clinical relevance, alterations of several UPR components including increased sXBP-1, increased phosphorylation of eIF2α and JNK, and activation of ATF4 and CHOP were further found in the livers of patients with NAFLD/NASH (Puri et al. 2008). Taken together, these reports highlight the importance of ER stress in the development of NAFLD/NASH.

How ER stress regulates the lipid metabolism in the liver is under intense investigations. Several lines of molecular evidence indicate that each of the three ER stress-sensing pathways (IRE1α/XBP1, PERK/eIF2α, and ATF6α) plays a direct role in regulating the development of microvesicular steatosis in the liver (Fig. 10.3). For the role of the IRE1α-XBP1 pathway in steatosis, a hepatocyte-specific deletion of IRE1α in mice has been demonstrated to repress expression of key metabolic transcriptional regulators, including C/EBPβ, C/EBPδ, and PPARγ and enzymes involved in triglyceride biosynthesis resulting in severe hepatic steatosis under ER stress (Zhang et al. 2011). IRE1α was required for efficient secretion of apolipoproteins upon disruption of ER homeostasis in the liver of IRE1α-deficient mice. XBP1 of this pathway regulates hepatic lipogenesis through directly binding to the promoters of lipogenic genes including SCD1, diacylglycerol acyltransferase-2 (DGAT2), and ACC2, which activates expression of these genes in hepatocytes (Lee et al. 2008). De novo lipid biosynthesis was reduced in the livers of mice with an

XBP1 deletion. A recent study demonstrated that the IRE1α-XBP1 pathway plays a critical role in the assembly and secretion of very low-density lipoprotein (VLDL) in the liver. Mice with the genetic ablation of IRE1α in hepatocytes demonstrated much severe hepatosteatosis upon metabolic challenges such as fasting and a high fructose diet (Wang et al. 2012), suggesting that the IRE1α-XBP1 pathway appears to be required for the maintenance of hepatic lipid homeostasis under ER stress conditions. For the role of the PERK-eIF2α-ATF4 pathway in steatosis, a deletion of PERK has been shown to inhibit the sustained expression of the lipogenic enzymes FAS, ATP citrate lyase, and SCD1 (Bobrovnikova-Marjon et al. 2008). ATF4 has been shown to activate C/EBPα and C/EBPβ regulating hepatic lipogenesis (Rahman et al. 2007; Millward et al. 2007). A heterozygous ATF4 deletion not only resulted in significantly decreased liver and white adipose tissue expression of lipogenic genes, such as PPARγ, SREBP-1c, ACC, and FAS, but also protected mice from diet-induced hepatic steatosis (Seo et al. 2009; Wang et al. 2010; Xiao et al. 2013). ATF4 also modulates CHOP functions regulating key genes involved in lipid metabolism (Chikka et al. 2013). Further, overexpression of GADD34 downstream of this pathway selectively dephosphorylated eIF2α, which compromised eIF2α-mediated signaling under ER stress conditions and resulted in diminished hepatosteatosis in animals fed a high-fat diet (Oyadomari et al. 2008). In addition, protein kinase-mediated p-eIF2α increases ATF4 translation. Thus, the PERK-eIF2α-ATF4 pathway plays an important role in promoting lipogenesis. For the role of the ATF6 pathway in steatosis, activated nuclear ATF6α has also been shown to interact with the nuclear form of SREBP2 thus inhibiting its transcriptional activity and reducing hepatic lipid accumulation (Zeng et al. 2004). In response to pharmacological ER stress, ATF6α-ablated mice developed hepatic steatosis as a result of reduced fatty acid β-oxidation and attenuated VLDL formation (Rutkowski et al. 2008; Yamamoto et al. 2010)). Moreover, in response to feeding of a high-fat diet, the ATF6α-deficient mice developed greater degrees of hepatic steatosis and glucose intolerance in association with increased expression of SREBP1c (Usui et al. 2012). Therefore, the ATF6 pathway also plays a direct role in the ER stress-induced lipid accumulation.

Under conditions of chronic ER stress, inflammatory response such as activation of NF-κB and JNK can be initiated generally by unfolded protein that caused ROS and oxidative stress and by PERK, ATF6, or the IRE1α-TRAF2 complex (Urano et al. 2000). More significantly, the inflammation can be contributed by lipotoxic members of the lipid family during NAFLD. Several lines of evidence indicate that excessive hepatic lipid accumulation interferes with ER stress response and promotes cell death and progression of steatosis to steatohepatitis. ER stress is reported to impair hepatic ABCA1 function, HDL biogenesis, and cholesterol efflux and synthesis in human hepatic cells (Röhrl et al. 2014). Excess cellular cholesterol loading increases cholesterol trafficking to ER membranes, depletes ER calcium stores, and activates the UPR and CHOP leading to apoptosis (Feng et al. 2003; Lluis et al. 2003). Studies have also shown that palmitate and other saturated fatty acids activate the UPR and apoptosis via the PERK/ATF4/CHOP pathways in human liver cells (Wei et al. 2006; Cao et al. 2012). Deletion of CHOP protects hepatocytes from palmitate-induced apoptosis in vitro via inhibition of JNK activation (Pfaffenbach et al. 2010; Zhang et al. 2015). Further, unsaturated fatty acids rescue palmitate-induced apoptosis by channeling palmitate into triglyceride pools and away from pathways leading to apoptosis (Wei et al. 2006). Particularly, palmitoleate can block ER stress-associated increases of the BH3-only proteins Bim and PUMA and inhibits lipoapoptosis in human and mouse primary hepatocytes and the Huh7 and Hep 3B cell lines (Cazanave et al. 2010), which is probably through protecting the hepatocytes against downstream death mediator Bax. Furthermore, saturated fatty acids and cholesterol have been shown to work synergistically to promote ER stress and cell death (Pineau et al. 2009), which could be attenuated by short-term exposure to unsaturated fatty acids that enables more efficient triglyceride synthesis and VLDL secretion (Listenberger et al. 2003; Fu et al. 2012). However, the lipotoxic effect of fatty acids and cholesterol on ER stress and cell death could not be rescued under chronic conditions as chronic exposure to oleic acid, a more physiologically relevant model of metabolic disease, induces ER stress and suppresses VLDL secretion (Ota et al. 2008). Under obese conditions, there are disruptions in ER calcium retention and shift in lipid composition (Fu et al. 2011), which could promote ER stress and result in an irreversible progression of NAFLD to NASH. Most interestingly, these lipotoxic effects resulting from chronic ER stress apparently occur in human patients with NASH (Lake et al. 2014).

In summary, the UPR signaling directly regulates lipid metabolism which contributes to lipid accumulation in the liver. Excessive lipids resulted from chronic/severe ER stress upregulate genes involved in apoptosis, which may further contribute to the liver transformation from simple steatosis to NASH.

10.5 ER Stress in Drug-Induced Hepatotoxicity

Many drugs target the liver, cellular enzymes of which such as cytochromes P450 can metabolize drugs generating numerous reactive electrophilic metabolites and free radicals that cause damages to the liver. Such drug-induced liver injury is a major concern for the well-being of both patients and pharmaceutical companies because it is one of the leading causes for drug failure, interruption of clinical trials, or withdrawal from the market after approval (Lee 2003; Kaplowitz 2005; Labbe et al. 2008; Senior 2014). To prevent or manage the drug-induced injury, it is fundamental to understand the molecular mechanism underlying the drug-induced hepatotoxicity. Multiple factors including binding of toxic drug metabolites to proteins/lipids/DNA, glutathione depletion, mitochondrial dysfunction, oxidative stress, lipid peroxidation, cell death, and impaired immune response are known to be involved in the liver toxicity (Gunawan and Kaplowitz 2007). In the recent years, ER stress has begun to be recognized playing an important role in the drug-induced liver toxicity, which has helped identify additional risk factors.

A number of pharmacological drugs induce ER response in the liver. First, nonsteroidal anti-inflammatory drugs (NSAIDs) such as cyclooxygenase (COX) inhibitors, diclofenac and indomethacin, are used in the treatment of acute and chronic inflammation. Apoptosis is associated with side effects of NSAIDs. Impaired UPR is found to be responsible for the NSAID-induced cell death in Huh7 cell lines (Franceschelli et al. 2011). Either diclofenac or indomethacin is able to activate only the PERK pathway that regulates the expression of the proapoptotic CHOP protein. The indomethacin-induced apoptosis could be suppressed in CHOP-deficient mice and in cultured guinea pig gastric mucosal cells by expression of the dominant-negative form of CHOP. These results indicate that ER stress-related CHOP may be involved in NSAID-induced apoptosis (Tsutsumi et al. 2004).

Second, thiazolidinediones (TZDs) such as ciglitazone and troglitazone are synthetic PPAR ligands that promote increased insulin sensitivity in type II diabetic patients. However, ciglitazone and troglitazone were found to induce CaMKII phosphorylation in liver epithelial cells, which correlated with induction of ER stress as indicated by eIF2α or PERK activation. A selective CaMKII inhibitor attenuated MKK3/6 and p38 as well as PERK and eIF2α phosphorylation (Gardner et al. 2005). Macelignan, a natural compound isolated from *Myristica fragrans*, reduced ER stress and JNK activation in the liver and adipose tissue of db/db mice and subsequently increased insulin signaling (Han et al. 2008). These data provide evidence supporting that TZDs likely deplete ER calcium, activate CaMKII, and phosphorylate PERK or eIF2α leading to ER stress response, which exerts antiproliferative effects and subsequent hepatic cell death.

Third, antalgic and antipyretic acetaminophen (APAP) is a popular drug for the management of pain and hyperthermia. However, APAP often induces hepatic injury, which is mediated by several mechanisms including mitochondrial alterations, ATP depletion, JNK activation, oxidative stress, and increased cytosolic calcium (Burcham and Harman 1988; Jaeschke 1990; Aubert et al. 2012; Hur et al. 2012). Recent in vivo and in vitro evidence indicates that APAP is also able to induce ER stress response that could contribute to hepatic cell death (Lorz et al. 2004; Nagy et al. 2007; Uzi et al. 2013; Kalinec et al. 2014). Mice with CHOP deficiency are resistant to a lethal dose of APAP (1 g/kg). However, at a lower dose (500 mg/kg), the APAP-induced ER stress was a relatively late event (Hur et al. 2012; Uzi et al. 2013), suggesting that the ER stress-mediated hepatotoxicity by APAP could be either significant at extremely high doses or secondary to mitochondrial dysfunctions at medium to high doses (Macanas-Pirard et al. 2005).

Fourth, drug resistance is common in HCC, which can be a major problem in HCC chemotherapy. One of the underlying mechanisms that mediate the drug resistance involves drug-induced ER stress and surviving effects of the UPR. For instance, anticancer cisplatin treatment triggers the UPR in HCC cells, which subsequently increases Hsp27-mediated autophagy activation and inhibits cisplatin-induced apoptosis (Chen et al. 2011). The cisplatin-induced ER stress is reported to mediate chemoresistance after living donor liver transplantation (LDLT) for patients, in which HCC tumor recurrence is an obstacle and anticancer treatment is often needed after the liver surgery. The cisplatin treatment has been shown to activate ATF6/GRP78 signaling and cause overexpression of interferon gamma-inducible protein 10 (IP10) that promoted HCC cell proliferation and tumor growth (Geng et al. 2015). Similarly, certain antioxidants are also shown to be inducers of hepatic ER stress that compromises the anti-HCC treatment (Schonthal 2012). Baicalein is a flavonoid that activates eIF2α and IRE1α and autophagy and downregulates the pro-survival Bcl-2 family, increases intracellular calcium, and activates JNK and CHOP. These mixed pro-survival and pro-death effects may limit baicalein's anti-HCC activity (Wang et al. 2014). Thus, combination of the anticancer drugs with other inhibitors against survival effects of UPR and autophagy may result in better therapeutic effects against HCC.

Fifth, antiretroviral drugs such as HIV protease inhibitors (HIV PIs) and non-nucleoside reverse transcriptase inhibitor (NNRTI) are used in the highly effective treatment for HIV-infected patients. However, antivirals singly or in combination increase the risk for HIV-associated liver disease (Núñez et al. 2001; Jones and Núñez 2012). There are reports indicating that some PIs such as atazanavir, lopinavir, and ritonavir and NNRTI such as efavirenz induce ER stress response, which is associated with altered lipid metabolism, inflammation, and apoptosis in hepatocytes and mouse liver (Cao et al. 2010; Apostolova et al. 2013; Taura et al. 2013). Moreover, the HIV PI-induced ER stress indicated by major ER stress markers can be worsened synergistically by alcohol exposure in mice and primary mouse and human hepatocytes (Kao et al. 2012). These reports support that ER stress is a pathogenic mechanism for the liver toxicities of some antiretroviral drugs. However, the precise mechanisms by which the PIs induce ER stress are still under investigations. One possible mechanism could be nonspecific inhibition of proteasome activity by the peptide mimetic PIs. Proteasome inhibition has been shown for atazanavir, lopinavir, and ritonavir (Schmidtke et al. 1999; Pajonk et al. 2002; Piccinini et al. 2002; Bono et al. 2012). Another possibility is that PI-induced oxidative stress could trigger the ER stress response (Touzet and Philips 2010; Taura et al. 2013). In addition, the PIs could interact directly with ER membrane proteins and the UPR components. For instance, inhibition of the ER-resident SERCA that regulates calcium homeostasis by lopinavir and ritonavir has been reported (Kao et al. 2012; Hu et al. 2015; Han et al. 2017).

10.6 ER Stress in Hepatitis Virus Infections

In hepatitis virus infections, viruses subvert the cellular translation machinery to produce large amounts of viral proteins, which leads to an overwhelming load of unfolded or misfolded proteins, causes ER stress, and triggers the UPR in the host hepatocytes. The UPR in the host interacts with the viruses, and the outcome can either be host cell protective, i.e., restricting viral protein production through translation attenuation, or host cell apoptosis and proviral, i.e., enabling viral protein production through chaperone production. The diverse role of ER stress and UPR in hepatitis virus infections and pathogenesis has often been demonstrated. In cirrhotic and carcinoma liver biopsies from HBV-infected patients, HBV infection was found to alter subcellular structures and induce a marked hypertrophy of the ER within hepatocytes (Hadziyannis et al. 1973). These hepatocytes termed "ground glass hepatocyte" (GGH) were found to have unusual expression and intracellular accumulation of HBV surface proteins (Hadziyannis et al. 1973; Popper 1975). Further, the accumulation of the viral proteins and the development of the GGH morphology were linked to mutations in the entire pre-S region that are secretion incompetent (Xu and Yen 1996; Tai et al. 2002). The accumulation of HBV surface proteins in host hepatocytes specifically activates ER stress signaling pathways as indicated by the upregulation of GRP78, GRP94, and ER-resident kinases (Wang et al. 2006a, b). These studies demonstrated that the ER stress induced by the pre-S mutants plays a key role in progression of HBV pathogenesis and the virus may have evolved through mutation to manipulate host UPR signaling pathways to promote its translation and persistence in infected cells.

Interactions of specific UPR components with unfolded viral proteins can trigger the UPR response during viral infection. It is reported that the chaperone GRP78 was associated transiently and non-covalently with the unfolded or immature glycoproteins during HCV infection, which were not transported to the plasma membrane but persist as complexes in the ER for a long period of time before degradation (Ng et al. 1989; Choukhi et al. 1998). On the other hand, in virus-infected cells, core and NS3 proteins of HCV are inserted into the ER and transported to cell surface and activate the expression of GRP78 (Liberman et al. 1999). HCV replicons expressing only nonstructural proteins are also capable of stimulating GRP78 expression (Tardif et al. 2002). These studies suggest that either the process of viral replication or the production of a specific viral protein in the ER is capable of inducing UPR response.

HCV can also promote its replication through suppressing the pro-survival UPR pathways and impairing antiviral pathways. It is reported that HCV replicon induces sXBP-1 but inhibits ERAD in Huh7 (Tardif et al. 2004; Zheng et al. 2005). HCV gene expression correlates the translocation of ATF6 cytoplasmic domain to the nuclei of cells that are expressing HCV subgenomic replicons. ATF6 activates the IRE1-XBP1 pathway by upregulating the transcription of XBP1. Both XBP1-spliced mRNA and sXBP1 protein were elevated in the HCV replicon-expressing cells (Tardif et al. 2004; Zheng et al. 2005), indicating that IRE1-catalyzed splicing is enhanced in the cells that host HCV gene expression. However, the downstream activity of the XBP1 is somehow repressed in the HCV-infected cells which prevents transcriptional induction of the ER degradation-enhancing mannosidase-like protein (EDEM) and inhibits ERAD (Tardif et al. 2004). Similarly, in Hep3B cells infected with HBV, expression of the multifunctional regulatory protein of HBV (HBx protein) alone is sufficient to activate both the ATF6 and IRE1-XBP1 pathways, and silencing HBx blocks their activation induced by the constitutive replication of HBV (Li et al. 2007). Thus, the protective UPR signaling in virus-infected cells is somehow interrupted and fails to further remove the excessive viral proteins. In addition, HCV has been reported to stimulate fatty acid synthesis for virus replication and maturation (Ke and Chen 2012; Targett-Adams et al. 2010). This has been speculated to be an adaptive mechanism employed by the viruses to use the

UPR signaling pathways for their own benefit given that activation of the IRE1/XBP1 pathway could increase the surface area of the ER to support the buildup of glycoproteins and enable virion budding or expand ER to reduce ER stress resulted from virus maturation. Thus, regulation of the host UPR by HCV and HBV may be an important pathophysiologic process utilized by the viruses.

The hepatitis virus can evade the host UPR resulting in pathological consequences. For instance, expression of HCV replicon has been reported to decrease glycosylation and prevented assembly of MHC-1, which allow continued virus production in host cells under severe ER stress and avoidance of immune clearance resulting in apoptosis and liver injury (Fournillier et al. 2001; Tardif and Siddiqui 2003; Fang et al. 2006; Lin et al. 2014). HCV or HBV protein expression in liver cells induces ER stress response resulting in Ca^{2+} release from the ER, which has been demonstrated to activate cyclic AMP response element-binding protein (CREB) (Benali-Furet et al. 2005, Christen et al. 2007). The activated CREB upregulates protein phosphatase 2Ac (PP2Ac), which impairs cell cycle regulation leading to apoptosis and hepatocarcinogenesis. In addition, expression of HCV core in Huh7 or HepG2 triggers hyper-expression of GRP78, GRP94, and calreticulin, CHOP induction, BAX translocation to mitochondria, cytochrome c release, and cleavages of caspase-3 and PARP (Benali-Furet et al. 2005). In the liver of transgenic mice, conditional expression of HCV structural proteins increased CHOP and cleaved caspase-3 and hepatic apoptosis (Sreejayan et al. 2008). These reports support that there is a direct link between hepatitis virus infection, ER stress, and liver injury.

10.7 ER Stress in Liver Cancer

During initiation, transformation, and establishment of liver cancer, the UPR can be activated by increased protein folding and assembly in the ER of liver cancer cells that rapidly grow and proliferate, by accumulations of mutant proteins that cannot be correctly folded, and by poor vascularization that creates hypoxia, glucose deprivation, and oxidative stress. The pro-survival capabilities of the UPR may provide chronic support for continuous proliferation and survival of liver cancer cells even under adverse conditions. Recent evidence indicates that nearly all the branches of the UPR signaling pathways are associated with liver cancer. In human HCC, elevated GRP78, ATF6, and IRE1 activations were observed (Shuda et al. 2003; Hetz et al. 2006; Al-Rawashdeh et al. 2010). Elevated Grp78 level has been reported to correlate well with higher pathologic grade, recurrence rate, and poor survival in patients with liver cancer (Luo and Lee 2013). Similarly, animal studies demonstrated that XBP1 is required for tumor growth in vivo. Knocking out Xbp1 blocked tumor formation in mice even though their growth rate and secretion of vascular endothelial growth factor (VEGF) were not decreased (Romero-Ramirez et al. 2004). Overexpression of XBP1s is associated with HCC in patients (Carrasco et al. 2007). Moreover, the contributory role of each of the UPR components to liver tumor initiation and development can be dynamic. For example, in a mouse model of HCC, increased expression of ER-resident chaperones was observed during tumor initiation and tumor progression (Vandewynckel et al. 2014). The IRE1 signaling pathway was activated only during tumor initiation, whereas the ATF6 pathway was activated only after tumor initiation. Interestingly, the PERK pathway was activated during tumor progression and continued to rise in the tumors, which make the PERK pathway the most promising target for HCC therapy (Vandewynckel 2014; Bu and Diehl 2016). In addition, ER stress can induce other antiapoptotic responses during tumorigenesis. Activated GSK3β by ER stress has been reported to phosphorylate p53, increase its degradation, and therefore protect cancer cells from p53-dependent apoptosis (Goldstein and Li 2009). NF-κB is activated during ER stress to regulate antiapoptotic responses (Kitamura 2011). Major oncogenes including Bcl-2, Bcl-XL, Mcl-1, Akt, and Ras and tumor suppressors such as p53, PTEN, and Beclin 1 directly regulate SERCA (Bittremieux et al. 2016), which alters Ca^{2+} signaling, promotes a high tolerance toward cell stress and damage, and favors malignancy.

Despite the above firm association of ER stress with liver cancer, the complete role of the UPR in liver carcinogenesis, especially how it mediates two conflicting outcomes, has yet to be defined. The dual role of GRP78/BiP in tumorigenesis is a typical example (Ji et al. 2011; Lau et al. 2013; Zhu and Lee 2015). BiP controls early tumor development through suppressive mechanisms such as the induction of cell cycle arrest, reprograming of lipid metabolism, or tumor dormancy via activating PERK (Weston and Puthalakath 2010). On the other hand, at more advanced stages of tumor progression, during which cells are exposed to more severe stressors, BiP suppresses caspase-7 activation and interacts with ER stress-induced chaperone proteins such as clusterin to promote cell survival and tumor development (Wang et al. 2013a, b). In addition, spontaneous hepatocellular adenomas (HCA) in aged female are found in mice with a liver-specific BiP deletion and under constitutive ER stress (Han et al. 2013; Lau et al. 2013). Active ATF6, CHOP, GSK3β, and Creld2 (cysteine rich with EGF-like domains 2) are increased in the knockout, indicative of continuous ER stress response. None of p53, HNF1α, or GP130 is significantly changed compared between wild type and knockout. Interestingly, cyclin D is reduced in the tumor portion of the knockout mice, which is specifically associated with overexpression of a variant of estrogen receptor-α (ERα). In contrast, there were no significant changes in the expression of ERβ, androgen receptor-α (ARα), cyclin E, or cyclin G. These findings indicate that inhibition of cyclin D and overexpression of an estrogen receptor variant promote the liver tumor development in the female BiP knockouts (Lau et al. 2013). Moreover, pathways of ERK1/2, Stat3, and p38 indicative of high malignancy are activated in these knockout mice treated with additional stresses such as high-fat diet or alcohol exposure (Han et al. 2013; Lau et al. 2013; Ji 2015). Therefore, the role of ER stress in liver cancer is more complex than initially anticipated, which may be determined by multiple endogenous and exogenous factors that change the delicate balance between the UPR-influenced oncogenes and tumor suppressors.

10.8 ER Stress in Inter-organelle Cross Talk

The ER, Golgi apparatus, and lysosomes are major components of the endomembrane system of hepatocytes, which play critical roles in cell homeostasis. Although each of the membranous organelles forms a single functional unit and their membranes cannot fuse with each other directly, they communicate through vesicle transport or inter-organellar connections (Daniele and Schiaffino 2014; Senft and Ronai 2015). The inter-organellar connections also exist between membranes of the ER, the nuclei, and the mitochondria. These connections enable inter-organellar information exchange, pathway coordination, or cross talk during stress responses, which leads to either recovery of cellular homeostasis or facilitation of disease development. There are several potential pathways for the inter-organelle cross talk. First, the activation of ATF6 requires both the ER and Golgi apparatus. Upon ER stress, phosphorylated ATF6 is transported to the Golgi in a COP-II vesicle and sequentially cleaved by S1P and S2P proteases, resulting in release of the cytoplasmic domain of ATF6 (Ye et al. 2000). The activated ATF6 mediates the UPR survival response involving an arrest of general protein synthesis and selected synthesis of chaperone proteins needed for ER homeostasis (Walter and Ron 2011; Wang and Kaufman 2012). Second, CHOP was initially found to mediate ER stress-caused cell death (Ji et al. 2005; Ron and Walter 2007; Hu et al. 2015). However, CHOP is also increased upon accumulation of unfolded proteins in the mitochondrial matrix, which mediates the mitochondrial UPR (UPR^{mt}) (Aldridge et al. 2007; Horibe and Hoogenraad 2007). Third, the expression of the ER-resident HSP47 can be induced by the Golgi dysfunction, which protects cells from Golgi stress-induced apoptosis (Miyata et al. 2013). Fourth, it is known that acute or chronic alcohol inhibits the ER-resident SERCA in human and mouse hepatocytes (Kao et al. 2012) and in neurons of mouse spinal cord and rat Purkinje (Dlugos 2014). Since there is a SERCA-mediated intricate relationship between the ER and mitochondria in intracellular calcium homeostasis, the alcoholic inhibition of SERCA could accommodate a massive transfer of calcium from the ER to mitochondria exacerbating oxidative stress. Similarly, palmitate treatment of primary hepatocytes resulted in calcium efflux from the ER, leading to mitochondrial dysfunction and oxidative stress (Egnatchik et al. 2014). In NASH, palmitate-induced ER stress also inhibits SERCA activity and causes ER calcium release into cytosol. High cytosolic levels of calcium in the hepatocytes inhibit autophagosome-lysosome fusion and protective autophagy (Czaja 2015; Dubois et al. 2016). Fifth, decreased mitochondrial O_2 consumption due to an altered metabolism can lead to an increase in reactive oxygen species (ROS) and reduction in ATP generation, both of which could worsen ER stress-caused cell death (Mohammad et al. 2012; Softic et al. 2016). In addition, the enzyme CYP2E1 is known to hydrolyze various small molecules such as fatty acid and alcohol into by-products, which is implicated in cellular stress-induced hepatocyte injury and progression to NASH by promoting oxidative stress, inflammation, and protein modification (Abdelmegeed et al. 2012). Recent evidence indicates that hepatic CYP2E1 undergoes transport from ER to mitochondria where it has distinctly different enzyme activities, which complicate pathological outcomes (Aubert et al. 2011; Hartman et al. 2015). Therefore, these inter-organellar interactions lead to the pleiotropic effects of ER stress that are often observed and potentially impact the interpretation and prediction of UPR signaling outputs.

10.9 Conclusions

The UPR in the liver is essentially a defense mechanism to attenuate ER stress and maintain the protein homeostasis. Under pathophysiological conditions, the molecular components of the UPR signaling have diverse biological functions in the hepatocytes and interact with other cellular signaling pathways, including oxidative stress, inflammation, autophagy, and mitochondrial function. Persistent ER stress overwhelms the distinct defensive capacity of the UPR and generally leads to hepatic cell death, which plays contributory roles in nearly all forms of liver disease. In alcohol-induced hepatic ER stress and injury, persistence of ER stress can be caused by alcohol metabolites/derivatives, elevated homocysteine, ceramides, increased SAH, and disruptions of calcium, iron, or zinc homeostasis. In NAFLD, lipid accumulation may initially trigger the ER stress response, which further regulates lipid metabolism and accumulation. Excessive lipids upregulate genes involved in hepatic cell death, which help transform the liver from simple steatosis to NASH. In drug-induced hepatotoxicity, ER stress inducers can be drugs and their derivatives that idiosyncratically interact with the ER components or the UPR pathways. In hepatitis virus infections, ER stress and the UPR in the host hepatocytes are most likely attributed by the persistent production of large amounts of viral proteins. In liver carcinogenesis, the UPR pro-survival signals may be preferentially activated in ER-stressed hepatocytes, which interact with multiple endogenous and exogenous factors that change the delicate balance between oncogenes and tumor suppressors. Therefore, the pathophysiological role of ER stress and the UPR demonstrated for the development of various forms of liver disease will provide us unique therapeutic opportunities. Drugs targeting the UPR signaling will be developed, and more clinical trials in humans toward treating the ER stress-induced liver diseases are anticipated.

Acknowledgments This work is supported by the US National Institutes of Health (NIH) grants R01AA018612 (NCE), R21AA023952, and R01DA042632 (to C. Ji). H. Han is a PhD student in Ji's lab. Dr. Ji is grateful to the graduate students and postdoctoral fellows who contributed to the studies in the lab. The USC Research Center for Liver Diseases (P30DK48522) and the Southern California Research Center for Alcoholic Liver and Pancreatic Diseases and Cirrhosis (P50AA011999-20) provided technical assistance for the studies.

References

Abdelmegeed MA, Banerjee A, Yoo SH, Jang S, Gonzalez FJ, Song BJ (2012) Critical role of cytochrome P450 2E1 (CYP2E1) in the development of high fat-induced non-alcoholic steatohepatitis. J Hepatol 57(4):860–866

Aldridge JE, Horibe T, Hoogenraad NJ (2007) Discovery of genes activated by the mitochondrial unfolded protein response (mtUPR) and cognate promoter elements. PLoS One 2(9):e874

Al-Rawashdeh FY, Scriven P, Cameron IC, Vergani PV, Wyld L (2010) Unfolded protein response activation contributes to chemoresistance in hepatocellular carcinoma. Eur J Gastroenterol Hepatol 22(9):1099–1105

Apostolova N, Gomez-Sucerquia LJ, Alegre F, Funes HA, Victor VM, Barrachina MD, Blas-Garcia A, Esplugues JV (2013) ER stress in human hepatic cells treated with Efavirenz: mitochondria again. J Hepatol 59(4):780–789

Aubert J, Begriche K, Knockaert L, Robin MA, Fromenty B (2011) Increased expression of cytochrome P450 2E1 in nonalcoholic fatty liver disease: mechanisms and pathophysiological role. Clin Res Hepatol Gastroenterol 35(10):630–637

Aubert J, Begriche K, Delannoy M, Morel I, Pajaud J, Ribault C, Lepage S, McGill MR, Lucas-Clerc C, Turlin B, Robin MA, Jaeschke H, Fromenty B (2012) Differences in early acetaminophen hepatotoxicity between obese ob/ob and db/db mice. J Pharmacol Exp Ther 342(3):676–687

Barak AJ, Beckenhauer HC, Tuma DJ, Badakhsh S (1987) Effects of prolonged ethanol feeding on methionine metabolism in rat liver. Biochem Cell Biol 65(3):230–233

Benali-Furet NL, Chami M, Houel L, De Giorgi F, Vernejoul F, Lagorce D, Buscail L, Bartenschlager R, Ichas F, Rizzuto R, Paterlini-Bréchot P (2005) Hepatitis C virus core triggers apoptosis in liver cells by inducing ER stress and ER calcium depletion. Oncogene 24(31):4921–4933

Bittremieux M, Parys JB, Pinton P, Bultynck G (2016) ER functions of oncogenes and tumor suppressors: modulators of intracellular Ca(2+) signaling. Biochim Biophys Acta 1863(6 Pt B):1364–1378

Blasco C, Caballería J, Deulofeu R, Lligoña A, Parés A, Lluis JM, Gual A, Rodés J (2005) Prevalence and mechanisms of hyperhomocysteinemia in chronic alcoholics. Alcohol Clin Exp Res 29(6):1044–1048

Bleich S, Lenz B, Ziegenbein M, Beutler S, Frieling H, Kornhuber J, Bönsch D (2006) Epigenetic DNA hypermethylation of the HERP gene promoter induces down-regulation of its mRNA expression in patients with alcohol dependence. Alcohol Clin Exp Res 30(4):587–591

Bobrovnikova-Marjon E, Hatzivassiliou G, Grigoriadou C, Romero M, Cavener DR, Thompson CB, Diehl JA (2008) PERK-dependent regulation of lipogenesis during mouse mammary gland development and adipocyte differentiation. Proc Natl Acad Sci 105(42):16314–16319

Bono C, Karlin L, Harel S, Mouly E, Labaume S, Galicier L, Apcher S, Sauvageon H, Fermand JP, Bories JC, Arnulf B (2012) The human immunodeficiency virus-1 protease inhibitor nelfinavir impairs proteasome activity and inhibits the proliferation of multiple myeloma cells in vitro and in vivo. Haematologica 97(7):1101–1109

Boukli NM, Saiyed ZM, Ricaurte M, Rodriguez JW, Ríos Olivares E, Cubano LA, Nair MP (2010) Implications of ER stress, the unfolded protein response, and pro- and anti-apoptotic protein fingerprints in human monocyte-derived dendritic cells treated with alcohol. Alcohol Clin Exp Res 34(12):2081–2088

Bu Y, Diehl JA (2016) PERK integrates oncogenic signaling and cell survival during cancer development. J Cell Physiol. doi:10.1002/jcp.25336

Bulleid NJ (2012) Disulfide bond formation in the mammalian endoplasmic reticulum. Cold Spring Harb Perspect Biol 4(11). pii: a013219

Burcham PC, Harman AW (1988) Effect of acetaminophen hepatotoxicity on hepatic mitochondrial and microsomal calcium contents in mice. Toxicol Lett 44(1–2):91–99

Cao R, Hu Y, Wang Y, Gurley EC, Studer EJ, Wang X, Hylemon PB, Pandak WM, Sanyal AJ, Zhang L, Zhou H (2010) Prevention of HIV protease inhibitor-induced dysregulation of

hepatic lipid metabolism by raltegravir via endoplasmic reticulum stress signaling pathways. J Pharmacol Exp Ther 334(2):530–539
Cao J, Dai DL, Yao L, Yu HH, Ning B, Zhang Q, Chen J, Cheng WH, Shen W, Yang ZX (2012) Saturated fatty acid induction of endoplasmic reticulum stress and apoptosis in human liver cells via the PERK/ATF4/CHOP signaling pathway. Mol Cell Biochem 364(1–2):115–129
Carrasco DR, Sukhdeo K, Protopopova M, Sinha R, Enos M, Carrasco DE, Zheng M, Mani M, Henderson J, Pinkus GS, Munshi N, Horner J, Ivanova EV, Protopopov A, Anderson KC, Tonon G, DePinho RA (2007) The differentiation and stress response factor XBP-1 drives multiple myeloma pathogenesis. Cancer Cell 11(4):349–360
Cazanave SC, Elmi NA, Akazawa Y, Bronk SF, Mott JL, Gores GJ (2010) CHOP and AP-1 cooperatively mediate PUMA expression during lipoapoptosis. Am J Physiol Gastrointest Liver Physiol 299(1):G236–G243
Cederbaum AI, Lu Y, Wu D (2009) Role of oxidative stress in alcohol-induced liver injury. Arch Toxicol 83(6):519–548
Chen R, Dai RY, Duan CY, Liu YP, Chen SK, Yan DM, Chen CN, Wei M, Li H (2011) Unfolded protein response suppresses cisplatin-induced apoptosis via autophagy regulation in human hepatocellular carcinoma cells. Folia Biol (Praha) 57(3):87–95
Chikka MR, McCabe DD, Tyra HM, Rutkowski DT (2013) C/EBP homologous protein (CHOP) contributes to suppression of metabolic genes during endoplasmic reticulum stress in the liver. J Biol Chem 288(6):4405–4415
Choukhi A, Ung S, Wychowski C, Dubuisson J (1998) Involvement of endoplasmic reticulum chaperones in the folding of hepatitis C virus glycoproteins. J Virol 72(5):3851–3858
Christen V, Treves S, Duong FH, Heim MH (2007) Activation of endoplasmic reticulum stress response by hepatitis viruses up-regulates protein phosphatase 2A. Hepatology 46(2):558–565
Cinti DL, Grundin R, Orrenius S (1973) The effect of ethanol on drug oxidations in vitro and the significance of ethanol-cytochrome P-450 interaction. Biochem J 134(2):367–375
Claude A (1943) The constitution of protoplasm. Science 97:451–456
Czaja MJ (2015) A new mechanism of lipotoxicity: calcium channel blockers as a treatment for nonalcoholic steatohepatitis? Hepatology 62(1):312–314
Daniele T, Schiaffino MV (2014) Organelle biogenesis and interorganellar connections: better in contact than in isolation. Commun Integr Biol 7:e29587
Dlugos CA (2014) ATF6 and caspase 12 expression in Purkinje neurons in acute slices from adult, ethanol-fed rats. Brain Res 1577:11–20
Dubois C, Prevarskaya N, Vanden Abeele F (2016) The calcium-signaling toolkit: updates needed. Biochim Biophys Acta 1863(6 Pt B):1337–1343
Egnatchik RA, Leamy AK, Jacobson DA, Shiota M, Young JD (2014) ER calcium release promotes mitochondrial dysfunction and hepatic cell lipotoxicity in response to palmitate overload. Mol Metab 3(5):544–553
Esfandiari F, Villanueva JA, Wong DH et al (2005) Chronic ethanol feeding and folate deficiency activate hepatic endoplasmic reticulum stress pathway in micropigs. Am J Physiol Gastrointest Liver Physiol 289(1):54–63
Esfandiari F, Medici V, Wong DH, Jose S, Dolatshahi M, Quinlivan E, Dayal S, Lentz SR, Tsukamoto H, Zhang YH, French SW, Halsted CH (2010) Epigenetic regulation of hepatic endoplasmic reticulum stress pathways in the ethanol-fed cystathionine beta synthase-deficient mouse. Hepatology 51(3):932–941
Fang C, Yi Z, Liu F, Lan S, Wang J, Lu H, Yang P, Yuan Z (2006) Proteome analysis of human liver carcinoma Huh7 cells harboring hepatitis C virus subgenomic replicon. Proteomics 6(2):519–527
Feng B, Yao PM, Li Y, Devlin CM, Zhang D, Harding HP, Sweeney M, Rong JX, Kuriakose G, Fisher EA, Marks AR, Ron D, Tabas I (2003) The endoplasmic reticulum is the site of cholesterol-induced cytotoxicity in macrophages. Nat Cell Biol 5(9):781–792
Fernandez A, Matias N, Fucho R, Ribas V, Von Montfort C, Nuño N, Baulies A, Martinez L, Tarrats N, Mari M, Colell A, Morales A, Dubuquoy L, Mathurin P, Bataller R, Caballeria J,

Elena M, Balsinde J, Kaplowitz N, Garcia-Ruiz C, Fernandez-Checa JC (2013) ASMase is required for chronic alcohol induced hepatic endoplasmic reticulum stress and mitochondrial cholesterol loading. J Hepatol 59(4):805–813

Fournillier A, Wychowski C, Boucreux D, Baumert TF, Meunier JC, Jacobs D, Muguet S, Depla E, Inchauspé G (2001) Induction of hepatitis C virus E1 envelope protein-specific immune response can be enhanced by mutation of N-glycosylation sites. J Virol 75(24):12088–12097

Franceschelli S, Moltedo O, Amodio G, Tajana G, Remondelli P (2011) In the Huh7 hepatoma cells diclofenac and indomethacin activate differently the unfolded protein response and induce ER stress apoptosis. Open Biochem J 5:45–51

Fu S, Yang L, Li P, Hofmann O, Dicker L, Hide W, Lin X, Watkins SM, Ivanov AR, Hotamisligil GS (2011) Aberrant lipid metabolism disrupts calcium homeostasis causing liver endoplasmic reticulum stress in obesity. Nature 473(7348):528–531

Fu S, Watkins SM, Hotamisligil GS (2012) The role of endoplasmic reticulum in hepatic lipid homeostasis and stress signaling. Cell Metab 15(5):623–634

Gardner OS, Shiau CW, Chen CS, Graves LM (2005) Peroxisome proliferator-activated receptor gamma-independent activation of p38 MAPK by thiazolidinediones involves calcium/calmodulin-dependent protein kinase II and protein kinase R: correlation with endoplasmic reticulum stress. J Biol Chem 280(11):10109–10118

Gaut JR, Hendershot LM (1993) The modification and assembly of proteins in the endoplasmic reticulum. Curr Opin Cell Biol 5(4):589–595

Geng W, Lo CM, Ng KT, Ling CC, Qi X, Li CX, Zhai Y, Liu XB, Ma YY, Man K (2015) Interferon-gamma inducible protein 10 (IP10) induced cisplatin resistance of HCC after liver transplantation through ER stress signaling pathway. Oncotarget 6(29):28042–28056

Goldstein MG, Li Z (2009) Heat-shock proteins in infection-mediated inflammation-induced tumorigenesis. J Hematol Oncol 2:5

Gunawan BK, Kaplowitz N (2007) Mechanisms of drug-induced liver disease. Clin liver dis 11:459–475, v

Hadziyannis S, Gerber MA, Vissoulis C, Popper H (1973) Cytoplasmic hepatitis B antigen in "ground-glass" hepatocytes of carriers. Arch Pathol 96(5):327–330

Hampton RY (2002) ER-associated degradation in protein quality control and cellular regulation. Curr Opin Cell Biol 14(4):476–482

Han KL, Choi JS, Lee JY, Song J, Joe MK, Jung MH, Hwang JK (2008) Therapeutic potential of peroxisome proliferators – activated receptor-alpha/gamma dual agonist with alleviation of endoplasmic reticulum stress for the treatment of diabetes. Diabetes 57(3):737–745

Han H, He Y, Hu J, Lau R, Lee H, Ji C (2017) Disrupted ER-to-Golgi trafficking underlies anti-HIV drugs and alcoholinduced cellular stress and hepatic injury. Hepatology Communications 1(2):122–139

Han H, Hu J, Lau MY, Feng M, Petrovic LM, Ji C (2013) Altered methylation and expression of ER-associated degradation factors in long-term alcohol and constitutive ER stress-induced murine hepatic tumors. Front Genet 4:224

Hartman JH, Martin HC, Caro AA, Pearce AR, Miller GP (2015) Subcellular localization of rat CYP2E1 impacts metabolic efficiency toward common substrates. Toxicology 338:47–58

Hetz C, Bernasconi P, Fisher J, Lee AH, Bassik MC, Antonsson B, Brandt GS, Iwakoshi NN, Schinzel A, Glimcher LH, Korsmeyer SJ (2006) Proapoptotic BAX and BAK modulate the unfolded protein response by a direct interaction with IRE1alpha. Science 312(5773):572–576

Hiroyuki T, Yuxia Z, Sangmin L, Rana S, Li W (2014) Circadian clock mediates ER stress signaling in alcoholic fatty liver. FASEB J 28:1116.12

Horibe T, Hoogenraad NJ (2007) The chop gene contains an element for the positive regulation of the mitochondrial unfolded protein response. PLoS One 2(9):e835

Hotamisligil GS (2010) Endoplasmic reticulum stress and the inflammatory basis of metabolic disease. Cell 140(6):900–917

Hu J, Han H, Lau MY, Lee H, MacVeigh-Aloni M, Ji C (2015) Effects of combined alcohol and anti-HIV drugs on cellular stress responses in primary hepatocytes and hepatic stellate and kupffer cells. Alcohol Clin Exp Res 39(1):11–20

Hultberg B, Berglund M, Andersson A, Frank A (1993) Elevated plasma homocysteine in alcoholics. Alcohol Clin Exp Res 17(3):687–689

Hur KY, So JS, Ruda V, Frank-Kamenetsky M, Fitzgerald K, Koteliansky V, Iwawaki T, Glimcher LH, Lee AH (2012) IRE1α activation protects mice against acetaminophen-induced hepatotoxicity. J Exp Med 209(2):307–318

Ient B, Edwards R, Mould R, Hannah M, Holden-Dye L, O'Connor V (2012) HSP-4 endoplasmic reticulum (ER) stress pathway is not activated in a *C. elegans* model of ethanol intoxication and withdrawal. Invertebr Neurosci 12(2):93–102

Jaeschke H (1990) Glutathione disulfide formation and oxidant stress during acetaminophen-induced hepatotoxicity in mice in vivo: the protective effect of allopurinol. J Pharmacol Exp Ther 255(3):935–941

Ji C (2015) Advances and new concepts in alcohol-induced organelle stress, unfolded protein responses and organ damage. Biomol Ther 5(2):1099–1121

Ji C, Kaplowitz N (2003) Betaine decreases hyperhomocysteinemia, endoplasmic reticulum stress, and liver injury in alcohol-fed mice. Gastroenterology 124(5):1488–99

Ji C, Kaplowitz N (2006) ER stress: can the liver cope? J Hepatol 45(2):321–333

Ji C, Deng Q, Kaplowitz N (2004) Role of TNF-alpha in ethanol-induced hyperhomocysteinemia and murine alcoholic liver injury. Hepatology 40(2):442–451

Ji C, Mehrian-Shai R, Chan C, Hsu YH, Kaplowitz N (2005) Role of CHOP in hepatic apoptosis in the murine model of intragastric ethanol feeding. Alcohol Clin Exp Res 29(8):1486–1503

Ji C, Chan C, Kaplowitz N (2006) Predominant role of sterol response element binding proteins (SREBP) lipogenic pathways in hepatic steatosis in the murine intragastric ethanol feeding model. J Hepatol 45(5):717–724

Ji C, Shinohara M, Vance D, Than TA, Ookhtens M, Chan C, Kaplowitz N (2008) Effect of transgenic extrahepatic expression of betaine-homocysteine methyltransferase on alcohol or homocysteine-induced fatty liver. Alcohol Clin Exp Res 32(6):1049–1058

Ji C, Kaplowitz N, Lau MY, Kao E, Petrovic LM, Lee AS (2011) Liver-specific loss of glucose-regulated protein 78 perturbs the unfolded protein response and exacerbates a spectrum of liver diseases in mice. Hepatology 54(1):229–239

Jones M, Núñez M (2012) Liver toxicity of antiretroviral drugs. Semin Liver Dis 32(2):167–176

Kalinec GM, Thein P, Parsa A, Yorgason J, Luxford W, Urrutia R, Kalinec F (2014) Acetaminophen and NAPQI are toxic to auditory cells via oxidative and endoplasmic reticulum stress-dependent pathways. Hear Res 313:26–37

Kao E, Shinohara M, Feng M, Lau MY, Ji C (2012) Human immunodeficiency virus protease inhibitors modulate Ca^{2+} homeostasis and potentiate alcoholic stress and injury in mice and primary mouse and human hepatocytes. Hepatology 56(2):594–604

Kaplowitz N (2005) Idiosyncratic drug hepatotoxicity. Nat Rev Drug Discov 4(6):489–499

Ke PY, Chen SS (2012) Hepatitis C virus and cellular stress response: implications to molecular pathogenesis of liver diseases. Virus 4(10):2251–2290

Kim S, Joe Y, Kim HJ, Kim YS, Jeong SO, Pae HO, Ryter SW, Surh YJ, Chung HT (2015) Endoplasmic reticulum stress-induced IRE1α activation mediates cross-talk of GSK-3β and XBP-1 to regulate inflammatory cytokine production. J Immunol 194(9):4498–4506

Kitamura M (2011) Control of NF-κB and inflammation by the unfolded protein response. Int Rev Immunol 30(1):4–15

Labbe G, Pessayre D, Fromenty B (2008) Drug-induced liver injury through mitochondrial dysfunction: mechanisms and detection during preclinical safety studies. Fundam Clin Pharmacol 22(4):335–353

Lake AD, Novak P, Hardwick RN, Flores-Keown B, Zhao F, Klimecki WT, Cherrington NJ (2014) The adaptive endoplasmic reticulum stress response to lipotoxicity in progressive human nonalcoholic fatty liver disease. Toxicol Sci 137(1):26–35

Lau MY, Han H, Hu J, Ji C (2013) Association of cyclin D and estrogen receptor α36 with hepatocellular adenomas of female mice under chronic endoplasmic reticulum stress. J Gastroenterol Hepatol 28(3):576–583

Lee WM (2003) Acute liver failure in the United States. Semin Liver Dis 23(3):217–226

Lee AH, Scapa EF, Cohen DE, Glimcher LH (2008) Regulation of hepatic lipogenesis by the transcription factor XBP1. Science 320(5882):1492–1496

Li B, Gao B, Ye L, Han X, Wang W, Kong L, Fang X, Zeng Y, Zheng H, Li S, Wu Z, Ye L (2007) Hepatitis B virus X protein (HBx) activates ATF6 and IRE1-XBP1 pathways of unfolded protein response. Virus Res 124(1–2):44–49

Liao W, Chan L (2001) Tunicamycin induces ubiquitination and degradation of apolipoprotein B in HepG2 cells. Biochem J 353(Pt 3):493–501

Liberman E, Fong YL, Selby MJ, Choo QL, Cousens L, Houghton M, Yen TS (1999) Activation of the grp78 and grp94 promoters by hepatitis C virus E2 envelope protein. J Virol 73(5):3718–3722

Lieber CS (1987) Microsomal ethanol-oxidizing system. Enzyme 37(1–2):45–56

Lin RJ, Chu JS, Chien HL, Tseng CH, Ko PC, Mei YY, Tang WC, Kao YT, Cheng HY, Liang YC, Lin SY (2014) MCPIP1 suppresses hepatitis C virus replication and negatively regulates virus-induced proinflammatory cytokine responses. J Immunol 193(8):4159–4168

Listenberger LL, Han X, Lewis SE, Cases S, Farese RV Jr, Ory DS, Schaffer JE (2003) Triglyceride accumulation protects against fatty acid-induced lipotoxicity. Proc Natl Acad Sci U S A 100(6):3077–3082

Lluis JM, Colell A, García-Ruiz C, Kaplowitz N, Fernández-Checa JC (2003) Acetaldehyde impairs mitochondrial glutathione transport in HepG2 cells through endoplasmic reticulum stress. Gastroenterology 124(3):708–724

Longato L, Ripp K, Setshedi M, Dostalek M, Akhlaghi F, Branda M, Wands JR, de la Monte SM (2012) Insulin resistance, ceramide accumulation, and endoplasmic reticulum stress in human chronic alcohol-related liver disease. Oxidative Med Cell Longev 2012:479348

Lorz C, Justo P, Sanz A, Subirá D, Egido J, Ortiz A (2004) Paracetamol-induced renal tubular injury: a role for ER stress. J Am Soc Nephrol 15(2):380–389

Luo B, Lee AS (2013) The critical roles of endoplasmic reticulum chaperones and unfolded protein response in tumorigenesis and anticancer therapies. Oncogene 32(7):805–818

Lutz UC (2008) Alterations in homocysteine metabolism among alcohol dependent patients – clinical, pathobiochemical and genetic aspects. Curr Drug Abuse Rev 1(1):47–55

Macanas-Pirard P, Yaacob NS, Lee PC, Holder JC, Hinton RH, Kass GE (2005) Glycogen synthase kinase-3 mediates acetaminophen-induced apoptosis in human hepatoma cells. J Pharmacol Exp Ther 313(2):780–789

Magne L, Blanc E, Legrand B, Lucas D, Barouki R, Rouach H, Garlatti M (2011) ATF4 and the integrated stress response are induced by ethanol and cytochrome P450 2E1 in human hepatocytes. J Hepatol 54(4):729–737

Marí M, Morales A, Colell A, García-Ruiz C, Fernández-Checa JC (2014) Mitochondrial cholesterol accumulation in alcoholic liver disease: role of ASMase and endoplasmic reticulum stress. Redox Biol 3:100–108

Maurel M, Chevet E, Tavernier J, Gerlo S (2014) Getting RIDD of RNA: IRE1 in cell fate regulation. Trends Biochem Sci 39(5):245–254

Millward CA, Heaney JD, Sinasac DS, Chu EC, Bederman IR, Gilge DA, Previs SF, Croniger CM (2007) Mice with a deletion in the gene for CCAAT/enhancer-binding protein beta are protected against diet-induced obesity. Diabetes 56(1):161–167

Miyata S, Mizuno T, Koyama Y, Katayama T, Tohyama M (2013) The endoplasmic reticulum-resident chaperone heat shock protein 47 protects the Golgi apparatus from the effects of O-glycosylation inhibition. PLoS One 8(7):e69732

Mohammad MK, Avila D, Zhang J, Barve S, Arteel G, McClain C, Joshi-Barve S (2012) Acrolein cytotoxicity in hepatocytes involves endoplasmic reticulum stress, mitochondrial dysfunction and oxidative stress. Toxicol Appl Pharmacol 265(1):73–82

Mori K (2000) Tripartite management of unfolded proteins in the endoplasmic reticulum. Cell 101(5):451–454

Mori K (2015) The unfolded protein response: the dawn of a new field. Proc Jpn Acad Ser B Phys Biol Sci 91(9):469–480

Nagy G, Kardon T, Wunderlich L, Szarka A, Kiss A, Schaff Z, Bánhegyi G, Mandl J (2007) Acetaminophen induces ER dependent signaling in mouse liver. Arch Biochem Biophys 459(2):273–279

Ng DT, Randall RE, Lamb RA (1989) Intracellular maturation and transport of the SV5 type II glycoprotein hemagglutinin-neuraminidase: specific and transient association with GRP78-BiP in the endoplasmic reticulum and extensive internalization from the cell surface. J Cell Biol 109(6 Pt 2):3273–3289

Núñez M, Lana R, Mendoza JL, Martín-Carbonero L, Soriano V (2001) Risk factors for severe hepatic injury after introduction of highly active antiretroviral therapy. J Acquir Immune Defic Syndr 27(5):426–431

Ogata M, Hino S, Saito A, Morikawa K, Kondo S, Kanemoto S, Murakami T, Taniguchi M, Tanii I, Yoshinaga K, Shiosaka S, Hammarback JA, Urano F, Imaizumi K (2006) Autophagy is activated for cell survival after endoplasmic reticulum stress. Mol Cell Biol 26(24):9220–9231

Ota T, Gayet C, Ginsberg HN (2008) Inhibition of apolipoprotein B100 secretion by lipid-induced hepatic endoplasmic reticulum stress in rodents. J Clin Invest 118(1):316–332

Oyadomari S, Mori M (2004) Roles of CHOP/GADD153 in endoplasmic reticulum stress. Cell Death Differ 11(4):381–389

Oyadomari S, Harding HP, Zhang Y, Oyadomari M, Ron D (2008) Dephosphorylation of translation initiation factor 2alpha enhances glucose tolerance and attenuates hepatosteatosis in mice. Cell Metab 7(6):520–532

Ozcan L, Tabas I (2012) Role of endoplasmic reticulum stress in metabolic disease and other disorders. Annu Rev Med 63:317–328

Ozcan U, Cao Q, Yilmaz E, Lee AH, Iwakoshi NN, Ozdelen E, Tuncman G, Görgün C, Glimcher LH, Hotamisligil GS (2004) Endoplasmic reticulum stress links obesity, insulin action, and type 2 diabetes. Science 306(5695):457–461

Ozcan U, Yilmaz E, Ozcan L, Furuhashi M, Vaillancourt E, Smith RO, Görgün CZ, Hotamisligil GS (2006) Chemical chaperones reduce ER stress and restore glucose homeostasis in a mouse model of type 2 diabetes. Science 313(5790):1137–1140

Pajonk F, Himmelsbach J, Riess K, Sommer A, McBride WH (2002) The human immunodeficiency virus (HIV)-1 protease inhibitor saquinavir inhibits proteasome function and causes apoptosis and radiosensitization in non-HIV-associated human cancer cells. Cancer Res 62(18):5230–5235

Palade GE (1995) The endoplasmic reticulum. J Biophys Biochem Cytol 2(4 Suppl):85–98

Palade GE, Siekevitz P (1956) Liver microsomes; an integrated morphological and biochemical study. J Biophys Biochem Cytol 2(2):171–200

Passeri MJ, Cinaroglu A, Gao C, Sadler KC (2009) Hepatic steatosis in response to acute alcohol exposure in zebrafish requires sterol regulatory element binding protein activation. Hepatology 49(2):443–452

Patil C, Walter P (2001) Intracellular signaling from the endoplasmic reticulum to the nucleus: the unfolded protein response in yeast and mammals. Curr Opin Cell Biol 13(3):349–355

Pfaffenbach KT, Gentile CL, Nivala AM, Wang D, Wei Y, Pagliassotti MJ (2010) Linking endoplasmic reticulum stress to cell death in hepatocytes: roles of C/EBP homologous protein and chemical chaperones in palmitate-mediated cell death. Am J Physiol Endocrinol Metab 298(5):E1027–E1035

Piccinini M, Rinaudo MT, Chiapello N, Ricotti E, Baldovino S, Mostert M, Tovo PA (2002) The human 26S proteasome is a target of antiretroviral agents. AIDS 16(5):693–700

Pineau L, Colas J, Dupont S, Beney L, Fleurat-Lessard P, Berjeaud JM, Bergès T, Ferreira T (2009) Lipid-induced ER stress: synergistic effects of sterols and saturated fatty acids. Traffic 10(6):673–690

Pinton P, Ferrari D, Rapizzi E, Di Virgilio F, Pozzan T, Rizzuto R (2001) The Ca2+ concentration of the endoplasmic reticulum is a key determinant of ceramide-induced apoptosis: significance for the molecular mechanism of Bcl-2 action. EMBO J 20(11):2690–2701

Popper H (1975) Clinical pathologic correlation in viral hepatitis. The effect of the virus on the liver. Am J Pathol 81(3):609–628

Porter KR, Claude A, Fullam EF (1945) A study of tissue culture cells by electron microscopy: methods and preliminary observations. J Exp Med 81(3):233–246

Puri P, Mirshahi F, Cheung O, Natarajan R, Maher JW, Kellum JM, Sanyal AJ (2008) Activation and dysregulation of the unfolded protein response in nonalcoholic fatty liver disease. Gastroenterology 134(2):568–576

Rahman SM, Schroeder-Gloeckler JM, Janssen RC, Jiang H, Qadri I, Maclean KN, Friedman JE (2007) CCAAT/enhancing binding protein beta deletion in mice attenuates inflammation, endoplasmic reticulum stress, and lipid accumulation in diet-induced nonalcoholic steatohepatitis. Hepatology 45(5):1108–1117

Ramirez T, Longato L, Dostalek M, Tong M, Wands JR, de la Monte SM (2013) Insulin resistance, ceramide accumulation and endoplasmic reticulum stress in experimental chronic alcohol-induced steatohepatitis. Alcohol Alcohol 48(1):39–52

Röhrl C, Eigner K, Winter K, Korbelius M, Obrowsky S, Kratky D, Kovacs WJ, Stangl H (2014) Endoplasmic reticulum stress impairs cholesterol efflux and synthesis in hepatic cells. J Lipid Res 55(1):94–103

Romero-Ramirez L, Cao H, Nelson D, Hammond E, Lee AH, Yoshida H, Mori K, Glimcher LH, Denko NC, Giaccia AJ, Le QT, Koong AC (2004) XBP1 is essential for survival under hypoxic conditions and is required for tumor growth. Cancer Res 64(17):5943–5947

Ron D, Hubbard SR (2013) How IRE1 reacts to ER stress. Cell 132(1):24–26

Ron D, Walter P (2007) Signal integration in the endoplasmic reticulum unfolded protein response. Nat Rev Mol Cell Biol 8(7):519–529

Ronis MJ, Korourian S, Blackburn ML, Badeaux J, Badger TM (2010) The role of ethanol metabolism in development of alcoholic steatohepatitis in the rat. Alcohol 44(2):157–169

Rutkowski DT, Wu J, Back SH, Callaghan MU, Ferris SP, Iqbal J, Clark R, Miao H, Hassler JR, Fornek J, Katze MG, Hussain MM, Song B, Swathirajan J, Wang J, Yau GD, Kaufman RJ (2008) UPR pathways combine to prevent hepatic steatosis caused by ER stress-mediated suppression of transcriptional master regulators. Dev Cell 15(6):829–840

Schmidtke G, Holzhütter HG, Bogyo M, Kairies N, Groll M, de Giuli R, Emch S, Groettrup M (1999) How an inhibitor of the HIV-I protease modulates proteasome activity. J Biol Chem 274(50):35734–35740

Schonthal AH (2012) Targeting endoplasmic reticulum stress for cancer therapy. Front Biosci (Schol Ed) 4:412–431

Schröder M, Kaufman RJ (2005) The mammalian unfolded protein response. Annu Rev Biochem 74:739–789

Senft D, Ronai ZA (2015) UPR, autophagy, and mitochondria crosstalk underlies the ER stress response. Trends Biochem Sci 40(3):141–148

Senior JR (2014) Evolution of the Food and Drug Administration approach to liver safety assessment for new drugs: current status and challenges. Drug Saf 37(Spl 1):S9–17

Seo J, Fortuno ES 3rd, Suh JM, Stenesen D, Tang W, Parks EJ, Adams CM, Townes T, Graff JM (2009) Atf4 regulates obesity, glucose homeostasis, and energy expenditure. Diabetes 58(11):2565–2573

Seth D, Leo PH (2003) McGuinness et al, "Gene expression profiling of alcoholic liver disease in the baboon (Papiohamadryas) and human liver". Am J Pathol 163(6):2303–2317

Shinohara M, Ji C, Kaplowitz N (2010) Differences in betaine-homocysteine methyltransferase expression, endoplasmic reticulum stress response, and liver injury between alcohol-fed mice and rats. Hepatology 51(3):796–805

Shuda M, Kondoh N, Imazeki N, Tanaka K, Okada T, Mori K, Hada A, Arai M, Wakatsuki T, Matsubara O, Yamamoto N, Yamamoto M (2003) Activation of the ATF6, XBP1 and grp78 genes in human hepatocellular carcinoma: a possible involvement of the ER stress pathway in hepatocarcinogenesis. J Hepatol 38(5):605–614

Softic S, Cohen DE, Kahn CR (2016) Role of dietary fructose and hepatic de novo lipogenesis in fatty liver disease. Dig Dis Sci 61(5):1282–1293

Sreejayan N, Dong F, Kandadi MR, Yang X, Ren J (2008) Chromium alleviates glucose intolerance, insulin resistance, and hepatic ER stress in obese mice. Obesity (Silver Spring) 16(6):1331–1337

Sun Q, Zhong W, Zhang W, Li Q, Sun X, Tan X, Sun X, Dong D, Zhou Z (2015) Zinc deficiency mediates alcohol-induced apoptotic cell death in the liver of rats through activating ER and mitochondrial cell death pathways. Am J Physiol Gastrointest Liver Physiol. doi:10.1152/ajpgi.00442.2014

Tabas I, Ron D (2011) Integrating the mechanisms of apoptosis induced by endoplasmic reticulum stress. Nat Cell Biol 13(3):184–190

Tai PC, Suk FM, Gerlich WH, Neurath AR, Shih C (2002) Hypermodification and immune escape of an internally deleted middle-envelope (M) protein of frequent and predominant hepatitis B virus variants. Virology 292(1):44–58

Tan TC, Crawford DH, Jaskowski LA, Subramaniam VN, Clouston AD, Crane DI, Bridle KR, Anderson GJ, Fletcher LM (2013) Excess iron modulates endoplasmic reticulum stress-associated pathways in a mouse model of alcohol and high-fat diet-induced liver injury. Lab Investig 93(12):1295–1312

Tardif KD, Siddiqui A (2003) Cell surface expression of major histocompatibility complex class I molecules is reduced in hepatitis C virus subgenomic replicon-expressing cells. J Virol 77(21):11644–11650

Tardif KD, Mori K, Siddiqui A (2002) Hepatitis C virus subgenomic replicons induce endoplasmic reticulum stress activating an intracellular signaling pathway. J Virol 76(15):7453–7459

Tardif KD, Mori K, Kaufman RJ, Siddiqui A (2004) Hepatitis C virus suppresses the IRE1-XBP1 pathway of the unfolded protein response. J Biol Chem 279(17):17158–17164

Targett-Adams P, Boulant S, Douglas MW, McLauchlan J (2010) Lipid metabolism and HCV infection. Virus 2(5):1195–1217

Taura M, Kariya R, Kudo E, Goto H, Iwawaki T, Amano M, Suico MA, Kai H, Mitsuya H, Okada S (2013) Comparative analysis of ER stress response into HIV protease inhibitors: lopinavir but not darunavir induces potent ER stress response via ROS/JNK pathway. Free Radic Biol Med 65:778–788

Tazi KA, Bièche I, Paradis V et al (2007) In vivo altered unfolded protein response and apoptosis in livers from lipopolysaccharide-challenged cirrhotic rats. J Hepatol 46(6):1075–1088

Thakur PC, Stuckenholz C, Rivera MR, Davison JM, Yao JK, Amsterdam A, Sadler KC, Bahary N (2011) Lack of de novo phosphatidylinositol synthesis leads to endoplasmic reticulum stress and hepatic steatosis in cdipt-deficient zebrafish. Hepatology 54(2):452–462

Touzet O, Philips A (2010) Resveratrol protects against protease inhibitor-induced reactive oxygen species production, reticulum stress and lipid raft perturbation. AIDS 24(10):1437–1447

Tsuchiya M, Ji C, Kosyk O, Shymonyak S, Melnyk S, Kono H, Tryndyak V, Muskhelishvili L, Pogribny IP, Kaplowitz N, Rusyn I (2012) Interstrain differences in liver injury and one-carbon metabolism in alcohol-fed mice. Hepatology 56(1):130–139

Tsutsumi S, Gotoh T, Tomisato W, Mima S, Hoshino T, Hwang HJ, Takenaka H, Tsuchiya T, Mori M, Mizushima T (2004) Endoplasmic reticulum stress response is involved in nonsteroidal anti-inflammatory drug-induced apoptosis. Cell Death Differ 11(9):1009–1016

Urano F, Wang X, Bertolotti A, Zhang Y, Chung P, Harding HP, Ron D (2000) Coupling of stress in the ER to activation of JNK protein kinases by transmembrane protein kinase IRE1. Science 287(5453):664–666

Usui M, Yamaguchi S, Tanji Y, Tominaga R, Ishigaki Y, Fukumoto M, Katagiri H, Mori K, Oka Y, Ishihara H (2012) Atf6α-null mice are glucose intolerant due to pancreatic β-cell failure on a high-fat diet but partially resistant to diet-induced insulin resistance. Metabolism 61(8):1118–1128

Uzi D, Barda L, Scaiewicz V, Mills M, Mueller T, Gonzalez-Rodriguez A, Valverde AM, Iwawaki T, Nahmias Y, Xavier R, Chung RT, Tirosh B, Shibolet O (2013) CHOP is a critical regulator of acetaminophen-induced hepatotoxicity. J Hepatol 59(3):495–503

Vandewynckel YP, Laukens D, Bogaerts E, Paridaens A, Van den Bussche A, Verhelst X, Van Steenkiste C, Descamps B, Vanhove C, Libbrecht L, De Rycke R, Lambrecht BN, Geerts A, Janssens S, Van Vlierberghe H (2014) Modulation of the unfolded protein response impedes tumor cell adaptation to proteotoxic stress: a PERK for hepatocellular carcinoma therapy. Hepatol Int 9(1):93–104

Walter P, Ron D (2011) The unfolded protein response: from stress pathway to homeostatic regulation. Science 334(6059):1081–1086
Wang S, Kaufman RJ (2012) The impact of the unfolded protein response on human disease. J Cell Biol 197(7):857–867
Wang D, Wei Y, Schmoll D, Maclean KN, Pagliassotti MJ (2006a) Endoplasmic reticulum stress increases glucose-6-phosphatase and glucose cycling in liver cells. Endocrinology 147(1):350–358
Wang HC, Huang W, Lai MD, Su IJ (2006b) Hepatitis B virus pre-S mutants, endoplasmic reticulum stress and hepatocarcinogenesis. Cancer Sci 97(8):683–688
Wang C, Huang Z, Du Y, Cheng Y, Chen S, Guo F (2010) ATF4 regulates lipid metabolism and thermogenesis. Cell Res 20(2):174–184
Wang S, Chen Z, Lam V, Han J, Hassler J, Finck BN, Davidson NO, Kaufman RJ (2012) IRE1α-XBP1s induces PDI expression to increase MTP activity for hepatic VLDL assembly and lipid homeostasis. Cell Metab 16(4):473–486
Wang Y, Zhang L, Wu X, Gurley EC, Kennedy E, Hylemon PB, Pandak WM, Sanyal AJ, Zhou H (2013a) The role of CCAAT enhancer-binding protein homologous protein in human immunodeficiency virus protease-inhibitor-induced hepatic lipotoxicity in mice. Hepatology 57(3):1005–1016
Wang C, Jiang K, Gao D, Kang X, Sun C, Zhang Q, Li Y, Sun L, Zhang S, Guo K, Liu Y (2013b) Clusterin protects hepatocellular carcinoma cells from endoplasmic reticulum stress induced apoptosis through GRP78. PLoS One 8(2):e55981
Wang Z, Jiang C, Chen W, Zhang G, Luo D, Cao Y, Wu J, Ding Y, Liu B (2014) Baicalein induces apoptosis and autophagy via endoplasmic reticulum stress in hepatocellular carcinoma cells. Biomed Res Int 2014:732516
Wei Y, Wang D, Topczewski F, Pagliassotti MJ (2006) Saturated fatty acids induce endoplasmic reticulum stress and apoptosis independently of ceramide in liver cells. Am J Physiol Endocrinol Metab 291(2):E275–E281
Weston RT, Puthalakath H (2010) Endoplasmic reticulum stress and BCL-2 family members. Adv Exp Med Biol 687:65–77
Xiao G, Zhang T, Yu S, Lee S, Calabuig-Navarro V, Yamauchi J, Ringquist S, Dong HH (2013) ATF4 protein deficiency protects against high fructose-induced hypertriglyceridemia in mice. J Biol Chem 288(35):25350–25361
Xie Q, Khaoustov VI, Chung CC, Sohn J, Krishnan B, Lewis DE, Yoffe B (2002) Effect of tauroursodeoxycholic acid on endoplasmic reticulum stress-induced caspase-12 activation. Hepatology 36(3):592–601
Xu Z, Yen TS (1996) Intracellular retention of surface protein by a hepatitis B virus mutant that releases virion particles. J Virol 70(1):133–140
Xu J, Lai KK, Verlinsky A et al (2011) Synergistic steatohepatitis by moderate obesity and alcohol in mice despite increased adiponectin and p-AMPK. J Hepatol 55(3):673–682
Yamamoto K, Takahara K, Oyadomari S, Okada T, Sato T, Harada A, Mori K (2010) Induction of liver steatosis and lipid droplet formation in ATF6alpha-knockout mice burdened with pharmacological endoplasmic reticulum stress. Mol Biol Cell 21(17):2975–2986
Yang L, Jhaveri R, Huang J, Qi Y, Diehl AM (2007) Endoplasmic reticulum stress, hepatocyte CD1d and NKT cell abnormalities in murine fatty livers. Lab Investig 87(9):927–937
Ye J, Rawson RB, Komuro R, Chen X, Davé UP, Prywes R, Brown MS, Goldstein JL (2000) ER stress induces cleavage of membrane-bound ATF6 by the same proteases that process SREBPs. Mol Cell 6(6):1355–1364
Yi Z, Fang C, Pan T, Wang J, Yang P, Yuan Z (2006) Subproteomic study of hepatitis C virus replicon reveals Ras-GTPase-activating protein binding protein 1 as potential HCV RC component. Biochem Biophys Res Commun 350(1):174–178
Zeng L, Lu M, Mori K, Luo S, Lee AS, Zhu Y, Shyy JY (2004) ATF6 modulates SREBP2-mediated lipogenesis. EMBO J 23(4):950–958

Zhang K, Wang S, Malhotra J, Hassler JR, Back SH, Wang G, Chang L, Xu W, Miao H, Leonardi R, Chen YE, Jackowski S, Kaufman RJ (2011) The unfolded protein response transducer IRE1α prevents ER stress-induced hepatic steatosis. EMBO J 30(7):1357–1375

Zhang J, Singh N, Robinson-Taylor KS, Dorsett-Martin WA, Morris MW Jr, Earl TM, Anderson CD (2015) Hepatocyte autophagy is linked to C/EBP-homologous protein, Bcl2-interacting mediator of cell death, and BH3-interacting domain death agonist gene expression. J Surg Res 195(2):588–595

Zheng Y, Gao B, Ye L, Kong L, Jing W, Yang X, Wu Z, Ye L (2005) Hepatitis C virus non-structural protein NS4B can modulate an unfolded protein response. J Microbiol 43(6):529–536

Zhu G, Lee AS (2015) Role of the unfolded protein response, GRP78 and GRP94 in organ homeostasis. J Cell Physiol 230(7):1413–1420

Chapter 11
The Impact of Gut Microbiota on Liver Injury

Wenke Feng and Craig McClain

Abbreviations

4-HNE	4-Hydroxynonenal
AC	Alcoholic cirrhosis
ACLF	Acute-on-chronic liver failure
ALD	Alcoholic liver disease
ALT	Alanine transaminase
AST	Aspartate transaminase
AXOS	Arabinoxylanoligosaccharide
BCAA	Branched chain amino acid
BCFA	Branched chain fatty acid
CDR	Cirrhosis dysbiosis ratio
CRAMP	Cathelicidin-related antimicrobial peptide
DCA	Deoxycholic acid
DEN	Diethylnitrosamine
FMT	Fecal microbiota transplantation
FOS	Fructooligosaccharide
FXR	Farnesoid X receptor
GLP-1	Glucagon-like peptide-1
GOS	Galactooligosaccharide
HCC	Hepatocellular carcinoma
HE	Hepatic encephalopathy
HFD	High-fat diet
HOMA	Homeostasis model assessment

W. Feng (✉) • C. McClain
Department of Medicine, University of Louisville School of Medicine, Louisville, KY, USA
e-mail: wenke.feng@louisville.edu

W.-X. Ding, X.-M. Yin (eds.), *Molecules, Systems and Signaling in Liver Injury*,
Cell Death in Biology and Diseases, DOI 10.1007/978-3-319-58106-4_11

X.-M. Yin and Z. Dong (Series eds.), *Cell Death in Biology and Diseases*

IFALD	Failure-associated liver disease
LAB	Lactic acid bacteria
LGG	*Lactobacillus rhamnosus* GG
LPS	Lipopolysaccharide
MCD	Methionine-choline deficiency
MDA	Malondialdehyde
MELD	Model for end-stage liver disease
MHE	Minimal HE
NAFLD	Nonalcoholic fatty liver disease
NASH	Nonalcoholic steatohepatitis
NF-kB	Nuclear factor kB
NLRP	Nod-like receptor protein
NOD2	Nucleotide-binding oligomerization domain 2
OFC	Oligofructose
PAMP	Pathogen-associated molecular pattern
PPARγ	Peroxisome proliferator-activated receptor-γ
sAH	Severe alcoholic hepatitis
SBP	Spontaneous bacterial peritonitis
SCFA	Short-chain fatty acid
SIBO	Small intestinal bacterial overgrowth
TJ	Tight junction
TLR	Toll-like receptor
γGT	Gamma glutamyl transferase

11.1 Dysbiotic Factors Contributing to Liver Disease, in General

The human intestinal tract harbors over 100 trillion bacteria in a diverse and complex community, which equates to tenfold more bacteria than human body cells. Studies of gut flora initially relied on culture-dependent techniques, but over 80% of the gut microbes are not culturable by standard methods. Because of recent advancements in high-throughput next-generation sequencing technology, we are beginning to gain more details about the gut microbiota. We now know that the collective microbiota genome (microbiota) has 100-fold more genes than the human genome, as determined by sequencing. Gut bacteria coevolve with human beings (symbiosis), and the changes of the bacterial population and structure can have important consequences, both beneficial and harmful, to the host. Bacterial metabolism and microbiota-host interactions influence the pathophysiology of the host. The gut microbiota composition is tightly regulated. Genetic factors, the mucosal immune system, dietary factors, and environmental factors are essential regulators of gut microbiota homeostasis.

Genetic factors It is clear that host genetics and gut microbiota both have an impact on metabolic phenotypes. Several twin studies revealed that genetic heritage has an influence on the human gut bacterial composition (Dicksved et al. 2008; Lee et al. 2011; Turnbaugh et al. 2009). A study including 10 monozygotic twin pairs with Crohn's disease and 8 healthy twin pairs demonstrates that fecal microbial communities are similar, and this similarity is disrupted when the twins are discordant for Crohn's disease. A larger twin study including 154 individuals revealed that the composition of the human gut microbiota is shared among family members. Testing core measurable microbiota in mice identified 18 host quantitative trait loci which are linked to bacterial abundance, suggesting that host genetics shape individual microbiota diversity (Turnbaugh et al. 2009). It has been found that an increased abundance of *Enterobacteriaceae* is strongly associated with the nucleotide-binding oligomerization domain 2 (NOD2) risk allele for inflammatory bowel disease (Knights et al. 2014). Intestinal inflammation causes intestinal barrier dysfunction and increases permeability leading to a translocation of bacteria and their products to the liver, and this is an important contributor to liver disease. In the liver, NOD2 variants are genetic risk factors for death and spontaneous bacterial peritonitis (SBP) in cirrhosis (Appenrodt et al. 2010; Bruns et al. 2012), and genetic polymorphisms of NOD2 are associated with increased mortality in liver transplant patients with nonalcoholic liver diseases (Saner et al. 2014). An alteration of gut microbiota noted in NOD-like receptor protein 6 (NLRP6) and NLRP3 inflammasome-deficient mice is associated with exacerbated hepatic steatosis and inflammation through regulation of TLR4 and TLR9 that drives NASH (nonalcoholic steatohepatitis) progression (Henao-Mejia et al. 2012). In a more specific aspect, a recent study showed that the *Christensenellaceae* family is a heritable taxon which forms a co-occurrence network with other heritable taxa (Goodrich et al. 2014). These bacteria families are enriched in individuals with low body mass index (Goodrich et al. 2014). Despite recent studies investigating genetic contributions to the gut microbiota changes, future carefully designed studies to determine whether known genetic risk alleles for liver disease have an impact on gut microbiota and how environmental factors interact with genetic factors are needed.

Environmental factor Environmental factors are important contributors shaping gut bacterial composition. People with similar cultural factors share similarities in microbiota. Diet, hygiene habits, antibiotic usage, and environmental exposure are among the most important contributors. In particular, emerging evidence suggests that environmental pollutants could interact with gut microbiota (Liu et al. 2014; Choi et al. 2013; Ribiere et al. 2016). Gut bacteria can metabolize environmental chemicals from various categories, either increasing or decreasing their toxicity to the host. Conversely, these chemicals may also influence the gut flora homeostasis and function. However, the relevance of the microbiota-xenobiotics/toxicants interactions in both experimental or human liver disease needs more detailed studies.

Dietary factors Energy from the diet is extracted by gut bacteria, and the nutrients are provided to both bacteria and the host. In newborns, their feeding pattern has a strong impact on gastrointestinal bacterial communities. *Bifidobacteria* dominate the microbiota in babies that are breast-fed until weaning, while babies that are formula-fed have a more diverse microbiota (Magne et al. 2006). Solid food induces changes in the microbiota, and a relatively stable, healthy-like microbiome assembles by the age of three (Scholtens et al. 2006). In adults, major macronutrients, including carbohydrate, protein, and fat, can have a significant impact on the microbial community composition.

There are three main classes of carbohydrate: resistant starches (RS), non-starch polysaccharides (NSP) and oligosaccharides, including di- and monosaccharides. Intake of high amounts of carbohydrate, especially fructose, has been considered as a contributing factor promoting nonalcoholic fatty liver disease (NAFLD) (Jin and Vos 2015). Changes in the amount and type of carbohydrate profoundly influence microbiota composition and the microbial metabolites in adult human volunteers. Recent studies revealed that high fructose consumption changed intestinal microbiota in humans and in experimental animals (Ferrere et al. 2016). Commonly used prebiotics including fructooligosaccharides (FOS), galactooligosaccharides (GOS), and lactulose are classified as carbohydrates. Many studies have shown that following dietary supplementation with GOS and FOS, there was an increase in *Bifidobacteria*, which is positively correlated with healthy status. In a humanized rat (inoculated with a human microbiota) model, arabinoxylanoligosaccharide (AXOS) supplementation increased numbers of *Roseburia/E. rectale* group, which have been recognized as butyrate-producing bacteria (Van den Abbeele et al. 2011). Butyrate is one of the major short-chain fatty acids (SCFAs), and it has beneficial effects in NAFLD and alcoholic liver disease (ALD) through multiple mechanisms. In an arsenic–/high-fat diet-induced liver injury mouse model, supplementation with oligofructose (OFC) showed a protective effect, partially through microbiota alteration (Massey et al. 2015). OFC feeding increased the relative abundance of *Bacteroidetes* with a concomitant decrease in *Firmicutes*. In addition, OFC restored the growth of *Tenericutes* which was lost in animals that received high-fat diet (Massey et al. 2015).

Protein malnutrition is strongly associated with more advanced liver disease. Protein turnover in the colon provides nitrogen and amino acids for certain bacterial growth, which in turn affects protein fermentation. High-protein feeding increased cecal branched chain amino acids (BCAAs) in conventional rats compared to germ-free rats (Lhoste et al. 1996). Most studies so far focus on the effects of altered fermentation products on microbiota composition, rather than dietary protein itself. Gut flora humanized rats fed meat diets had lower levels of SCFAs and higher levels of branched chain fatty acids (BCFAs), greater ammonia, and higher pH in cecal contents compared to pea protein-fed animals (Lhoste et al. 1998; Scott et al. 2013).

It is well-established that high-fat (HF) consumption is a major contributing factor to NAFLD and ALD. HF diet feeding significantly reduced *Bifidobacteria* counts and fecal SCFAs, including butyrate, concentrations compared to low-fat (LF) feeding in human subjects (Brinkworth et al. 2009). HF diet-induced microbiota alterations seem to be reversible. In mice, a 12-week HF diet induced obesity and changed microbiota composition, while the mouse phenotype became indistinguishable from control diet-fed animals after switching the HF diet to a normal diet for additional 10 weeks (Zhang et al. 2012a). This reversibility of microbiota in obesity provides a rationale for developing approaches to influence microbiota in order to treat metabolic diseases, including NASH and obesity, if in fact it plays an etiologic role in these diseases (Zhang et al. 2012a).

11.2 Microbiota and Alcoholic Liver Disease

ALD encompasses hepatic steatosis, steatohepatitis, and more serious forms including fibrosis, cirrhosis, and liver cancer (Gao and Bataller 2011). Consumption of alcohol may induce small and large intestinal bacterial overgrowth, particularly Gram-negative bacteria and alterations in bacterial composition. Alcohol metabolism in the intestine also causes intestinal epithelial barrier dysfunction resulting in the increased translocation of bacterial products, including endotoxin, bacterial DNA, and pathogen-associated molecular patterns (PAMPs), from the gut lumen to the liver. Those toxic microbial-related products trigger the inflammatory cascade and the induction of reactive oxygen species and other pro-inflammatory/toxic factors in Kupffer cells, hepatocytes, and other liver cells. Clinical studies show that human subjects with alcoholic cirrhosis have higher levels of serum bacterial products than healthy controls (Parlesak et al. 2000; Bajaj et al. 2014b).

Intestinal dysbiosis and bacterial overgrowth have long been observed in human alcoholic subjects. Alcohol abuse-induced intestinal bacterial overgrowth in humans was documented over three decades ago (Bode et al. 1984). Anaerobic and aerobic cultures of samples from jejunal juice displayed significantly higher bacterial counts in alcoholics compared to healthy controls, indicating small intestinal bacterial overgrowth might contribute to functional abnormalities of the small intestine in patients with chronic alcohol abuse. This same group of the investigators further showed a higher prevalence of small intestine bacterial overgrowth in chronic alcoholics compared to controls using a breath test (Bode et al. 1993). The observation was later confirmed by other groups in alcoholic cirrhosis patients (Morencos et al. 1995; Kirpich et al. 2008; Bhonchal et al. 2007). Recently, Bajaj et al. studied intestinal bacterial composition in 244 alcoholic cirrhotic patients using a novel index, namely, cirrhosis dysbiosis ratio (CDR, a low number indicating dysbiosis). The authors found that intestinal dysbiosis was more severe in decompensated cirrhotics compared to compensated cirrhotics (Bajaj et al. 2013). Gut microbiota are not only affected by excessive alcohol consumption; moderate drinking has also been shown to be a strong risk factor for small intestine bacterial overgrowth (Gabbard et al. 2014). Mutlu and colleagues investigated the mucosa-associated colon microbiota in alcoholics with and without cirrhosis and in controls (Mutlu et al. 2012). Pyrosequencing analysis of colon biopsy samples revealed that mucosa-associated bacteria were persistently altered in a subset of alcoholics, and this was correlated with endotoxemia (Mutlu et al. 2012). Although studies on bacterial overgrowth and dysbiosis are increasing, as yet there is no specific intestinal bacterial pattern identified to have a proven etiologic role in the development of ALD. However, the fact that alcohol consumption causes bacterial overgrowth and dysbiosis provides an opportunity for the treatment and/or prevention of ALD by targeting intestinal microbiota.

Another human study showed that alcohol-dependent subjects developed gut leakiness after 3 weeks of alcohol abstinence together with fecal dysbiosis and changes in colonic microbiota (Leclercq et al. 2014). The functional changes in microbiota have also been studied. Alcohol-dependent subjects with high intestinal permeability showed higher phenol and lower 4-methyl-phenol concentration in feces than the subjects with low intestinal permeability, while the indole and 3-methyl-indole concentrations were oppositely changed. These metabolite alterations could be associated with liver injury from alcohol abuse.

Compared with clinical observations, studies in experimental animal models of ALD provide more detailed information. In an intragastric alcohol-feeding mouse model, Yan et al. (2011) showed that 3 weeks of alcohol ingestion led to bacterial overgrowth in the proximal small intestine and to dysbiosis, which was associated with the suppression of antimicrobial peptides, Reg3b and Reg3g. Alcohol-fed mice displayed an increase in *Bacteroidetes* and *Verrucomicrobia* abundance and a decrease in *Firmicutes* level compared to pair-fed animals. Interestingly, an overgrowth of *Akkermansia muciniphila* was observed in alcohol-fed mice, and this is believed to be responsible for mucin degradation. Moreover, the population of *Lactobacilli* was depleted in alcohol-fed mice, and *Lactobacilli* are generally considered to be a beneficial bacterial group.

In a study using the Lieber-DeCarli alcohol-feeding mouse model (Bull-Otterson et al. 2013), metagenomics analysis demonstrated a decline in the abundance of both *Bacteroidetes* and *Firmicutes* phyla, with a proportional increase in the Gram-negative *Proteobacteria* and Gram-positive *Actinobacteria* phyla. Genera analysis showed the greatest expansion in Gram-negative alkaline-tolerant *Alcaligenes* and Gram-positive *Corynebacterium*. These alterations were accompanied by the changes in colonic pH and liver steatosis (Bull-Otterson et al. 2013).

Alcohol exposure-associated gut bacterial alterations were also studied in mice with various genetic backgrounds. Wild-type mice have higher alcohol-induced hepatic steatosis and injury than mucin knockout mice ($Muc2^{-/-}$) (Hartmann et al. 2013), and mucin plays an important role in intestinal mucus homeostasis. $Muc2^{-/-}$ mice are protected from intestinal bacterial overgrowth after alcohol feeding and have a higher expression of antimicrobial peptides, Reg3b and Reg3g. Interestingly, overexpression of Reg3g in the intestinal epithelial cells restricts bacterial colonization of the mucosal surface, reduces bacterial translocation, and protects mice from alcohol-induced steatohepatitis (Wang et al. 2016b). Although wild-type and $Reg3b^{-/-}$ mice had a similar increase in gut microbiota after alcohol feeding, $Reg3b^{-/-}$ mice showed significantly higher numbers of mucosa-associated bacteria in the mucus and epithelial layer of the small intestine than wild-type mice. Alcohol-induced bacterial colonization in the intestine was also demonstrated by another group who observed significant changes in the mucosa-associated microbiota in the colon in mice after a 10-week alcohol-feeding regimen (Mutlu et al. 2012). The authors concluded that alcohol appears to impair control of the mucosa-associated microbiota, and the subsequent breach of the mucosal barrier facilitates progression of alcoholic liver disease.

Germ-free mice were used to study the impact of alcohol on gut microbiota homeostasis with somewhat conflicting results. Surprisingly, the absence of the microbiota (germ-free) in mice did not protect but, in fact, exacerbated ALD (Chen et al. 2015a), suggesting that microbiota are required for hepatic homeostasis including ethanol metabolism, oxidative stress, and bile acid metabolism. Germ-free mice exhibited dysregulation of ethanol-metabolizing enzymes, as a small dose of alcohol could induce liver damage (Chen et al. 2015a). The requirement of microbiota for liver homeostasis has also been demonstrated in other types of liver disease (Tabibian et al. 2016; Mazagova et al. 2015). A study of ALD using germ-free mice showed that 7-day alcohol feeding in the drinking water caused less inflammation in the liver and did not cause liver injury in germ-free mice compared with alcohol-fed conventional mice. Interestingly, fecal transplantation from conventional mice fed alcohol to germ-free mice induced injury and inflammation in both liver and intestine, suggesting that alcohol consumption caused dysbiosis. A more recent work extended the study using germ-free mice. Humanization of mice using human intestinal microbiota transplants from alcoholic patients showed significant differences between germ-free and conventional mice (Llopis et al. 2016). Two patient donors had distinct microbiota compositions, with a large amount of *Bifidobacteria* and *Streptococci* in the patient with severe alcoholic hepatitis (sAH) compared to the patient without severe alcoholic hepatitis (noAH). The germ-free mice receiving the intestinal microbiota from a patient with sAH developed more severe liver inflammation, higher liver necrosis, greater intestinal permeability, and higher translocation of bacteria than mice harboring the intestinal microbiota from an alcoholic patient with noAH. Interestingly, in conventional mice humanized with the intestinal microbiota from an sAH patient, a second subsequent transfer of intestinal microbiota from the noAH patient improved alcohol-induced liver lesions.

Investigators at two nearby institutions in France incurred a reproducibility problem in creating experimentally-induced ALD (Ferrere et al. 2017). The mice were from the same vendor, and the investigators used the same Lieber-DeCarli liquid alcohol diet at both facilities. The group of alcohol-sensitive animals developed classic ALD and disruption of gut barrier function, while the alcohol-resistant mice from the other institution did not develop liver disease. The investigators evaluated why they could not generate the same liver injury at both institutions, and their final conclusion was gut microbiota. They ultimately performed fecal transplantation of fresh feces from alcohol-resistant donor mice to alcohol-sensitive mice three times a week. Another group of mice received dietary pectin during the entire alcohol-feeding period. Both of these interventions prevented liver steatosis/inflammation and improved gut homeostasis in the sensitive mice. Analysis of the intestinal microbiota showed that the proportion of *Bacteroides* was significantly lower in the alcohol-sensitive mice. Principal coordinate analysis showed major differences between the intestinal microbiota of sensitive and resistant mice. These data demonstrate the importance of gut flora and the gut-liver axis in the development and reproducibility of experimental ALD. If these results can be extrapolated to humans, it may also help explain why some people who drink heavily develop ALD.

The effect of the types of dietary fat on gut microbiota homeostasis has been evaluated in mice with ALD. Saturated fat (medium chain triglycerides enriched) feeding reduced *Proteobacteria* and *Actinobacteria* and increased *Bacteroidetes* in feces of mice exposed to 8 weeks of alcohol feeding (Kirpich et al. 2016). A reduction in endotoxemia and liver steatosis and injury was seen in animals fed the saturated fat diet. In addition, supplementation of saturated long-chain fatty acids maintained intestinal eubiosis and reduced alcohol-induced liver injury (Chen et al. 2015b).

11.3 Microbiota and Nonalcoholic Liver Disease

NAFLD is pathologically similar to alcohol-induced fatty liver disease in subjects without an alcohol drinking history. It is regarded as the most common liver disease worldwide. The prevalence of NAFLD is greatly increased in patients with the metabolic syndrome-obesity, type 2 diabetes, and hyperlipidemia. In NAFLD patients, approximately 25% may progress to NASH (nonalcoholic steatohepatitis) which may result in liver fibrosis and cirrhosis. Patients with liver cirrhosis are at increased risk for hepatocellular carcinoma.

The importance of microbiota in hepatic steatosis was elucidated clinically as early as three decades ago when Drenick et al. (1982) studied patients undergoing gastric bypass surgery. The authors found that hepatic steatosis coincided with bacterial overgrowth, and treatment with antibiotics in these patients reduced hepatic steatosis. Later on, it was demonstrated that small intestine bacterial overgrowth is more prevalent in patients with NASH than in healthy controls (Wigg et al. 2001). Since then, increasing evidence in animals and humans has improved our understanding of the role of microbiota homeostasis in the development of NAFLD (Abu-Shanab and Quigley 2010).

Several clinical studies have shown that alterations of the gut microbial community are closely associated with the prevalence and development of NAFLD. Significant shifts of the intestinal microbiota community, such as over-representation of selected members of *Firmicutes* and *Lactobacillus* species, and decreased bacteria of phylum *Firmicutes* were found in NAFLD patients (Raman et al. 2013). Microbiota samples from patients with NAFLD or NASH exhibited a lower abundance of Ruminococcaceae family members. Obese children with NASH have a significantly disproportionate *Escherichia* in their intestinal microbiota. Adult NAFLD patients had a significant lower percentage of *Clostridium coccoides* than NASH patients in their gut, indicating a possible role for *Clostridium coccoides* in the development of NAFLD to NASH.

In a 6-month longitudinal study involving 16 NASH patients and 22 healthy controls (Wong et al. 2013a), fecal samples displayed a lower abundance of *Faecalibacterium* and *Anaerosporobacter* but higher abundance of *Parabacteroides* and *Allisonella* in the patients. In a cross-sectional study involving 11 biopsy-proven NALFD, 22 NASH, and 17 living liver donors as controls (Mouzaki et al. 2013), PCR measurement showed a lower abundance of *Bacteroides* and *Prevotella* and higher fecal *Clostridium coccoides* in NASH patients compared with patients with simple steatosis. It is concluded that there is an inverse and diet–/BMI-independent association between the *Bacteroidetes/Prevotella* bacteria and the presence of NASH. Similarly, a recent study of 57 patients with biopsy-proven NAFLD revealed that NAFLD patients with fibrosis had a significant increase in *Bacteroides* and *Ruminococcus* and a decrease in *Prevotella* (Boursier et al. 2016). It appears that *Bacteroides* are independently associated with NASH, and *Ruminococcus* are independently associated with significant fibrosis.

NAFLD often occurs in obese patients, but it is not rare in non-obese adults, especially in Asian populations. A study (Wang et al. 2016a) aimed at understanding whether abnormal intestinal microbiota are associated with NAFLD in non-obese patients found that the NAFLD patients had lower diversity and changes in phylum levels in fecal microbiota. Specifically, 20% more of phylum *Bacteroidetes* and 24% less of *Firmicutes* were found in fecal samples of non-obese NAFLD patients compared to non-obese healthy controls. In addition, four families and their eight genera within *Firmicutes*, which were SCFA-producing and 7α-dehydroxylating bacteria, were significantly decreased. Furthermore, dysbiosis-associated metabolic markers were found to be correlated with non-obese NAFLD.

In a recent pediatric NAFLD study, 61 children diagnosed with NAFLD, NASH, or obesity and 54 healthy children were recruited (Del Chierico et al. 2017). A multivariate analysis of the stool microbiota and volatile organic compounds found that a combination of *Oscillospira*, *Rikenellaceae*, *Parabacteroides*, *Bacteroides fragilis*, *Sutterella*, *Lachnospiraceae*, 4-methyl-2-pentanone, 1-butanol, and 2-butanone could serve as a marker to classify NAFLD patients and healthy subjects.

However, studies comparing intestinal microbiota in NAFLD and NASH patients have produced inconsistent and even contradictory results. To better determine whether patients with NAFLD and NASH have distinct microbiota profiles, studies with larger and better characterized patient cohorts are needed. Identification of specific microbial alterations of NAFLD and NASH patients will promote our understanding of the role of gut contents in liver pathophysiology and could lead to biomarker discovery and strategies for the treatment of NAFLD and NASH.

The role of gut microbiota in high-fat diet (HFD)-induced NAFLD has been widely studied in animal models. HFD feeding increases obesity, but not all obese subjects develop NAFLD. A 10-week HFD feeding increased body weight, hepatic fat, inflammatory cell infiltration, and inducible nitric oxide synthase concentration in mice along with an increase in fecal *Lactobacillus* species (Zeng et al. 2013). The increased lactobacillus may have an effect on lipid metabolism via bile acid metabolism, which contributes to the fatty liver formation. In a mouse model (Le Roy et al. 2013), germ-free mice were colonized with intestinal microbiota from two distinct mice fed HFD. One of the donor mice developed hyperglycemia with higher concentrations of pro-inflammatory cytokines (responder), while another mouse was normoglycemic and had less systemic inflammation (non-responder). The germ-free mice that received microbiota from the responder developed hyperglycemia and hepatic steatosis, while the germ-free mice that received microbiota from the non-responder remained normoglycemic and did not develop NAFLD. The microbial composition showed distinct differences at the phylum, genus, and species levels between the recipients from the responder and the recipients from the non-responder. This study demonstrates that NAFLD is transmissible to germ-free mice, and the gut microbiota contribute to the development of NAFLD independent of obesity.

In children with NAFLD or NASH, fecal microbiota were compared to healthy controls (Zhu et al. 2013). Blood alcohol concentrations were similar between healthy subjects and NAFLD children, but higher in NASH patients with increased abundance of alcohol-producing bacteria. The authors postulate that distinct composition of gut microbiota in NASH, obese and healthy subjects could offer targets for intervention in the disease. In a mouse study (Cope et al. 2000), the authors showed that endogenous alcohol concentration was increased in obese mice without exogenous alcohol exposure. Neomycin treatment reduced blood alcohol concentrations together with inducing an improvement of NALFD, suggesting a role for the microbiota in the production of endogenous alcohol and the development of NAFLD.

Inflammasome-mediated gut microbiota homeostasis in NAFLD and obesity progression has been studied in a mouse model (Henao-Mejia et al. 2012). The NLRP6 and NLRP3 inflammasome-deficient mouse model revealed that changes in the composition of the gut microbiota are associated with exacerbated hepatic steatosis and inflammation. Gut microbiota-associated influx of bacterial products into portal circulation act as TLR4 and TLR9 agonists leading to enhanced hepatic TNF-α expression that drives NASH progression. Co-housing of inflammasome-deficient mice with wild-type mice, or transfer of the dominant microbiota from the deficient mice to wild-type mice, resulted in exacerbation of hepatic steatosis and the progression of NAFLD. This study highlights the central role of the microbiota in the pathogenesis of unrelated systemic auto-inflammatory and metabolic disorders.

Results from human studies suggest that in addition to high-fat diets, the intake of sugar, particularly fructose, may be a risk factor for the development of NAFLD (Jin and Vos 2015). Mice were either fed with 30% fructose in drinking water/tap water with or without antibiotics for 8 weeks (Wagnerberger et al. 2012). Hepatic steatosis found in the livers of fructose-fed mice was associated with a significant induction of TLR1-4 and TLR6-8. The deleterious effects of fructose were attenuated in antibiotic-treated mice, indicating that microbiota homeostasis and increased intestinal translocation may be involved in the onset of fructose-induced NAFLD. Another mouse study showed that fructose feeding caused steatosis and dyslipidemia, with significantly decreased *Bifidobacterium* and *Lactobacillus* and increased endotoxemia (Jegatheesan et al. 2016).

11.4 Microbiota and Liver Cancer

Hepatocellular carcinoma (HCC) is the most common primary malignancy of the liver and one of leading causes of cancer mortality. Approximately 80% of the liver cancer patients also have liver cirrhosis. Among the various etiologies, alcohol abuse and hepatitis C/B virus are the most common and most studied risk factors. Patients with liver cirrhosis and HCC exhibited a significant increase in serum endotoxin levels (Zhang et al. 2012b), indicating a correlation of gut homeostasis with cirrhosis and HCC. The association of gut microbiota and HCC has been described in recent animal studies (Li et al. 2016b; Xie et al. 2016a, b). These studies have implications for human liver cancer risk assessment and prevention.

In a rat model (Zhang et al. 2012b), chronic diethylnitrosamine (DEN) administration was associated with an imbalance of subpopulations of gut microbiota. A significant suppression of *Lactobacillus* species, *Bifidobacterium* species, and *Enterococcus* species was found in feces of the rats with DEN-induced HCC. The changes in gut microbiota were associated with intestinal inflammation. Induction of enteric dysbiosis or intestinal inflammation by penicillin or dextran sulfate sodium, respectively, also significantly promoted tumor formation.

Obesity is increasingly recognized as a major risk factor for several common types of cancer, including HCC. In a murine model, Yoshimoto and colleagues (2013) showed that dietary or genetic obesity induced gut dysbiosis which led to the increase of deoxycholic acid (DCA) levels, a gut bacterial metabolite known to cause DNA damage. DCA in the circulation provokes the senescence-associated secretory phenotype in hepatic stellate cells, which, in turn, produce inflammatory factors in the liver, thus facilitating HCC development in mice after exposure to the chemical carcinogen. Blocking DCA production or reducing gut bacteria prevents HCC development in obese mice. Gut microbiota alterations and dysregulated bile acids were also described in another murine model of NASH-HCC (Xie et al. 2016a), in which dysbiosis was closely correlated with altered bile acid concentrations and the progression of liver disease from NASH to HCC (Xie et al. 2016b).

The role of specific intestinal microbiota in HCC development has been studied (Fox et al. 2010). Helicobacter-free C3H/HeN mice were inoculated with the foodborne contaminant, aflatoxin B1 (AFB), and/or *Helicobacter* hepaticus. Intestinal colonization by *H. hepaticus* was sufficient to promote aflatoxin- and HCV transgene-induced HCC. Neither bacterial translocation to the liver nor induction of hepatitis was necessary. In summary, enteric microbiota directly relates to HCC risk in mice exposed to carcinogenic chemicals or hepatitis virus transgenes.

In another murine study, Dapito et al. demonstrated that intestinal microbiota exerted a profound influence on HCC promotion through the LPS (lipopolysaccharide)-TLR4 pathway (Dapito et al. 2012). HCC development in chronically injured livers depended on the intestinal microbiota and TLR4 activation in non-bone-marrow-derived resident liver cells, which is distinct from previous studies in which bone-marrow-derived cells such as macrophages were believed to be responsible for inflammatory cytokine production and genotoxic HCC. This study suggests that the intestinal microbiota and TLR4 represent potential therapeutic targets for HCC prevention in advanced liver disease.

Grat et al. (2016) determined the gut microbiota profile in 15 patients with HCC and 15 non-HCC patients matched according to etiology of cirrhosis and for model for end-stage liver disease (MELD) scores, all of whom underwent liver transplantations. Gut microbiota analysis showed that the presence of HCC was associated with significantly increased fecal counts of *Escherichia coli* (*E. coli*). Intestinal overgrowth of *E. coli* may contribute to hepatocarcinogenesis.

11.5 Microbiota and Other Liver Diseases

Gut flora and bacterial translocation play an important role in the pathogenesis of the cirrhosis and its complications, including hepatic encephalopathy (HE) and spontaneous bacterial peritonitis (SBP, the most common type of infection). Bacterial infections promote dysregulated intestinal immunity and bacterial overgrowth/dysbiosis, contributing to acute-on-chronic liver failure (ACLF) and death in cirrhosis.

Bacterial infection is a severe complication in the course of cirrhosis. Increased endotoxemia has been shown in patients with advanced stage cirrhosis, and this is associated with the activation of the cytokine cascade in SBP. To study the alterations in gut microbiota in cirrhosis and their longitudinal changes with decompensation, stool samples from 244 age-matched compensated/decompensated cirrhotic patients and controls were analyzed (Bajaj et al. 2014b). The CDR (low score indicates more severe disease) was highest in controls followed by compensated and decompensated patients. These worsening dysbiosis was correlated with the increasing endotoxin levels.

Hepatic encephalopathy (HE) is a common neurocognitive impairment in cirrhosis (Bajaj et al. 2009) that is mechanistically associated with gut bacteria and colonic mucosal microbiota composition and inflammation in the setting of intestinal barrier dysfunction (Bajaj et al. 2012b). Dysbiosis has been seen in cirrhotic patients with HE with a higher abundance of potentially pathogenic bacteria and a reduction of commensal bacteria (Bajaj et al. 2012c). A study involving 69 cirrhotic patients with minimal HE (MHE) showed substantial derangements in the gut microecology (Liu et al. 2004), with significant fecal overgrowth of potentially pathogenic *E. coli* and Staphylococcal species compared to cirrhotic patients without HE. Synbiotic treatment (combination of pre- and probiotics) modified the gut flora and was associated with a significant reduction in blood ammonia levels and endotoxemia. Another MHE study (Zhang et al. 2013) showed an overrepresentation of two bacterial families, *Streptococcaceae* and *Veillonellaceae*, in cirrhotic patients with and without MHE, compared with normal individuals. These studies suggest that targeting the gut microbiota with novel agents may be an alternative form of management of minimal HE in patients with cirrhosis.

The salivary microbiota have also been studied in cirrhotic patients with HE (Bajaj et al. 2015). Cirrhotic patients with HE-associated gut dysbiosis have a reduction in autochthonous bacteria and changes in bacterial function, which are also present in saliva. Dysbiosis in saliva is associated with a higher risk of further hospitalization owing to liver-related complications.

11.6 Mechanisms Linking Gut Microbiota and Liver Disease

Intestinal integrity It is becoming increasingly clear that intestinal barrier dysfunction contributes to most of the liver diseases studied thus far. The integrated intestinal barrier has a vital role in preventing translocation of harmful bacteria and their products into the portal circulation. The epithelial cells form a lining with the paracellular space sealed by tight junctions (TJ) and adherens junctions (Rao 2009), and this is covered by a protective mucus layer that physically blocks most particles from direct contact with the epithelial cells (Turner 2009). Intestinal IgA, a major class of antibody secreted by the gut mucosa, is an important contributor to gut barrier function (Conley and Delacroix 1987). Dietary factors in ALD and NAFLD-caused dysbiosis directly affect the gut intestinal barrier at multiple levels including tight junctions, production and stabilization of mucin, recruitment and activation of inflammatory cells, and reduction of IgA secretion leading to the increased intestinal permeability and the exposure of mucosal and liver cells to potentially harmful bacteria and products (Hooper and Macpherson 2010; Wei et al. 2011). However, a recent study demonstrates that absence of IgA does not affect the development of ALD, likely due to a compensated increase in IgM, which limits bacterial translocation in ALD (Inamine et al. 2016).

Increased intestinal oxidative stress in ALD and NAFLD subjects could damage intestinal barrier function. Although the majority of ingested alcohol is metabolized in the liver, a significant amount of alcohol is also metabolized in the intestine, leading to increased ROS mediated by Cyp2E1 (Forsyth et al. 2014). In addition, gut bacteria-produced endogenous alcohol in NAFLD subjects contributes to ROS production causing barrier dysfunction.

Intestinal mucosa experiences profound fluctuations in blood flow and metabolism (Glover et al. 2016). Intestinal tissue hypoxia induces protective mechanisms including increased hypoxia-inducible factor (HIF) expression, and HIF is a master transcription factor regulating many genes in intestinal homeostasis (Saeedi et al. 2015). Studies have shown that the expression of several HIF-targeting genes, including intestinal trefoil factor, p-glycoprotein, antimicrobial peptides [Reg3 and cathelicidin-related antimicrobial peptide (CRAMP)], and tight junction claudin-1, is reduced in the intestines in ALD (Wang et al. 2011, 2012) and NAFLD subjects. Deleting HIF in the intestine causes dysbiosis and exacerbates liver injury (Shao et al. 2016), suggesting that HIF signaling is important for gut microbiota homeostasis and intestinal barrier function.

The tight junction protein, occludin, was decreased by both alcohol and high-fructose diet due to dysbiosis. It was shown that intestinal inflammation-induced upregulation of TNF-α increased microRNA 122a (miR122a), which regulates occludin expression (Ye et al. 2011). Overexpression of miR122a decreased occludin protein levels and increased permeability in epithelial Caco-2 cells (Zhao et al. 2015).

Clinical and animal studies have shown that liver steatosis and injury induced by alcohol and high-fat diet are associated with dysbiosis, increased permeability, and circulating endotoxins. The effect of LPS is not limited to the liver but also affects other tissues/organs, such as adipose and pancreas, that contribute to hepatic fat accumulation. Several mechanisms have been proposed to be responsible for the translocated bacteria and products contributing to the pathogenesis of fatty liver disease. The TLR pathway is a well-known target for the effects of increased LPS, which activates TLR4 in hepatic Kupffer cells and stellate cells to stimulate pro-inflammatory and profibrotic pathways (Seki and Brenner 2008). Mucosal TLR activation might also contribute to hepatic steatosis via intestinal epithelial MYD88 and inflammasome activation (Everard et al. 2014).

Short-chain fatty acids In the intestine, SCFAs are produced in the distal small intestine and colon where nondigestible carbohydrates such as resistant starch, dietary fiber, and other low-digestible polysaccharides are fermented by saccharolytic bacteria including the phyla *Bacteroidetes*, *Firmicutes*, and *Actinobacteria*. Acetate and propionate are the main products of the *Bacteroidetes* phylum, and butyrate is mainly produced by the *Firmicutes* phylum. As an energy precursor, SCFAs are implicated in the pathogenesis of NAFLD because of their possible contribution to obesity. The first evidence regarding SCFAs was from a study by Turnbaugh et al. (2006) showing that the cecum in ob/ob mice has an increased concentration of SCFAs and that transplantation of germ-free mice with the gut microbiota from ob/ob mice caused greater fat gain than from transplants from lean animals. In humans, increased production of SCFAs by the gut microbiota was also observed in overweight and obese people, compared to lean subjects (Schwiertz et al. 2010). Using metagenomics analysis, the majority of studies showed that ob/ob mice (Ley et al. 2005) and obese patients (Ley et al. 2006) exhibit reduced abundance of *Bacteroidetes* and proportionally increased abundance of *Firmicutes*. However, how these ratio changes affect energy imbalance leading to obesity and its complications including NAFLD needs further study. In fact, generally, SCFAs have more beneficial effects than obesity-causing effects (den Besten et al. 2013). Beneficial effects of SCFAs are through several pathways, including immunoregulation, enhanced intestinal barrier function, acting as a histone deacetylase 1 (HDAC) inhibitor, decreased expression of lipogenic genes, increased carnitine palmitoyltransferase 1A expression (den Besten et al. 2013), and a peroxisome proliferator-activated receptor-γ (PPARγ)-dependent mechanism, thereby shifting metabolism in adipose and liver tissue from lipogenesis to fatty acid oxidation (den Besten et al. 2015).

Bile acids Gut microbiota produce numerous metabolites to regulate host metabolism through binding to their cognate receptors. Bile acids are produced in the liver and metabolized in the intestine by gut bacteria. The role of bile acids in ALD and NAFLD is complex (Rinella and Sanyal 2015). Farnesoid X receptor (FXR) and G protein-coupled receptor 5 (Zhu et al. 2016; Yuan and Bambha 2015; Yamada et al. 2016; McMahan et al. 2011, 2013) are considered as the major regulators of bile acid metabolism, as they are involved in numerous metabolic pathways in the host.

Bile acids can regulate gut microbiota through activation of innate immune genes in the intestine (Wahlstrom et al. 2016). Hepatic FXR activation by bile acids suppresses NF-κB signaling and decreases hepatic inflammation in ALD (Manley and Ding 2015), NAFLD (Carr and Reid 2015), and fibrosis (Verbeke et al. 2016). Accumulating data demonstrate that both alcohol and high-fat diet consumption inhibit FXR activation. Intestinal FXR activation by bile acids inhibits bacterial overgrowth and mucosal damage (Inagaki et al. 2006). In contrast, deletion of FXR promotes bacterial overgrowth (Swennes et al. 2014) and intestinal epithelial barrier dysfunction (Inagaki et al. 2006). A recent clinical trial using an FXR ligand, obeticholic acid, in patients with non-cirrhotic NASH (FLINT trial) showed that activation of FXR by obeticholic acid improved histological NASH (Neuschwander-Tetri et al. 2015). However, another experimental NASH study showed that FXR antagonism might also be beneficial for NASH (Jiang et al. 2015). Manipulation of the gut microbiota resulted in an alteration of intestinal bile acid composition leading to intestinal FXR inactivation. This FXR antagonism reduced ceramide synthesis and de novo lipogenesis in the liver. Therefore, FXR regulation is complex with a broad range of effects, and gut microbiota could regulate FXR activity through bile acid metabolism. It should be noted that human and animal bile acid metabolism are different. Caution should be taken when attempting to explain experimental results that are in conflict.

Endogenous ethanol ALD and NAFLD share a high similarity in histological features and might have some similar pathways in their pathogenesis. Although dietary alcohol is the primary cause of liver injury in ALD, the bacterial-produced ethanol cannot be overlooked, due to its substantial production. Several studies suggest that endogenous alcohol might be involved in the development of NAFLD (Cope et al. 2000; Zhu et al. 2013). *Proteobacteria* (especially *E. coli* and other *Enterobacteriaceae*) is the major phyla of gut bacteria producing alcohol (Zhu et al. 2013). For example, 1 g of *E. coli* produces 0.8 g of ethanol per hour under anaerobic conditions (Dawes and Foster 1956). In NASH patients, those bacteria have been found to be increased in the gut, as shown by an increase in breath ethanol concentration (Nair et al. 2001) and an upregulation of alcohol metabolizing capacity (Baker et al. 2010). Gut bacteria-produced ethanol might contribute to increased intestinal permeability and endotoxemia, leading to TLR and inflammasome activation. Absorbed endogenous ethanol also may have a direct toxic effect on the liver.

11.7 Treatment Targeting Microbiota in Liver Disease

Manipulating gut microbiota to counteract the deleterious effects of pathogenic bacteria has been widely used in the management of liver disease. Probiotics, prebiotics, and synbiotics have been shown to be beneficial to the health of intestinal microbiota.

Probiotics Probiotics are defined as live microorganisms which, when administered in adequate amounts, confer a health benefit on the host, according to the FAO/WHO definition (Fooladi et al. 2013). The beneficial effects of probiotics have been widely investigated in multiple animal models and clinical studies of a variety of disease conditions in the gastrointestinal system such as inflammatory bowel disease, NASH, cirrhosis, and ALD (O'Mahony et al. 2005; Kirpich and McClain 2012). Ideal probiotic strains for this kind of application should be resistant to bile, hydrochloric acid, and pancreatic juice; be able to tolerate stomach and duodenum conditions and gastric transport; and have the ability to stimulate the immune system, thereby improving intestinal function via adhering to and colonizing the intestinal epithelium. In addition, probiotic strains must be able to survive during manufacture and storage in order to exert considerable healthful outcomes (Lin et al. 2006). Currently, the most often used probiotics are *Bifidobacteria*, lactic acid bacteria (LAB), *Propionibacteria*, yeasts (*Saccharomyces boulardii*), and the Gram-negative *Escherichia coli* strain (Nissle 1917). *Lactobacilli*, major contributors to the LAB group, also are frequently used probiotics. Various species and strains of *Lactobacilli* have been used in animals and humans, including *Lactobacillus acidophilus*, *Lactobacillus casei*, *Lactobacillus rhamnosus*, and *Lactobacillus helveticus*. Most of these species belong to phylum *Firmicutes*. *Bifidobacterium*, which produces lactic acid, is another commonly used probiotic genus and belongs to the *Actinobacteria* phylum. To date, a large number of probiotics have been reported to be suitable for the treatment of a variety of diseases, and this number is still growing.

Prebiotics Unlike probiotics, prebiotics, which have also been frequently used for disease treatment, are not live bacteria but rather nondigestible carbohydrates. Prebiotics serve as an energy source for"good" bacteria and stimulate the growth and activities of specific bacteria in the gut (Yan et al. 2011). SCFAs, in particular, acetate (Perry et al. 2016) and butyrate (Berni Canani et al. 2012), have been recognized as beneficial metabolites associated with many biological functions in the gut. One of the important functions of butyrate is its ability to regulate gene expression through epigenetic mechanisms (Berni Canani et al. 2012). Butyrate enhances cell proliferation and inhibits cell apoptosis in normal cells but not in the transformed cells (Hass et al. 1997).

Synbiotics Using prebiotics and probiotics in combination is usually described as synbiotics. The net health effect of synbiotics is usually synergistic (Pineiro et al. 2008). Synbiotics practically have been used in a large body of clinical and experimental studies on liver disease.

Probiotic treatment/prevention of ALD Probiotics have been used in experimental animals and, to some extent, in humans, to modulate gut microbial homeostasis and to manage ALD. Probiotics have been used in various experimental ALD animal models, including chronic alcohol exposure, single-dose acute alcohol exposure, multiple-dose alcohol exposure, and alcohol exposure plus LPS challenge. A variety of probiotic strains have been used, such as *Lactobacillus rhamnosus* GG,

Lactobacillus acidophilus, *Lactobacillus helveticus*, *Bifidobacterium*, VSL#3, heat-killed *Lactobacillus brevis* SBC8803, and *Lactobacillus rhamnosus* GG supernatant.

Among those, *Lactobacillus rhamnosus* GG (LGG) is the most frequently used strain. LGG is a Gram-positive bacterial strain of the *Lactobacillus rhamnosus* species that was isolated in 1983 by Barry R. Goldin and Sherwood L. Gorbach. In several models of ALD in rats and mice, LGG administration showed significant protective effects. LGG reduced plasma endotoxin level, improved liver enzymes alanine transaminase (ALT) and aspartate transaminase (AST), and reduced hepatic steatosis and injury.

Nanji and coworkers were one of the earliest groups demonstrating the effectiveness of LGG in experimental ALD (Nanji et al. 1994). LGG was administrated to Wistar rats at 10^{10} CFU (colony forming units), and this probiotic reduced alcohol-induced endotoxemia and liver injury. In another study, a combination treatment using *Lactobacillus acidophilus*, *Lactobacillus helveticus*, and *Bifidobacterium* in rats with alcohol pancreatitis-related liver damage effectively protected against endotoxin/bacterial translocation, as well as liver damage in the course of acute pancreatitis and concomitant heavy alcohol consumption (Marotta et al. 2005). Additional studies using LGG in rats demonstrated reduced alcohol-induced gut leakiness, oxidative stress, and inflammation in both intestine and liver (Forsyth et al. 2009) kкand improved intestinal dysbiosis (Mutlu et al. 2009). Another frequently used probiotic mixture, VSL#3, was shown to be effective in modulating gut microbiota and protecting against alcohol-induced intestinal barrier dysfunction in rats (Chang et al. 2013).

Our group fed mice with the Lieber-DeCarli liquid diet containing 5% alcohol for 8 weeks to produce hepatic fatty liver and injury. These mice were treated with LGG culture broth at 10^9 CFU for the final 2 weeks along with continued chronic alcohol administration. LGG supplementation reversed established alcoholic hepatic steatosis and injury (Wang et al. 2011). This beneficial effect was associated with a reduction in circulating LPS and improved intestinal barrier function mediated, at least in part, by intestinal HIF-modulated mucus layer regulation.

While many reports have studied the effects of probiotics in experimental ALD, clinical trials are limited. Stadlbauer and coworkers evaluated the effectiveness of the probiotic *Lactobacillus casei* Shirota on alcoholic cirrhosis (AC) patients ($n = 12$) and healthy controls ($n = 13$) in a small open-labeled study (Stadlbauer et al. 2008). Compared to the baseline before treatment, cirrhotic patients who received the probiotics for 4 weeks had a significantly lower TLR4 expression and IL-10, sTNFR1 (soluble TNF receptor), and sTNFR2 levels, along with a restored neutrophil phagocytic activity, suggesting that the probiotic is safe and may be effective in the treatment of patients with defective immunity.

In a brief report, Loguercio et al. (2002) showed that treatment of 10 alcoholic cirrhosis patients (who were all persistent alcohol users with a median daily intake of pure ethanol of 150 g) with a synbiotic mixture of different bacteria strains and a prebiotic significantly improved liver damage and function compared to basal values. Patients were treated with the synbiotics for 2 months, followed by 1 month of a washout period. The ALT and γGT (gamma glutamyl transferase) levels were slightly, but not significantly, increased after the washout period. These results indicate that the effects of synbiotic treatment partially persisted beyond the end of treatment. The same group (Loguercio et al. 2005) also reported that a commonly used probiotics mixture, VSL#3, was beneficial in liver disease. This open-labeled study involved 22 NAFLD and 20 alcoholic cirrhosis (AC) patients and 36 hepatitis C virus (HCV)-positive patients with and without liver cirrhosis for comparison. VSL#3 treatment significantly improved plasma levels of malondialdehyde (MDA) and 4-hydroxynonenal (4-HNE) in NAFLD and AC patients, but cytokines (TNF-α, IL-6, and IL-10) improved only in AC patients.

More recently, Dhiman et al. (2014) reported that probiotic VSL#3 treatment reduced liver disease severity and hospitalization in a double-blind trial in patients with cirrhosis including AC ($n = 89$, 46 probiotics, 43 placebo; patients who had an alcohol use history in the previous 6 weeks were excluded). Lata and colleagues (2007) in a double-blind, randomized study in 34 cirrhosis patients (19 on probiotics; 15 on placebo), who had an alcoholic etiology of their cirrhosis, showed that treatment with the probiotic *Escherichia coli* Nissle for 42 days improved colonic colonization and liver function with probiotic treatment. In an open-labeled, randomized study which involved 66 patients who were diagnosed with alcoholic psychosis and liver disease as well as 24 matched healthy controls, Kirpich et al. (2008) demonstrated that after 5 days of treatment with *Bifidobacterium bifidum* and *Lactobacillus plantarum* 8PA3, mild alcoholic hepatitis patients had a significant end-of-treatment reduction of ALT and AST, lactate dehydrogenase, and total bilirubin. Compared to standard therapy, probiotic treatment significantly reduced serum ALT. This liver function improvement was associated with changes in the fecal commensal bacteria *Bifidobacteria* and *Lactobacilli*.

Taken together, clinical studies suggest that targeting the gut-liver axis through the use of probiotics may have a therapeutic role in the treatment of patients ranging from those with mild alcoholic hepatitis to those with severe alcoholic cirrhosis. As noted, further rigorous studies with larger sample sizes for testing the effects of probiotics on ALD are needed. Developing novel probiotic strains and related products (including isolating new probiotic bacteria with improved potency for inhibiting pathogenic bacterial growth, strengthening intestinal barrier function, and improving immunoregulation and engineered probiotic bacteria producing specific metabolites) will provide more selectivity for treating ALD patients at different disease stages (Li et al. 2016a).

Accumulating evidence demonstrates the protective effect of probiotics on multiple pathological disorders. However, these treatments are not always effective because, in many cases, live bacteria must colonize the gut to confer their beneficial effects. The spectrum of pathogenic bacteria varies from patient to patient. Drugs, in particular antibiotics, used by patients may be harmful to live probiotics leading to an unstable and variable effect of the live probiotics. Moreover, the clinically recommended dose of probiotics usually consists of billions of live bacteria. Generally, probiotics are considered safe, but several reports have raised safety concerns about ingesting such large amounts of bacteria, especially when the intestinal function and/or the patient's immune response are compromised (Pereg et al. 2011; Tandon et al. 2009; Wiest et al. 2003; Bauer et al. 2002). In an attempt to address these potential concerns, soluble factors secreted from live probiotics and dead probiotics have been used in the treatment of several diseases/conditions, such as inflammatory bowel disease, colitis, and arthritis (Nowak et al. 2012; Zakostelska et al. 2011; Yan and Polk 2012). Yan et al. demonstrated that soluble proteins produced by probiotic bacteria regulate intestinal epithelial cell survival and growth (Yan et al. 2013). Interestingly, the beneficial effects of probiotics on ALD do not appear to be restricted to viable probiotic bacteria. Segawa and colleagues demonstrated that oral administration of heat-killed *Lactobacillus brevis* SBC8803 induced the expression of cytoprotective heat shock proteins and improvement of intestinal barrier function leading to amelioration of experimental ALD (Segawa et al. 2008). Recently, we evaluated the effectiveness of LGG culture supernatant in the prevention of acute and chronic alcohol-induced hepatic steatosis and liver injury (Zhang et al. 2015; Zhao et al. 2015; Wang et al. 2012). Pretreatment with LGG supernatant (LGG-s) reduced hepatic fat accumulation in mice subsequently exposed to acute-binge alcohol (Wang et al. 2012). Furthermore, co-administration of LGG supernatant with alcohol in the Lieber-DeCarli liquid diet for 4 weeks significantly prevented alcohol-induced intestinal barrier dysfunction, endotoxemia, fatty liver, and inflammation in mice (Zhang et al. 2015; Zhao et al. 2015). The use of probiotic culture supernatant opens a new avenue for the probiotic use. Further characterization of the LGG-s active components will enhance our understanding of the protective effect of probiotics in ALD and advance the development of new therapeutic strategies for ALD.

Probiotic treatment/prevention of NAFDL/NASH Evidence of the beneficial effects of probiotics and symbiotics on NAFLD is extensively derived from rodent models. In a pioneering study of an animal model of NAFLD (Li et al. 2003), VSL#3 probiotic mixture (a multi-strain preparation composed of *Bifidobacterium longum*, *Bifidobacterium breve*, *Bifidobacterium infants*, *Lactobacillus casei*, *Lactobacillus plantarum*, *Lactobacillus acidophilus*, *Lactobacillus delbrueckii* subsp. bulgaricus, and *Streptococcus thermophilus*) was used to treat high-fat diet-induced NAFLD. Treatment with VSL#3 improved liver histology, reduced hepatic total fatty acid content, and decreased serum ALT levels. Mechanistic analysis showed that VSL#3 treatment reduced the activity of Jun N-terminal kinase (JNK), a TNF-regulated kinase that promotes insulin resistance, and decreased the DNA binding activity of NF-κB, the target of IKKβ, another TNF-regulated enzyme that causes insulin resistance. Consistent with treatment-related improvements in hepatic insulin resistance, fatty acid β-oxidation and uncoupling protein (UCP)-2 expression decreased after treatment with VSL#3. The beneficial effects of VSL#3 treatment were comparable to anti-TNF antibody treatment. In a possibly more relevant mouse model (Ma et al. 2008) to diet-induced obesity in humans, the beneficial effects of VSL#3 supplementation on the prevention of the development of hepatic steatosis was also attributed to the regulation of hepatic natural killer T cells (NKT cells) and suppression of the TNF-α/IKK-β signaling pathway. Moreover, VSL#3 exhibited anti-inflammatory activity through a reduction of nuclear factor κB (NF-κB) activation, decreasing hepatic TNF-α production, and cyclooxygenase and inducible nitric oxide synthase expression (Ma et al. 2008). In a rat model of NASH, 4 weeks of VSL#3 treatment attenuated high-fat diet-induced oxidative and inflammatory liver damage (Esposito et al. 2009). In a methionine-choline deficiency (MCD) diet-induced steatohepatitis model, VSL#3 attenuated fibrosis, but did not reduce fatty liver and steatohepatitis (Velayudham et al. 2009).

Several strains of *Lactobacillus* and *Bifidobacterium* have shown protective effects on NAFLD. Administration of *Lactobacillus rhamnosus* PL60, a probiotic bacterium of human origin which produces conjugated linoleic acid, showed anti-obesity effects and improved liver steatosis in a mouse model of diet-induced obesity (Lee et al. 2006). *Lactobacillus acidophilus* and *Lactobacillus casei* administration for 8 weeks demonstrated antioxidant effects in the liver and pancreatic tissues in high-fructose diet-fed mice (Yadav et al 2007). Recent studies also showed that *Lactobacillus* probiotic treatments prevented fructose excess-induced (Wagnerberger et al. 2013; Yadav et al. 2007; Ritze et al. 2014) or choline deficiency-induced hepatic steatosis in animals (Okubo et al. 2013). The beneficial effects of probiotics in NAFLD have been linked to restoration of gut microbiota, gut barrier integrity, and attenuation of inflammatory cytokine expression in the liver (Ritze et al. 2014; Xu et al. 2012). *Lactobacillus* probiotics have also been used to prevent hepatic cholesterol and triglyceride accumulation in rats fed high-cholesterol diet (Wang et al. 2009).

Lactobacillus paracasei F19 significantly attenuated liver injury induced by ischemia-reperfusion and a methionine-restricted/choline-deficient diet in rats by restoring gut microbiota and reducing inflammation and steatosis (Nardone et al. 2010). Of note, administration of different strains of *Lactobacillus* and *Bifidobacterium* in an acute liver injury rat model has shown different effects on bacterial translocation and hepatocellular damage (Adawi et al. 2001). In addition, in a study comparing *Bifidobacterium longum* and *Lactobacillus acidophilus*, Xu et al. (2012) showed that *Bifidobacterium longum* was more effective in preventing NAFLD development, further supporting the importance of careful evaluation of probiotic strains in therapeutic use.

Prebiotic preparations have also been used to treat NAFLD in animals. ob/ob mice treated with prebiotic with non-prebiotic carbohydrate as control showed that the beneficial effect of the prebiotic was mediated by attenuation of liver inflammation in obese mice through a glucagon-like peptide-2 (GLP2)-dependent effect on the gut barrier (Cani et al. 2009). In another mouse model, prebiotic FOS administration reduced liver fat accumulation associated with cecal bacterial alterations (Pachikian et al. 2013).

Despite a large body of preclinical studies on probiotics and NAFLD, high-quality clinical studies are still limited. A double-blind clinical trial (Vajro et al. 2011) using *Lactobacillus* GG in obese children with NAFLD showed significant decreases in serum ALT values and in anti-peptidoglycan-polysaccharide antibodies, which are used as an indirect indicator of intestinal bacterial overgrowth and translocation. Loguercio et al. (2005) demonstrated that VSL#3 supplementation in patients affected by several types of chronic liver diseases, including NAFLD, may reduce liver damage and improve serum levels of various biomarkers. A NAFLD-related study endpoint was a reduction of ALT and AST accompanied by a reduction of pro-inflammatory cytokines.

A small, randomized, placebo-controlled study of NAFLD in children (Alisi et al. 2014) showed that supplementation with probiotics significantly improved their liver disease. They showed that 4-month supplementation of VSL#3 improved both fatty liver and BMI. A previous study (Yadav et al. 2013) showed that the same probiotic preparation suppressed body weight gain and insulin resistance in mice. Interestingly, VSL#3 supplementation in NAFLD children showed a clear trend toward increasing concentrations of total and activated glucagon-like peptide-1 (GLP-1), suggesting that the VSL#3-dependent GLP-1 increase could be responsible, at least in part, for these beneficial effects. A recent randomized triple-blind trial study (Famouri et al. 2017) including 64 obese children with NAFLD showed that 12-week supplementation with a probioitic mixture (*Lactobacillus acidophilus* ATCC B3208; *Bifidobacterium lactis* DSMZ 32269; *B. bifidum* ATCC SD6576; *L. rhamnosus* DSMZ 21690) significantly reduced serum levels of ALT, AST, cholesterol, LDL-C, and triglycerides, further suggesting that this probiotic mixture can be effective in improving pediatric NAFLD.

A randomized study on the use of a synbiotic that contains five probiotics (*L. plantarum*, *L. delbrueckii* spp. *bulgaricus*, *L. acidophilus*, *L. rhamnosus*, *B. bifidum*) and a prebiotic (inulin) over 6 months in adults with NASH produced a significant decrease in hepatic steatosis (Wong et al. 2013b). Another clinical study showed that synbiotic treatment (*L. casei*, *L. rhamnosus*, *S. thermophilus*, *B. breve*, *L. acidophilus*, *B. longum*, *L. bulgaricus*, and FOS) of NAFLD patients over 28 weeks inhibited NF-κB-induced TNF-α production (Eslamparast et al. 2014).

Several meta-analyses on the effectiveness of probiotics in NAFLD have been published. A meta-analysis (Ma et al. 2013) of four randomized controlled trials (Wong et al. 2013b; Malaguarnera et al. 2012; Vajro et al. 2011; Aller et al. 2011), selected from 475 items published from 1996 to 2013, on probiotics in patients with NAFLD/NASH concluded that probiotic therapies can reduce liver aminotransferase levels, serum cholesterol, and TNF-α and improve insulin resistance in patients with NAFLD. This suggests that modulation of the gut microbiota using probiotics may represent a unique method of treating or preventing NAFLD. Although this meta-analysis introduces difficulty in obtaining unequivocal correlations between biochemical changes and pharmacological treatment or lifestyle modifications, such as the inclusion of probiotics in diet, it is informative since it demonstrates the complexity of probiotic treatment of NAFLD.

A recent meta-analysis including nine international randomized controlled trials of probioitcs in childhood and adult NAFLD showed that probiotics provided improvements in the outcomes of homeostasis model assessment (HOMA), total cholesterol, high-density lipoprotein, and TNF-α in any NAFLD patient and in triglycerides in Italian and Spanish patients, but there were no improvements in the outcomes of BMI, glucose, or insulin in adult NAFLD patients (Gao et al. 2016).

A more comprehensive meta-analysis summarized the effects of probiotics and synbiotics on obesity, insulin resistance syndrome, type 2 diabetes, and NAFLD (Saez-Lara et al. 2016). Current scientific evidence shows that effects of probiotics in patients with NAFLD are primarily an improvement in liver tests and metabolic parameters. Some clinical results reported beneficial effects using both probiotic and synbiotic supplementation. However, there are studies that did not show any significant effect of probiotic administration on these chronic diseases. These contradictory effects might be related to inappropriate design such as diversity, the use of several strains, and the small number of individuals receiving some interventions. The authors suggest that further studies to evaluate the best dose-response effect of probiotics and synbiotics are needed.

Treatment of cirrhosis The first-line treatment of the cirrhotic with PSE is lactulose. However, treatment with lactulose showed only minimal impact on microbiota composition (Bajaj et al. 2012a). The microbial functionality may be more important than composition in HE. A recent study using LGG in patients with cirrhosis showed promising results (Bajaj et al. 2014a). In this randomized placebo-controlled double-blind Phase I trial involving 30 patients with MHE, LGG treatment reduced gut dysbiosis and endotoxemia with improved gut microbiota-metabolome linkage. In an animal study, VSL#3 has been shown to decrease bacterial translocation and improve intestinal permeability in rats with cirrhosis (Adawi et al. 2001).

It is obvious that antibiotics decrease gut bacterial growth and changes the composition. The effects on antibiotic alterations on gut microbiota appear to be more sustained than previously recognized (Cox and Blaser 2015). In animal studies, antibiotics corrected liver function induced by multiple pathogenic conditions. In clinical studies, however, antibiotics have been primarily employed in the treatment and prevention of several complications of liver disease (Rimola et al. 1985). Prophylactic antibiotics have been used mainly on bacterial infections which target the most common pathogenic microorganisms in cirrhosis (Wong et al. 2005). There is no research thus far to support the use of antibiotics as the primary therapy of ALD and NAFLD or related liver diseases such as intestinal failure-associated liver disease (IFALD) where small intestinal bacterial overgrowth (SIBO) is an important pathogenic factor (Barclay et al. 2011).

Fecal microbiota transplantation Fecal microbiota transplantation (FMT) is a technique in which intestinal microbiota are transferred from a healthy donor to the patient to introduce or restore healthy microbiota in the gut. Numerous studies have demonstrated that FMT is an effective therapeutic strategy for *Clostridium difficile* infection. There also is strong evidence that FMT can be used to treat NAFLD and ALD in animal models. In a NAFLD murine model study (Le Roy et al. 2013), the microbiota from mice that developed NAFLD from HFD feeding were transferred to germ-free mice, which subsequently developed steatosis and harbored a large number of *Barnesiella* and *Roseburia*. However, the mice which received the microbiota from the HFD mice that did not develop NAFLD did not develop steatosis. This indicates that FMT might be useful in NAFLD treatment. Similar results have also been found in ALD mice (Ferrere et al. 2017). Although there are promising data from rodents, a clinical study of FMT for liver disease is lacking. We need a better understanding of the host-intestine interactions to determine the optimal intestinal preparation and environment for FMT. Appropriate standardization and characterization of the FMT preparation and a simplified regulatory mechanism for utilizing FMT need to be established for it to become an alternative therapy for liver diseases associated with gut microbiota.

11.8 Conclusions

There is increasing evidence supporting the concept that gut microbiota play a central role in liver diseases such as NAFLD, ALD, cirrhosis, and liver cancer. Microbiota can impact the onset and progression of these diseases. Many of these studies were conducted in preclinical animal models in which the disease phenotypes could be altered by gut microbiota manipulation and antibiotics could reverse the disease. Further human study is urgently needed. A major difficulty of studying the human microbiota is the variation in the microbiota composition and functionality from individual to individual. In the past, multiple procedures have been used to assess the microbiota homeostasis, and they might produce different results. There is need to develop standardized and more sophisticated techniques to analyze microbiota and their functional activity.

It is clear that many factors affect gut microbiota and liver. In particular, dietary factors such as lipid type, fructose, and alcohol, as well as lifestyle factors such as exercise and sleep, have been studied. Manipulation of these factors may have preventive/therapeutic potential for liver disease. Probiotics, prebiotics, synbiotics, and FMT can have specific gut effects on microbiota and liver pathophysiology that impact on development of liver disease and offer therapeutic opportunities.

Many elegant studies have identified a number of regulatory mechanisms, such as bile acid metabolism, SCFA production, endogenous alcohol generation, increased intestinal barrier dysfunction, and dysregulated immune function in the gut and liver. These factors might be simultaneously involved as "multiple hits," but the individuals and disease types may determine the relative importance of those mechanisms. Furthermore, it is important to emphasize that the metabolic effects of dysbiosis play a central role. A combination of microbiota metagenomics and metabolomics, and possibly proteomics, genomics, lipidomics, and other high-throughput omics techniques, will provide insights into mechanisms and novel therapeutic strategies. Such a comprehensive study may enable a precision medicine approach based on patient's prognostic biomarkers, clinical observations, and pathogenic pathways.

Further studies should also focus on the identification of specific pathogenic microbiota species, complete description of the microbiota, microbiota functionality, and the interaction with diets, environmental factors, and lifestyles via longitudinal studies. Such an approach could reveal how microbiota influence disease phenotype and enable us to design specific therapeutic approaches targeting individual microbiota for liver diseases.

References

Abu-Shanab A, Quigley EM (2010) The role of the gut microbiota in nonalcoholic fatty liver disease. Nat Rev Gastroenterol Hepatol 7:691–701
Adawi D, Ahrne S, Molin G (2001) Effects of different probiotic strains of Lactobacillus and Bifidobacterium on bacterial translocation and liver injury in an acute liver injury model. Int J Food Microbiol 70:213–220
Alisi A, Bedogni G, Baviera G, Giorgio V, Porro E, Paris C, Giammaria P, Reali L, Anania F, Nobili V (2014) Randomised clinical trial: the beneficial effects of VSL#3 in obese children with non-alcoholic steatohepatitis. Aliment Pharmacol Ther 39:1276–1285
Aller R, De Luis DA, Izaola O, Conde R, Gonzalez Sagrado M, Primo D, De La Fuente B, Gonzalez J (2011) Effect of a probiotic on liver aminotransferases in nonalcoholic fatty liver disease patients: a double blind randomized clinical trial. Eur Rev Med Pharmacol Sci 15:1090–1095
Appenrodt B, Grunhage F, Gentemann MG, Thyssen L, Sauerbruch T, Lammert F (2010) Nucleotide-binding oligomerization domain containing 2 (NOD2) variants are genetic risk factors for death and spontaneous bacterial peritonitis in liver cirrhosis. Hepatology 51:1327–1333
Bajaj JS, Wade JB, Sanyal AJ (2009) Spectrum of neurocognitive impairment in cirrhosis: implications for the assessment of hepatic encephalopathy. Hepatology 50:2014–2021
Bajaj JS, Gillevet PM, Patel NR, Ahluwalia V, Ridlon JM, Kettenmann B, Schubert CM, Sikaroodi M, Heuman DM, Crossey MM, Bell DE, Hylemon PB, Fatouros PP, Taylor-Robinson SD (2012a) A longitudinal systems biology analysis of lactulose withdrawal in hepatic encephalopathy. Metab Brain Dis 27:205–215
Bajaj JS, Hylemon PB, Ridlon JM, Heuman DM, Daita K, White MB, Monteith P, Noble NA, Sikaroodi M, Gillevet PM (2012b) Colonic mucosal microbiome differs from stool microbiome in cirrhosis and hepatic encephalopathy and is linked to cognition and inflammation. Am J Physiol Gastrointest Liver Physiol 303:G675–G685
Bajaj JS, Ridlon JM, Hylemon PB, Thacker LR, Heuman DM, Smith S, Sikaroodi M, Gillevet PM (2012c) Linkage of gut microbiome with cognition in hepatic encephalopathy. Am J Physiol Gastrointest Liver Physiol 302:G168–G175
Bajaj JS, Hylemon P, Heuman DM, Sanyal AJ, Gillevet PM (2013) The cirrhosis dysbiosis ratio provides insight into gut microbiome changes across the spectrum of cirrhosis: a prospective study of 250 patients. Hepatology 58:274a
Bajaj JS, Heuman DM, Hylemon PB, Sanyal AJ, Puri P, Sterling RK, Luketic V, Stravitz RT, Siddiqui MS, Fuchs M, Thacker LR, Wade JB, Daita K, Sistrun S, White MB, Noble NA, Thorpe C, Kakiyama G, Pandak WM, Sikaroodi M, Gillevet PM (2014a) Randomised clinical trial: lactobacillus GG modulates gut microbiome, metabolome and endotoxemia in patients with cirrhosis. Aliment Pharm Ther 39:1113–1125
Bajaj JS, Heuman DM, Hylemon PB, Sanyal AJ, White MB, Monteith P, Noble NA, Unser AB, Daita K, Fisher AR, Sikaroodi M, Gillevet PM (2014b) Altered profile of human gut microbiome is associated with cirrhosis and its complications. J Hepatol 60:940–947
Bajaj JS, Betrapally NS, Hylemon PB, Heuman DM, Daita K, White MB, Unser A, Thacker LR, Sanyal AJ, Kang DJ, Sikaroodi M, Gillevet PM (2015) Salivary microbiota reflects changes in gut microbiota in cirrhosis with hepatic encephalopathy. Hepatology 62:1260–1271
Baker SS, Baker RD, Liu W, Nowak NJ, Zhu L (2010) Role of alcohol metabolism in non-alcoholic steatohepatitis. PLoS One 5:e9570
Barclay AR, Beattie LM, Weaver LT, Wilson DC (2011) Systematic review: medical and nutritional interventions for the management of intestinal failure and its resultant complications in children. Aliment Pharmacol Ther 33:175–184
Bauer TM, Fernandez J, Navasa M, Vila J, Rodes J (2002) Failure of lactobacillus spp. to prevent bacterial translocation in a rat model of experimental cirrhosis. J Hepatol 36:501–506
Berni Canani R, Di Costanzo M, Leone L (2012) The epigenetic effects of butyrate: potential therapeutic implications for clinical practice. Clin Epigenetics 4:4

Bhonchal S, Nain CK, Taneja N, Sharma M, Sharma AK, Sinha SK, Singh K (2007) Modification of small bowel microflora in chronic alcoholics with alcoholic liver disease. Trop Gastroenterol 28:64–66

Bode JC, Bode C, Heidelbach R, Durr HK, Martini GA (1984) Jejunal Microflora in patients with chronic alcohol-abuse. Hepato-Gastroenterol 31:30–34

Bode C, Kolepke R, Schafer K, Bode JC (1993) Breath hydrogen excretion in patients with alcoholic liver-disease – evidence of small intestinal bacterial overgrowth. Zeitschrift Fur Gastroenterologie 31:3–7

Boursier J, Mueller O, Barret M, Machado M, Fizanne L, Araujo-Perez F, Guy CD, Seed PC, Rawls JF, David LA, Hunault G, Oberti F, Cales P, Diehl AM (2016) The severity of nonalcoholic fatty liver disease is associated with gut dysbiosis and shift in the metabolic function of the gut Microbiota. Hepatology 63:764–775

Brinkworth GD, Noakes M, Clifton PM, Bird AR (2009) Comparative effects of very low-carbohydrate, high-fat and high-carbohydrate, low-fat weight-loss diets on bowel habit and faecal short-chain fatty acids and bacterial populations. Br J Nutr 101:1493–1502

Bruns T, Peter J, Reuken PA, Grabe DH, Schuldes SR, Brenmoehl J, Scholmerich J, Wiest R, Stallmach A (2012) NOD2 gene variants are a risk factor for culture-positive spontaneous bacterial peritonitis and monomicrobial bacterascites in cirrhosis. Liver Int 32:223–230

Bull-Otterson L, Feng W, Kirpich I, Wang Y, Qin X, Liu Y, Gobejishvili L, Joshi-Barve S, Ayvaz T, Petrosino J, Kong M, Barker D, McClain C, Barve S (2013) Metagenomic analyses of alcohol induced pathogenic alterations in the intestinal microbiome and the effect of lactobacillus rhamnosus GG treatment. PLoS One 8:e53028

Cani PD, Possemiers S, Van de Wiele T, Guiot Y, Everard A, Rottier O, Geurts L, Naslain D, Neyrinck A, Lambert DM, Muccioli GG, Delzenne NM (2009) Changes in gut microbiota control inflammation in obese mice through a mechanism involving GLP-2-driven improvement of gut permeability. Gut 58:1091–1103

Carr RM, Reid AE (2015) FXR agonists as therapeutic agents for non-alcoholic fatty liver disease. Curr Atheroscler Rep 17:16

Chang B, Sang L, Wang Y, Tong J, Zhang D, Wang B (2013) The protective effect of VSL#3 on intestinal permeability in a rat model of alcoholic intestinal injury. BMC Gastroenterol 13:151

Chen P, Miyamoto Y, Mazagova M, Lee KC, Eckmann L, Schnabl B (2015a) Microbiota protects mice against acute alcohol-induced liver injury. Alcohol Clin Exp Res 39:2313–2323

Chen P, Torralba M, Tan J, Embree M, Zengler K, Starkel P, van Pijkeren JP, DePew J, Loomba R, Ho SB, Bajaj JS, Mutlu EA, Keshavarzian A, Tsukamoto H, Nelson KE, Fouts DE, Schnabl B (2015b) Supplementation of saturated long-chain fatty acids maintains intestinal eubiosis and reduces ethanol-induced liver injury in mice. Gastroenterology 148:203–U345

Choi JJ, Eum SY, Rampersaud E, Daunert S, Abreu MT, Toborek M (2013) Exercise attenuates PCB-induced changes in the mouse gut microbiome. Environ Health Perspect 121:725–730

Conley ME, Delacroix DL (1987) Intravascular and mucosal immunoglobulin A: two separate but related systems of immune defense? Ann Intern Med 106:892–899

Cope K, Risby T, Diehl AM (2000) Increased gastrointestinal ethanol production in obese mice: implications for fatty liver disease pathogenesis. Gastroenterology 119:1340–1347

Cox LM, Blaser MJ (2015) Antibiotics in early life and obesity. Nat Rev Endocrinol 11:182–190

Dapito DH, Mencin A, Gwak GY, Pradere JP, Jang MK, Mederacke I, Caviglia JM, Khiabanian H, Adeyemi A, Bataller R, Lefkowitch JH, Bower M, Friedman R, Sartor RB, Rabadan R, Schwabe RF (2012) Promotion of hepatocellular carcinoma by the intestinal microbiota and TLR4. Cancer Cell 21:504–516

Dawes EA, Foster SM (1956) The formation of ethanol in *Escherichia coli*. Biochim Biophys Acta 22:253–265

Del Chierico F, Nobili V, Vernocchi P, Russo A, Stefanis C, Gnani D, Furlanello C, Zandona A, Paci P, Capuani G, Dallapiccola B, Miccheli A, Alisi A, Putignani L (2017) Gut microbiota profiling of pediatric nonalcoholic fatty liver disease and obese patients unveiled by an integrated meta-omics-based approach. Hepatology 65:451–464 den Besten G, van Eunen K, Groen AK, Venema K, Reijngoud DJ, Bakker BM (2013) The role of short-chain fatty acids in the interplay between diet, gut microbiota, and host energy metabolism. J Lipid Res 54:2325–2340

den Besten G, Bleeker A, Gerding A, van Eunen K, Havinga R, van Dijk TH, Oosterveer MH, Jonker JW, Groen AK, Reijngoud DJ, Bakker BM (2015) Short-chain fatty acids protect against high-fat diet-induced obesity via a PPARgamma-dependent switch from lipogenesis to fat oxidation. Diabetes 64:2398–2408

Dhiman RK, Rana B, Agrawal S, Garg A, Chopra M, Thumburu KK, Khattri A, Malhotra S, Duseja A, Chawla YK (2014) Probiotic VSL#3 reduces liver disease severity and hospitalization in patients with cirrhosis: a randomized, controlled trial. Gastroenterology 147:1327–1337.e1323

Dicksved J, Halfvarson J, Rosenquist M, Jarnerot G, Tysk C, Apajalahti J, Engstrand L, Jansson JK (2008) Molecular analysis of the gut microbiota of identical twins with Crohn's disease. ISME J 2:716–727

Drenick EJ, Fisler J, Johnson D (1982) Hepatic steatosis after intestinal bypass – prevention and reversal by metronidazole, irrespective of protein-calorie malnutrition. Gastroenterology 82:535–548

Eslamparast T, Poustchi H, Zamani F, Sharafkhah M, Malekzadeh R, Hekmatdoost A (2014) Synbiotic supplementation in nonalcoholic fatty liver disease: a randomized, double-blind, placebo-controlled pilot study. Am J Clin Nutr 99:535–542

Esposito E, Iacono A, Bianco G, Autore G, Cuzzocrea S, Vajro P, Canani RB, Calignano A, Raso GM, Meli R (2009) Probiotics reduce the inflammatory response induced by a high-fat diet in the liver of young rats. J Nutr 139:905–911

Everard A, Geurts L, Caesar R, Van Hul M, Matamoros S, Duparc T, Denis RGP, Cochez P, Pierard F, Castel J, Bindels LB, Plovier H, Robine S, Muccioli GG, Renauld JC, Dumoutier L, Delzenne NM, Luquet S, Bakhed F, Cani PD (2014) Intestinal epithelial MyD88 is a sensor switching host metabolism towards obesity according to nutritional status. Nat Commun 5:5648

Famouri F, Shariat Z, Hashemipour M, Keikha M, Kelishadi R (2017) Effects of probiotics on nonalcoholic fatty liver disease in obese children and adolescents. J Pediatr Gastroenterol Nutr 64:413–417

Ferrere G, Leroux A, Wrzosek L, Puchois V, Gaudin F, Ciocan D, Renoud ML, Naveau S, Perlemuter G, Cassard AM (2016) Activation of Kupffer cells is associated with a specific dysbiosis induced by fructose or high fat diet in mice. PLoS One 11:e0146177

Ferrere G, Wrzosek L, Cailleux F, Turpin W, Puchois V, Spatz M, Ciocan D, Rainteau D, Humbert L, Hugot C, Gaudin F, Noordine ML, Robert V, Berrebi D, Thomas M, Naveau S, Perlemuter G, Cassard AM (2017) Fecal microbiota manipulation prevents dysbiosis and alcohol-induced liver injury in mice. J Hepatol 66:806–815

Fooladi AAI, Hosseini HM, Nourani MR, Khani S, Alavian SM (2013) Probiotic as a novel treatment strategy against liver disease. Hepat Mon 13:e7521

Forsyth CB, Farhadi A, Jakate SM, Tang Y, Shaikh M, Keshavarzian A (2009) Lactobacillus GG treatment ameliorates alcohol-induced intestinal oxidative stress, gut leakiness, and liver injury in a rat model of alcoholic steatohepatitis. Alcohol 43:163–172

Forsyth CB, Voigt RM, Keshavarzian A (2014) Intestinal CYP2E1: a mediator of alcohol-induced gut leakiness. Redox Biol 3:40–46

Fox JG, Feng Y, Theve EJ, Raczynski AR, Fiala JLA, Doernte AL, Williams M, McFaline JL, Essigmann JM, Schauer DB, Tannenbaum SR, Dedon PC, Weinman SA, Lemon SM, Fry RC, Rogers AB (2010) Gut microbes define liver cancer risk in mice exposed to chemical and viral transgenic hepatocarcinogens. Gut 59:88–97

Gabbard SL, Lacy BE, Levine GM, Crowell MD (2014) The impact of alcohol consumption and cholecystectomy on small intestinal bacterial overgrowth. Digest Dis Sci 59:638–644

Gao B, Bataller R (2011) Alcoholic liver disease: pathogenesis and new therapeutic targets. Gastroenterology 141:1572–1585

Gao X, Zhu Y, Wen Y, Liu G, Wan C (2016) Efficacy of probiotics in non-alcoholic fatty liver disease in adult and children: a meta-analysis of randomized controlled trials. Hepatol Res 46:1226–1233

Glover LE, Lee JS, Colgan SP (2016) Oxygen metabolism and barrier regulation in the intestinal mucosa. J Clin Invest 126:3680–3688

Goodrich JK, Waters JL, Poole AC, Sutter JL, Koren O, Blekhman R, Beaumont M, Van Treuren W, Knight R, Bell JT, Spector TD, Clark AG, Ley RE (2014) Human genetics shape the gut microbiome. Cell 159:789–799

Grat M, Wronka KM, Krasnodebski M, Masior K, Lewandowski Z, Kosinska I, Grat K, Stypulkowski J, Rejowski S, Wasilewicz M, Galecka M, Szachta P, Krawczyk M (2016) Profile of gut microbiota associated with the presence of hepatocellular cancer in patients with liver cirrhosis. Transpl P 48:1687–1691

Hartmann P, Chen P, Wang HJ, Wang LR, McCole DF, Brandl K, Starkel P, Belzer C, Hellerbrand C, Tsukamoto H, Ho SB, Schnabl B (2013) Deficiency of intestinal mucin-2 ameliorates experimental alcoholic liver disease in mice. Hepatology 58:108–119

Hass R, Busche R, Luciano L, Reale E, vonEngelhardt W (1997) Lack of butyrate is associated with induction of Bax and subsequent apoptosis in the proximal colon of guinea pig. Gastroenterology 112:875–881

Henao-Mejia J, Elinav E, Jin C, Hao L, Mehal WZ, Strowig T, Thaiss CA, Kau AL, Eisenbarth SC, Jurczak MJ, Camporez JP, Shulman GI, Gordon JI, Hoffman HM, Flavell RA (2012) Inflammasome-mediated dysbiosis regulates progression of NAFLD and obesity. Nature 482:179–185

Hooper LV, Macpherson AJ (2010) Immune adaptations that maintain homeostasis with the intestinal microbiota. Nat Rev Immunol 10:159–169

Inagaki T, Moschetta A, Lee YK, Peng L, Zhao GX, Downes M, Yu RT, Shelton JM, Richardson JA, Repa JJ, Mangelsdorf DJ, Kliewer SA (2006) Regulation of antibacterial defense in the small intestine by the nuclear bile acid receptor. Proc Natl Acad Sci U S A 103:3920–3925

Inamine T, Yang A, Wang L, Lee K, Llorente C, Schnabl B (2016) Genetic loss of immunoglobulin a does not influence development of alcoholic Steatohepatitis in mice. Aclcohol Clin Exp Res 40:2604–2613

Jegatheesan P, Beutheu S, Ventura G, Sarfati G, Nubret E, Kapel N, Waligora-Dupriet AJ, Bergheim I, Cynober L, De-Bandt JP (2016) Effect of specific amino acids on hepatic lipid metabolism in fructose-induced non-alcoholic fatty liver disease. Clin Nutr 35:175–182

Jiang C, Xie C, Li F, Zhang L, Nichols RG, Krausz KW, Cai J, Qi Y, Fang ZZ, Takahashi S, Tanaka N, Desai D, Amin SG, Albert I, Patterson AD, Gonzalez FJ (2015) Intestinal farnesoid X receptor signaling promotes nonalcoholic fatty liver disease. J Clin Invest 125:386–402

Jin R, Vos MB (2015) Fructose and liver function – is this behind nonalcoholic liver disease? Curr Opin Clin Nutr 18:490–495

Kirpich IA, McClain CJ (2012) Probiotics in the treatment of the liver diseases. J Am Coll Nutr 31:14–23

Kirpich IA, Solovieva NV, Leikhter SN, Shidakova NA, Lebedeva OV, Sidorov PI, Bazhukova TA, Soloviev AG, Barve SS, McClain CJ, Cave M (2008) Probiotics restore bowel flora and improve liver enzymes in human alcohol-induced liver injury: a pilot study. Alcohol 42:675–682

Kirpich IA, Petrosino J, Ajami N, Feng W, Wang Y, Liu Y, Beier JI, Barve SS, Yin X, Wei X, Zhang X, McClain CJ (2016) Saturated and unsaturated dietary fats differentially modulate ethanol-induced changes in gut microbiome and metabolome in a mouse model of alcoholic liver disease. Am J Pathol 186:765–776

Knights D, Silverberg MS, Weersma RK, Gevers D, Dijkstra G, Huang H, Tyler AD, van Sommeren S, Imhann F, Stempak JM, Huang H, Vangay P, Al-Ghalith GA, Russell C, Sauk J, Knight J, Daly MJ, Huttenhower C, Xavier RJ (2014) Complex host genetics influence the microbiome in inflammatory bowel disease. Genome Med 6:107

Lata J, Novotny I, Pribramska V, Jurankova J, Fric P, Kroupa R, Stiburek O (2007) The effect of probiotics on gut flora, level of endotoxin and Child-Pugh score in cirrhotic patients: results of a double-blind randomized study. Eur J Gastroenterol Hepatol 19:1111–1113

Le Roy T, Llopis M, Lepage P, Bruneau A, Rabot S, Bevilacqua C, Martin P, Philippe C, Walker F, Bado A, Perlemuter G, Cassard-Doulcier AM, Gerard P (2013) Intestinal microbiota determines development of non-alcoholic fatty liver disease in mice. Gut 62:1787–1794

Leclercq S, Matamoros S, Cani PD, Neyrinck AM, Jamar F, Starkel P, Windey K, Tremaroli V, Backhed F, Verbeke K, de Timary P, Delzenne NM (2014) Intestinal permeability, gut-bacterial dysbiosis, and behavioral markers of alcohol-dependence severity. Proc Natl Acad Sci U S A 111:E4485–E4493

Lee HY, Park JH, Seok SH, Baek MW, Kim DJ, Lee KE, Paek KS, Lee Y, Park JH (2006) Human originated bacteria, lactobacillus rhamnosus PL60, produce conjugated linoleic acid and show anti-obesity effects in diet-induced obese mice. BBA-Mol Cell Biol L 1761:736–744

Lee S, Sung J, Lee J, Ko G (2011) Comparison of the gut microbiotas of healthy adult twins living in South Korea and the United States. Appl Environ Microbiol 77:7433–7437

Ley RE, Backhed F, Turnbaugh P, Lozupone CA, Knight RD, Gordon JI (2005) Obesity alters gut microbial ecology. Proc Natl Acad Sci U S A 102:11070–11075

Ley RE, Turnbaugh PJ, Klein S, Gordon JI (2006) Microbial ecology: human gut microbes associated with obesity. Nature 444:1022–1023

Lhoste EF, Catala I, Fiszlewicz M, Gueugneau AM, Popot F, Vaissade P, Corring T, Szylit O (1996) Influence of caecal microflora and of two dietary protein levels on the adaptation of the exocrine pancreas: comparative study in germ-free and conventional rats. Br J Nutr 75:433–444

Lhoste EF, Mouzon B, Andrieux C, Gueugneau AM, Fiszlewicz M, Corring T, Szylit O (1998) Physiological effects of a pea protein isolate in gnotobiotic rats: comparison with a soybean isolate and meat. Ann Nutr Metab 42:44–54

Li Z, Yang S, Lin H, Huang J, Watkins PA, Moser AB, Desimone C, Song XY, Diehl AM (2003) Probiotics and antibodies to TNF inhibit inflammatory activity and improve nonalcoholic fatty liver disease. Hepatology 37:343–350

Li F, Duan K, Wang C, McClain C, Feng W (2016a) Probiotics and alcoholic liver disease: treatment and potential mechanisms. Gastroenterol Res Pract 2016:5491465

Li J, Sung CY, Lee N, Ni Y, Pihlajamaki J, Panagiotou G, El-Nezami H (2016b) Probiotics modulated gut microbiota suppresses hepatocellular carcinoma growth in mice. Proc Natl Acad Sci U S A 113:E1306–E1315

Lin WH, Hwang CF, Chen LW, Tsen HY (2006) Viable counts, characteristic evaluation for commercial lactic acid bacteria products. Food Microbiol 23:74–81

Liu Q, Duan ZP, Ha DK, Bengmark S, Kurtovic J, Riordan SM (2004) Synbiotic modulation of gut flora: effect on minimal hepatic encephalopathy in patients with cirrhosis. Hepatology 39:1441–1449

Liu Y, Li Y, Liu K, Shen J (2014) Exposing to cadmium stress cause profound toxic effect on microbiota of the mice intestinal tract. PLoS One 9:e85323

Llopis M, Cassard AM, Wrzosek L, Boschat L, Bruneau A, Ferrere G, Puchois V, Martin JC, Lepage P, Le Roy T, Lefevre L, Langelier B, Cailleux F, Gonzalez-Castro AM, Rabot S, Gaudin F, Agostini H, Prevot S, Berrebi D, Ciocan D, Jousse C, Naveau S, Gerard P, Perlemuter G (2016) Intestinal microbiota contributes to individual susceptibility to alcoholic liver disease. Gut 65:830–839

Loguercio C, De Simone T, Federico A, Terracciano F, Tuccillo C, Di Chicco M, Carteni M (2002) Gut-liver axis: a new point of attack to treat chronic liver damage? Am J Gastroenterol 97:2144–2146

Loguercio C, Federico A, Tuccillo C, Terracciano F, D'Auria MV, De Simone C, Del Vecchio BC (2005) Beneficial effects of a probiotic VSL#3 on parameters of liver dysfunction in chronic liver diseases. J Clin Gastroenterol 39:540–543

Ma X, Hua J, Li ZP (2008) Probiotics improve high fat diet-induced hepatic steatosis and insulin resistance by increasing hepatic NKT cells. J Hepatol 49:821–830

Ma YY, Li L, Yu CH, Shen Z, Chen LH, Li YM (2013) Effects of probiotics on nonalcoholic fatty liver disease: a meta-analysis. World J Gastroenterol 19:6911–6918

Magne F, Hachelaf W, Suau A, Boudraa G, Mangin I, Touhami M, Bouziane-Nedjadi K, Pochart P (2006) A longitudinal study of infant faecal microbiota during weaning. FEMS Microbiol Ecol 58:563–571

Malaguarnera M, Vacante M, Antic T, Giordano M, Chisari G, Acquaviva R, Mastrojeni S, Malaguarnera G, Mistretta A, Li Volti G, Galvano F (2012) Bifidobacterium longum with fructo-oligosaccharides in patients with non alcoholic steatohepatitis. Dig Dis Sci 57:545–553

Manley S, Ding WX (2015) Role of farnesoid X receptor and bile acids in alcoholic liver disease. Acta Pharmacol Sin B 5:158–167

Marotta F, Barreto R, Wu CC, Naito Y, Gelosa F, Lorenzetti A, Yoshioka M, Fesce E (2005) Experimental acute alcohol pancreatitis-related liver damage and endotoxemia: synbiotics but not metronidazole have a protective effect. Chin J Dig Dis 6:193–197

Massey VL, Stocke KS, Schmidt RH, Tan M, Ajami N, Neal RE, Petrosino JF, Barve S, Arteel GE (2015) Oligofructose protects against arsenic-induced liver injury in a model of environment/obesity interaction. Toxicol Appl Pharmacol 284:304–314

Mazagova M, Wang LR, Anfora AT, Wissmueller M, Lesley SA, Miyamoto Y, Eckmann L, Dhungana S, Pathmasiri W, Sumner S, Westwater C, Brenner DA, Schnabl B (2015) Commensal microbiota is hepatoprotective and prevents liver fibrosis in mice. FASEB J 29:1043–1055

McMahan R, Wang XX, Nash R, Pruzanski M, Adorini L, Golden-Mason L, Levi M, Rosen HR (2011) Bile-acid receptor activation shifts hepatic monocytes/macrophages towards an anti-inflammatory phenotype and improves Nonalcoholic fatty liver disease. Hepatology 54:389a

McMahan RH, Wang XXX, Cheng LL, Krisko T, Smith M, El Kasmi K, Pruzanski M, Adorini L, Golden-Mason L, Levi M, Rosen HR (2013) Bile acid receptor activation modulates hepatic monocyte activity and improves nonalcoholic fatty liver disease. J Biol Chem 288:11761–11770

Morencos FC, Castano GDLH, Ramos LM, Arias MJL, Ledesma F, Romero FP (1995) Small-bowel bacterial overgrowth in patients with alcoholic cirrhosis. Digest Dis Sci 40:1252–1256

Mouzaki M, Comelli EM, Arendt BM, Bonengel J, Fung SK, Fischer SE, McGilvray ID, Allard JP (2013) Intestinal microbiota in patients with nonalcoholic fatty liver disease. Hepatology 58:120–127

Mutlu E, Keshavarzian A, Engen P, Forsyth CB, Sikaroodi M, Gillevet P (2009) Intestinal dysbiosis: a possible mechanism of alcohol-induced endotoxemia and alcoholic steatohepatitis in rats. Alcohol Clin Exp Res 33:1836–1846

Mutlu EA, Gillevet PM, Rangwala H, Sikaroodi M, Naqvi A, Engen PA, Kwasny M, Lau CK, Keshavarzian A (2012) Colonic microbiome is altered in alcoholism. Am J Physiol-Gastr L 302:G966–G978

Nair S, Cope K, Risby TH, Diehl AM (2001) Obesity and female gender increase breath ethanol concentration: potential implications for the pathogenesis of nonalcoholic steatohepatitis. Am J Gastroenterol 96:1200–1204

Nanji AA, Khettry U, Sadrzadeh SMH (1994) Lactobacillus feeding reduces endotoxemia and severity of experimental alcoholic liver (disease). Proc Soc Exp Biol Med 205:243–247

Nardone G, Compare D, Liguori E, Di Mauro V, Rocco A, Barone M, Napoli A, Lapi D, Iovene MR, Colantuoni A (2010) Protective effects of lactobacillus paracasei F19 in a rat model of oxidative and metabolic hepatic injury. Am J Physiol Gastrointest Liver Physiol 299: G669–G676

Neuschwander-Tetri BA, Loomba R, Sanyal AJ, Lavine JE, Van Natta ML, Abdelmalek MF, Chalasani N, Dasarathy S, Diehl AM, Hameed B, Kowdley KV, McCullough A, Terrault N, Clark JM, Tonascia J, Brunt EM, Kleiner DE, Doo E, Network NCR (2015) Farnesoid X nuclear receptor ligand obeticholic acid for non-cirrhotic, non-alcoholic steatohepatitis (FLINT): a multicentre, randomised, placebo-controlled trial. Lancet 385:956–965

Nowak B, Ciszek-Lenda M, Srottek M, Gamian A, Kontny E, Gorska-Fraczek S, Marcinkiewicz J (2012) Lactobacillus rhamnosus exopolysaccharide ameliorates arthritis induced by the systemic injection of collagen and lipopolysaccharide in DBA/1 mice. Arch Immunol Ther Exp 60:211–220

Okubo H, Sakoda H, Kushiyama A, Fujishiro M, Nakatsu Y, Fukushima T, Matsunaga Y, Kamata H, Asahara T, Yoshida Y, Chonan O, Iwashita M, Nishimura F, Asano T (2013) Lactobacillus casei strain Shirota protects against nonalcoholic steatohepatitis development in a rodent model. Am J Physiol-Gastr L 305:G911–G918

O'Mahony L, McCarthy J, Kelly P, Hurley G, Luo FY, Chen KS, O'Sullivan GC, Kiely B, Collins JK, Shanahan F, Quigley EMM (2005) Lactobacillus and Bifidobacterium in irritable bowel syndrome: symptom responses and relationship to cytokine profiles. Gastroenterology 128:541–551

Pachikian BD, Essaghir A, Demoulin JB, Catry E, Neyrinck AM, Dewulf EM, Sohet FM, Portois L, Clerbaux LA, Carpentier YA, Possemiers S, Bommer GT, Cani PD, Delzenne NM (2013) Prebiotic approach alleviates hepatic steatosis: implication of fatty acid oxidative and cholesterol synthesis pathways. Mol Nutr Food Res 57:347–359

Parlesak A, Schafer C, Schutz T, Bode JC, Bode C (2000) Increased intestinal permeability to macromolecules and endotoxemia in patients with chronic alcohol abuse in different stages of alcohol-induced liver disease. J Hepatol 32:742–747

Pereg D, Kotliroff A, Gadoth N, Hadary R, Lishner M, Kitay-Cohen Y (2011) Probiotics for patients with compensated liver cirrhosis: a double-blind placebo-controlled study. Nutrition 27:177–181

Perry RJ, Peng L, Barry NA, Cline GW, Zhang D, Cardone RL, Petersen KF, Kibbey RG, Goodman AL, Shulman GI (2016) Acetate mediates a microbiome-brain-beta-cell axis to promote metabolic syndrome. Nature 534:213–217

Pineiro M, Asp NG, Reid G, Macfarlane S, Morelli L, Brunser O, Tuohy K (2008) FAO technical meeting on prebiotics. J Clin Gastroenterol 42(Suppl 3 Pt 2):S156–S159

Raman M, Ahmed I, Gillevet PM, Probert CS, Ratcliffe NM, Smith S, Greenwood R, Sikaroodi M, Lam V, Crotty P, Bailey J, Myers RP, Rioux KP (2013) Fecal microbiome and volatile organic compound metabolome in obese humans with nonalcoholic fatty liver disease. Clin Gastroenterol Hepatol 11:868–875.e861–863

Rao R (2009) Endotoxemia and gut barrier dysfunction in alcoholic liver disease. Hepatology 50:638–644

Ribiere C, Peyret P, Parisot N, Darcha C, Dechelotte PJ, Barnich N, Peyretaillade E, Boucher D (2016) Oral exposure to environmental pollutant benzo[a]pyrene impacts the intestinal epithelium and induces gut microbial shifts in murine model. Sci Rep 6:31027

Rimola A, Bory F, Teres J, Perez-Ayuso RM, Arroyo V, Rodes J (1985) Oral, nonabsorbable antibiotics prevent infection in cirrhotics with gastrointestinal hemorrhage. Hepatology 5:463–467

Rinella ME, Sanyal AJ (2015) NAFLD in 2014: genetics, diagnostics and therapeutic advances in NAFLD. Nat Rev Gastroenterol Hepatol 12:65–66

Ritze Y, Bardos G, Claus A, Ehrmann V, Bergheim I, Schwiertz A, Bischoff SC (2014) Lactobacillus rhamnosus GG protects against non-alcoholic fatty liver disease in mice. PLoS One 9:e80169

Saeedi BJ, Kao DJ, Kitzenberg DA, Dobrinskikh E, Schwisow KD, Masterson JC, Kendrick AA, Kelly CJ, Bayless AJ, Kominsky DJ, Campbell EL, Kuhn KA, Furuta GT, Colgan SP, Glover LE (2015) HIF-dependent regulation of claudin-1 is central to intestinal epithelial tight junction integrity. Mol Biol Cell 26:2252–2262

Saez-Lara MJ, Robles-Sanchez C, Ruiz-Ojeda FJ, Plaza-Diaz J, Gil A (2016) Effects of probiotics and synbiotics on obesity, insulin resistance syndrome, type 2 diabetes and non-alcoholic fatty liver disease: a review of human clinical trials. Int J Mol Sci 17:928

Saner FH, Nowak K, Hoyer D, Rath P, Canbay A, Paul A, Koldehoff M, Elmaagacli A (2014) A non-interventional study of the genetic polymorphisms of NOD2 associated with increased mortality in non-alcoholic liver transplant patients. BMC Gastroenterol 14:4

Scholtens PA, Alles MS, Bindels JG, van der Linde EG, Tolboom JJ, Knol J (2006) Bifidogenic effects of solid weaning foods with added prebiotic oligosaccharides: a randomised controlled clinical trial. J Pediatr Gastroenterol Nutr 42:553–559

Schwiertz A, Taras D, Schafer K, Beijer S, Bos NA, Donus C, Hardt PD (2010) Microbiota and SCFA in lean and overweight healthy subjects. Obesity (Silver Spring) 18:190–195

Scott KP, Gratz SW, Sheridan PO, Flint HJ, Duncan SH (2013) The influence of diet on the gut microbiota. Pharmacol Res 69:52–60

Segawa S, Wakita Y, Hirata H, Watari J (2008) Oral administration of heat-killed lactobacillus brevis SBC8803 ameliorates alcoholic liver disease in ethanol-containing diet-fed C57BL/6N mice. Int J Food Microbiol 128:371–377

Seki E, Brenner DA (2008) Toll-like receptors and adaptor molecules in liver disease: update. Hepatology 48:322–335

Shao T, Liu L, Zhang L, McClain C, Feng W (2016) Intestinal HIF-1alpha deletion exacerbates alcohol-induced hepatic steatosis mediated by intestinal dysbiosis and barrier dysfunction in Series Intestinal HIF-1alpha deletion exacerbates alcohol-induced hepatic steatosis mediated by intestinal dysbiosis and barrier dysfunction. Alcohol Clin Exp Res 40:136A

Stadlbauer V, Mookerjee RP, Hodges S, Wright GA, Davies NA, Jalan R (2008) Effect of probiotic treatment on deranged neutrophil function and cytokine responses in patients with compensated alcoholic cirrhosis. J Hepatol 48:945–951

Swennes AG, Sheh A, Parry NM, Muthupalani S, Lertpiriyapong K, Garcia A, Fox JG (2014) Helicobacter hepaticus infection promotes hepatitis and preneoplastic foci in farnesoid X receptor (FXR) deficient mice. PLoS One 9:e106764

Tabibian JH, O'Hara SP, Trussoni CE, Tietz PS, Splinter PL, Mounajjed T, Hagey LR, LaRusso NF (2016) Absence of the intestinal microbiota exacerbates hepatobiliary disease in a murine model of primary sclerosing cholangitis. Hepatology 63:185–196

Tandon P, Moncrief K, Madsen K, Arrieta MC, Owen RJ, Bain VG, Wong WW, Ma MM (2009) Effects of probiotic therapy on portal pressure in patients with cirrhosis: a pilot study. Liver Int 29:1110–1115

Turnbaugh PJ, Ley RE, Mahowald MA, Magrini V, Mardis ER, Gordon JI (2006) An obesity-associated gut microbiome with increased capacity for energy harvest. Nature 444:1027–1031

Turnbaugh PJ, Hamady M, Yatsunenko T, Cantarel BL, Duncan A, Ley RE, Sogin ML, Jones WJ, Roe BA, Affourtit JP, Egholm M, Henrissat B, Heath AC, Knight R, Gordon JI (2009) A core gut microbiome in obese and lean twins. Nature 457:480–484

Turner JR (2009) Intestinal mucosal barrier function in health and disease. Nat Rev Immunol 9:799–809

Vajro P, Mandato C, Licenziati MR, Franzese A, Vitale DF, Lenta S, Caropreso M, Vallone G, Meli R (2011) Effects of lactobacillus rhamnosus strain GG in pediatric obesity-related liver disease. J Pediatr Gastroenterol Nutr 52:740–743

Van den Abbeele P, Gerard P, Rabot S, Bruneau A, El Aidy S, Derrien M, Kleerebezem M, Zoetendal EG, Smidt H, Verstraete W, Van de Wiele T, Possemiers S (2011) Arabinoxylans and inulin differentially modulate the mucosal and luminal gut microbiota and mucin-degradation in humanized rats. Environ Microbiol 13:2667–2680

Velayudham A, Dolganiuc A, Ellis M, Petrasek J, Kodys K, Mandrekar P, Szabo G (2009) VSL#3 probiotic treatment attenuates fibrosis without changes in steatohepatitis in a diet-induced non-alcoholic steatohepatitis model in mice. Hepatology 49:989–997

Verbeke L, Mannaerts I, Schierwagen R, Govaere O, Klein S, Elst IV, Windmolders P, Farre R, Wenes M, Mazzone M, Nevens F, van Grunsven LA, Trebicka J, Laleman W (2016) FXR agonist obeticholic acid reduces hepatic inflammation and fibrosis in a rat model of toxic cirrhosis. Sci Rep Uk 6:33453

Wagnerberger S, Spruss A, Kanuri G, Volynets V, Stahl C, Bischoff SC, Bergheim I (2012) Toll-like receptors 1-9 are elevated in livers with fructose-induced hepatic steatosis. Br J Nutr 107:1727–1738

Wagnerberger S, Spruss A, Kanuri G, Stahl C, Schroder M, Vetter W, Bischoff SC, Bergheim I (2013) Lactobacillus casei Shirota protects from fructose-induced liver steatosis: a mouse model. J Nutr Biochem 24:531–538

Wahlstrom A, Sayin SI, Marschall HU, Backhed F (2016) Intestinal crosstalk between bile acids and microbiota and its impact on host metabolism. Cell Metab 24:41–50

Wang Y, Xu N, Xi A, Ahmed Z, Zhang B, Bai X (2009) Effects of lactobacillus plantarum MA2 isolated from Tibet kefir on lipid metabolism and intestinal microflora of rats fed on high-cholesterol diet. Appl Microbiol Biotechnol 84:341–347

Wang Y, Kirpich I, Liu Y, Ma Z, Barve S, McClain CJ, Feng W (2011) Lactobacillus rhamnosus GG treatment potentiates intestinal hypoxia-inducible factor, promotes intestinal integrity and ameliorates alcohol-induced liver injury. Am J Pathol 179:2866–2875

Wang Y, Liu Y, Sidhu A, Ma Z, McClain C, Feng W (2012) Lactobacillus rhamnosus GG culture supernatant ameliorates acute alcohol-induced intestinal permeability and liver injury. Am J Physiol Gastrointest Liver Physiol 303:G32–G41

Wang B, Jiang X, Cao M, Ge J, Bao Q, Tang L, Chen Y, Li L (2016a) Altered fecal microbiota correlates with liver biochemistry in nonobese patients with non-alcoholic fatty liver disease. Sci Rep 6:32002

Wang LR, Fouts DE, Starkel P, Hartmann P, Chen P, Llorente C, DePew J, Moncera K, Ho SB, Brenner DA, Hooper LV, Schnabl B (2016b) Intestinal REG3 lectins protect against alcoholic steatohepatitis by reducing mucosa-associated microbiota and preventing bacterial translocation. Cell Host Microbe 19:227–239

Wei M, Shinkura R, Doi Y, Maruya M, Fagarasan S, Honjo T (2011) Mice carrying a knock-in mutation of Aicda resulting in a defect in somatic hypermutation have impaired gut homeostasis and compromised mucosal defense. Nat Immunol 12:264–U102

Wiest R, Chen F, Cadelina G, Groszmann RJ, Garcia-Tsao G (2003) Effect of lactobacillus-fermented diets on bacterial translocation and intestinal flora in experimental prehepatic portal hypertension. Dig Dis Sci 48:1136–1141

Wigg AJ, Roberts-Thomson IC, Dymock RB, McCarthy PJ, Grose RH, Cummins AG (2001) The role of small intestinal bacterial overgrowth, intestinal permeability, endotoxaemia, and tumour necrosis factor alpha in the pathogenesis of non-alcoholic steatohepatitis. Gut 48:206–211

Wong F, Bernardi M, Balk R, Christman B, Moreau R, Garcia-Tsao G, Patch D, Soriano G, Hoefs J, Navasa M, International Ascites Club (2005) Sepsis in cirrhosis: report on the 7th meeting of the International Ascites Club. Gut 54:718–725

Wong VW, Tse CH, Lam TT, Wong GL, Chim AM, Chu WC, Yeung DK, Law PT, Kwan HS, Yu J, Sung JJ, Chan HL (2013a) Molecular characterization of the fecal microbiota in patients with nonalcoholic steatohepatitis – a longitudinal study. PLoS One 8:e62885

Wong VW, Won GL, Chim AM, Chu WC, Yeung DK, Li KC, Chan HL (2013b) Treatment of nonalcoholic steatohepatitis with probiotics. A proof-of-concept study. Ann Hepatol 12:256–262

Xie G, Wang X, Huang F, Zhao A, Chen W, Yan J, Zhang Y, Lei S, Ge K, Zheng X, Liu J, Su M, Liu P, Jia W (2016a) Dysregulated hepatic bile acids collaboratively promote liver carcinogenesis. Int J Cancer 139:1764–1775

Xie G, Wang X, Liu P, Wei R, Chen W, Rajani C, Hernandez BY, Alegado R, Dong B, Li D, Jia W (2016b) Distinctly altered gut microbiota in the progression of liver disease. Oncotarget 7:19355–19366

Xu RY, Wan YP, Fang QY, Lu W, Cai W (2012) Supplementation with probiotics modifies gut flora and attenuates liver fat accumulation in rat nonalcoholic fatty liver disease model. J Clin Biochem Nutr 50:72–77

Yadav H, Jain S, Sinha PR (2007) Antidiabetic effect of probiotic dahi containing lactobacillus acidophilus and lactobacillus casei in high fructose fed rats. Nutrition 23:62–68

Yadav H, Lee JH, Lloyd J, Walter P, Rane SG (2013) Beneficial metabolic effects of a probiotic via butyrate-induced GLP-1 hormone secretion. J Biol Chem 288:25088–25097

Yamada S, Guo X, Wang KY, Tanimoto A, Sasaguri Y (2016) Novel function of histamine signaling via histamine receptors in cholesterol and bile acid metabolism: histamine H2 receptor protects against nonalcoholic fatty liver disease. Pathol Int 66:376–385

Yan F, Polk DB (2012) Characterization of a probiotic-derived soluble protein which reveals a mechanism of preventive and treatment effects of probiotics on intestinal inflammatory diseases. Gut Microbes 3:25–28

Yan AW, Fouts DE, Brandl J, Starkel P, Torralba M, Schott E, Tsukamoto H, Nelson KE, Brenner DA, Schnabl B (2011) Enteric dysbiosis associated with a mouse model of alcoholic liver disease. Hepatology 53:96–105

Yan F, Liu L, Dempsey PJ, Tsai YH, Raines EW, Wilson CL, Cao H, Cao Z, Liu L, Polk DB (2013) A lactobacillus rhamnosus GG-derived soluble protein, p40, stimulates ligand release from intestinal epithelial cells to transactivate epidermal growth factor receptor. J Biol Chem 288:30742–30751

Ye D, Guo S, Al-Sadi R, Ma TY (2011) MicroRNA regulation of intestinal epithelial tight junction permeability. Gastroenterology 141:1323–1333

Yoshimoto S, Loo TM, Atarashi K, Kanda H, Sato S, Oyadomari S, Iwakura Y, Oshima K, Morita H, Hattori M, Honda K, Ishikawa Y, Hara E, Ohtani N (2013) Obesity-induced gut microbial metabolite promotes liver cancer through senescence secretome. Nature 499:97–101

Yuan L, Bambha K (2015) Bile acid receptors and nonalcoholic fatty liver disease. World J Hepatol 7:2811–2818

Zakostelska Z, Kverka M, Klimesova K, Rossmann P, Mrazek J, Kopecny J, Hornova M, Srutkova D, Hudcovic T, Ridl J, Tlaskalova-Hogenova H (2011) Lysate of probiotic Lactobacillus casei DN-114 001 ameliorates colitis by strengthening the gut barrier function and changing the gut microenvironment. PLoS One 6:e27961

Zeng H, Liu J, Jackson MI, Zhao FQ, Yan L, Combs GF Jr (2013) Fatty liver accompanies an increase in lactobacillus species in the hind gut of C57BL/6 mice fed a high-fat diet. J Nutr 143:627–631

Zhang C, Zhang M, Pang X, Zhao Y, Wang L, Zhao L (2012a) Structural resilience of the gut microbiota in adult mice under high-fat dietary perturbations. ISME J 6:1848–1857

Zhang HL, Yu LX, Yang W, Tang L, Lin Y, Wu H, Zhai B, Tan YX, Shan L, Liu Q, Chen HY, Dai RY, Qiu BJ, He YQ, Wang C, Zheng LY, Li YQ, Wu FQ, Li Z, Yan HX, Wang HY (2012b) Profound impact of gut homeostasis on chemically-induced pro-tumorigenic inflammation and hepatocarcinogenesis in rats. J Hepatol 57:803–812

Zhang ZG, Zhai HQ, Geng JW, Yu R, Ren HQ, Fan H, Shi P (2013) Large-scale survey of gut microbiota associated with MHE via 16S rRNA-based pyrosequencing. Am J Gastroenterol 108:1601–1611

Zhang M, Wang C, Wang C, Zhao H, Zhao C, Chen Y, Wang Y, McClain C, Feng W (2015) Enhanced AMPK phosphorylation contributes to the beneficial effects of Lactobacillus rhamnosus GG supernatant on chronic-alcohol-induced fatty liver disease. J Nutr Biochem 26:337–344

Zhao H, Zhao C, Dong Y, Zhang M, Wang Y, Li F, Li X, McClain C, Yang S, Feng W (2015) Inhibition of miR122a by Lactobacillus rhamnosus GG culture supernatant increases intestinal occludin expression and protects mice from alcoholic liver disease. Toxicol Lett 234:194–200

Zhu L, Baker SS, Gill C, Liu W, Alkhouri R, Baker RD, Gill SR (2013) Characterization of gut microbiomes in nonalcoholic steatohepatitis (NASH) patients: a connection between endogenous alcohol and NASH. Hepatology 57:601–609

Zhu Y, Liu HX, Zhang M, Guo GL (2016) Fatty liver diseases, bile acids, and FXR. Acta Pharm Sin B 6:409–412

Index

W.-X. Ding, X.-M. Yin (eds.), *Molecules, Systems and Signaling in Liver Injury*, Cell Death in Biology and Diseases, DOI 10.1007/978-3-319-58106-4

X.-M. Yin and Z. Dong (Series eds.), *Cell Death in Biology and Diseases*

I

K

L

M

N

MIX
Papier aus verantwortungsvollen Quellen
Paper from responsible sources
FSC® C105338

If you have any concerns about our products,
you can contact us on
ProductSafety@springernature.com

In case Publisher is established outside the EU,
the EU authorized representative is:
Springer Nature Customer Service Center GmbH
Europaplatz 3, 69115 Heidelberg, Germany

Printed by Libri Plureos GmbH
in Hamburg, Germany